ALLE · ZEIT · WACH · 1842

W. R. Ullrich C. Rigano
A. Fuggi P. J. Aparicio (Eds.)

Inorganic Nitrogen in Plants and Microorganisms

Uptake and Metabolism

With 122 Figures

Springer-Verlag Berlin Heidelberg New York
London Paris Tokyo Hong Kong Barcelona

Professor WOLFRAM R. ULLRICH
Institut für Botanik
Technische Hochschule
Schnittspahnstraße 3
6100 Darmstadt, FRG

Professor CARMELO RIGANO
Dipartimento di Biologia Vegetale
Università di Napoli
Via Foria 223
80139 Napoli, Italy

Professor AMODIO FUGGI
Dipartimento di Biologia Vegetale
Università di Napoli
Via Foria 223
80139 Napoli, Italy

Professor PEDRO J. APARICIO
Centro de Investigaciones Biológicas
CSIC Velazquez 144
28006 Madrid, Spain

ISBN-13:978-3-642-75814-0 e-ISBN-13:978-3-642-75812-6
DOI: 10.1007/978-3-642-75812-6

Softcover reprint of the hardcover 1st edition 1990

2131/3145(3011)-543210 – Printed on acid-free paper

Preface

Three years after the first European Meeting on Inorganic Nitrogen Metabolism in Jarandilla de la Vera, Spain, it was time to meet again and exchange new results and ideas among the scientists involved in this field. The new meeting was held in Seiano di Vico Equense, Italy, in June 1989. It was also primarily European, but again several guests from abroad were invited as speakers. Already during the preparation of this "Second Advanced Course" we saw there had been such a development in research that we arranged with Springer-Verlag to publish the contributions under the cover of a new book. In order to extend the available space without a substantial increase in costs of producing a well-printed volume, we asked the contributors to send us their articles on diskettes. Many kindly responded to this request. We then converted all the corrected texts to the same font for laser printing.

The aim of this book is to give a survey of the progress within the last few years in research on inorganic nitrogen by the main experts. The contributions are in part original papers, but generally they are of a review character, leading the reader to the actual state of development. Thus, within this volume, the reader will find a rather wide spectrum of mainly physiological, biochemical and genetic approaches to "Inorganic Nitrogen Uptake and Metabolism", providing a sort of cross-section of present scientific research.

The main parts cover six topics, starting from some general or basic subjects in Part 1: The history of discoveries in inorganic nitrogen metabolism as elucidated in an interesting personal retrospective by one of the pioneers, Richard Hageman, followed by two articles with recent, and some novel, aspects of energetics in plant metabolism and, less related to the former but of great public interest, a critical review on nitrogen and the quality of human food. Parts 2 and 3 are closer to classical plant physiology, reporting progress in research on uptake and assimilation of inorganic nitrogen. Part 3 deals more particularly with the interface between nitrogen and carbon metabolism. Parts 4 and 5 are the mainly biochemical parts concerning the structure of enzymes and metabolic regulation, on the one hand, and genetics and molecular biology, on the other. The separation of these two parts is indistinct since presently most work on metabolic regulation includes molecular biology and genetics and vice versa. Finally, Part 6 is again dedicated to whole organisms, to storage and partitioning of inorganic nitrogen, ecology, and stress physiology with regard to uptake and metabolism.

Within the individual parts of the volume, the general line is always drawn from higher plants to eukaryotic algae and fungi, and then to prokaryotes. As to the sequence of metabolic steps or compounds, nitrate has been placed in the first position, followed by nitrite, nitric and nitrous oxides, ammonium, and dinitrogen.

It should be mentioned that besides the articles represented in this book there were 76 posters exhibited at the meeting that could not be included in this volume for reasons of space and the arrangements with Springer-Verlag. However, everyone knows the outstanding value these posters often have in contributing new ideas and data.

The "Second Advanced Course on Inorganic Nitrogen Metabolism" took place in the "Moon Valley Hotel" in Seiano di Vico Equense, a beautiful site south of Naples with a full view of the Gulf of Naples and Vesuvio. To Carmelo Rigano and Amodio Fuggi, together with other members of the Department of Plant Physiology of the Università di Napoli, fell the extensive task of organizing the whole meeting, and they chose this highly appropriate and well-equipped place. Also, the stimulating historical environment, with its Greek and Roman tradition, contributed to the excellent atmosphere and success of the meeting. Although antiquity had no idea of nitrogen as an element, Gaius Plinius Secundus, the well-known Roman writer who died during the disastrous eruption of Vesuvio in 79 A.D., reported on long experience with leguminous plants for fertilizing crop fields. In his *Naturalis Historia* (*liber* XVIII) he wrote: "*(Faba) solum in quo sata est laetificat stercoris vice. Ideo circa Macedoniam Thessaliamque, cum florere coepit, vertunt arva. ... Et vicia pinguescunt arva, nec ipsa agricolis operosa. Uno sulco sata non saritur, non stercoratur. ...*" [(Broad bean) improves the soil in which it is sown like a fertilizer. Thus in the area of Macedonia and Thessalia, when it has begun to flower, they turn the soil. ... Also from the vetch the fields become richer, and it does not require much work for the farmers. Once sown in the furrow it is not weeded, not fertilized...]

The "Advanced Course" was held under the patronage of several institutions to which we are highly indebted: Amministrazione Provinciale di Napoli, Assessorato Pubblica Istruzione della Regione Campania, Federation of European Societies of Plant Physiology, and Società Italiana di Fisiologia Vegetale.

The warm welcome by the President of "Amministrazione Provinciale di Napoli", Dr. Salvatore Piccolo, and his appreciation of the meeting and its subject during the opening session were especially appreciated as a sign of the high regard paid by political administrators to scientific research.

The financial support by various Italian and local institutions as well as by the respective national organizations of the participants is gratefully acknowledged. These were: Amministrazione Provinciale di Napoli, Commune di Vico Equense, Consiglio Nazionale delle Ricerche, Dipartimento di Biologia Vegetale, Federation of European Societies of Plant Physiology, Società Italiana di Fisiologia Vegetale, and Università di Napoli. With respect to the national institutions, support from the Consejo Superior de Investigaciones Científicas to the Spanish and from the Deutsche Forschungsgemeinschaft to the German participants for their bilateral contacts is acknowledged among others with gratitude. Without all this financial support a meeting of this size and scientific level would not have been possible.

As to the preparation of this volume, the editors wish to express their special thanks to Helmut Schipp von Branitz, Darmstadt, for his intensive and, indeed, very essential help in preparing manuscripts and diskettes for final publication.

Thanks are finally due to Springer-Verlag for accepting the task of the final printing and publishing of the volume with their usual high quality standard. May this book serve as a source of information on new data and recent ideas to all those who are involved in research in the field of nitrogen metabolism and to those who need and want an overview in its recent development.

Wolfram R. Ullrich
Carmelo Rigano
Amodio Fuggi
Pedro J. Aparicio

Contents

Part 1 History and General Topics

Part 2 Transport and Assimilation of Inorganic Nitrogen

Part 6 Distribution of Storage and Metabolism, Environmental Stress

List of Contributors

Addresses available at the beginning of the respective contributions

Part 1 History and General Topics

HISTORICAL PERSPECTIVES OF THE ENZYMES OF NITRATE ASSIMILATION BY CROP PLANTS AND POTENTIAL FOR BIOTECHNOLOGICAL APPLICATION

R.H. Hageman [1]

Nitrate Reductase

Early history. Early (1880-1948) literature reviews report on the effect of NO_3^- supply, experimental manipulations and environment, especially light, on levels of NO_3^-, NO_2^- and organic-N in various organs of several plant species. As early as 1904 it was noted that cell free extracts of potato tubers reduced NO_3^-. Subsequently, similar observations were made with other plant species. By 1952, knowledge was emerging on the characteristics of nitrate reductase (NR) extracted from *E. coli* and *Neurospora* (McElroy and Glass 1956). In 1953, Evans and Nason extracted and partially purified a NR from the primary leaves of 8-day-old soybean plants. In order of decreasing efficiency, the enzyme used reduced tri- or diphosphopyridine nucleotide (TPNH or DPNH) or flavin adenine dinucleotide ($FADH_2$) as electron donors. With TPNH, the NR reduced NO_3^- to NO_2^-, stiochiometrically. The optimum pH for assay was 6.5. Sulfhydryl group(s), FAD and at least one metal were essential components. *In vitro*, illuminated grana could reduce TPN for NO_3^- reduction. At that time, this provided an explanation for the role of light in NO_3^- assimilation in green leaves. In a survey of seven plant species, low levels of NR activity (1 to 7 nmol mg^{-1} dry weight h^{-1}) were found in crude extracts of shoots and roots. For four of these species, the enzyme could utilize either TPNH or DPNH. For wheat, corn and tomato, NR was found in the roots but not the shoots. With 17-day-old soybean seedlings, the primary leaves had no NR activity and activity of other leaves was very low.

Developments. Subsequently, numerous studies were made on NR of higher plants. However, other investigators as well as Evans and Nason (1953) were unable to detect or found only traces of NR and many other enzymes in extracts of mature leaves of field grown crop plants, especially cereals. Waygood found that additions of casein hydrolysate to the extraction buffers was superior to an "ammonia extraction" of triosephosphate dehydrogenase (TPD) from wheat. However, the results with casein were variable. Shortly thereafter, Hageman and Waygood (1959) noted that addition of cysteine (0.01 M) to the extraction media permitted the successful extraction of TPD, NR and 10 other enzymes, from mature leaves of cereals. The amount of cys-

[1] Professor, Emeritus, Department of Agronomy, 1102 South Goodwin Ave., University of Illinois, Urbana, Illinois 61801, USA

teine required for extraction of optimum activity varied with enzyme, plant species, tissue, plant age, and mode of cell disruption. With cysteine as a protectant, activities of NR as high as 27 and 17 μmol g^{-1} fresh weight h^{-1} were found by my students in extracts of leaves of seedling and mature corn plants. Although Streeter and Bosler (1972) found dithiothreitol (5mM) was superior (25%) to cysteine as a protectant of NR during extraction and assay of mature soybean leaves, NR activities were low (2 - 6 μmol g^{-1} fresh weight h^{-1}). When similar leaf samples were assayed by the *in vivo* procedure, NR activities were 5-fold greater than for the *in vitro* assay. Later work in my laboratory showed that when FAD, in addition to cysteine, was added to the extraction medium, NR activities of 16 and 17 μmol g^{-1} fresh weight h^{-1} for DPNH and TPNH, respectively, were obtained from the leaves of young soybean seedlings. Additional improvements (freezing the tissue with liquid N_2 before extraction with a medium containing 1 mM cysteine and 25 mM FAD and stopping the reaction by boiling and addition of phenazine methosulfate) improved recoveries of NR. *In vitro* activities as high as 47 μmol g^{-1} fresh weight h^{-1} for NAD-NR were obtained from laves of 40-day-old soybean plants. These *in vitro* activities were higher (up to 2-fold) than *in vivo* activities with similar tissue.

By 1974, it had been shown that NR extracted from many plant species had a pH optimum of 7.5 and a strong preference, but not an absolute specificity, for NADH. However, other workers had confirmed the initial report of Evans and Nason (1953) that NR from soybean had a pH optimum of 6.25 - 6.5, and utilized either NADH or NADPH (Beevers and Hageman 1969, 1980). This anomaly was clarified by Jolly et al. (1976) by partially separating a NAD-NR and a NADP-NR from extracts of wild type soybean leaves. Harper and coworkers identified a mutant soybean plant, designated nr_1, that was devoid of NADPH-NR. The mutant nr_1 type had no detectable NR when the plants were grown on urea. Upon the addition of NO_3^- to the growing medium, an inducible NAD-NR with a pH optimum of 7.5 could be extracted from the nr_1 plants. The wild type plants, when grown on a medium devoid of NO_3^- had a "constitutive" NADP-NR with a pH optimum of 6.5. Addition of NO_3^- to the growing medium caused, in time, a near doubling of total (constitutive and inducible) activity. The constitutive NR from urea grown wild type plants was separated into two forms that had the following characteristics: (a) NADP-NR; pH 6.5; K_m values for NO_3^- 5.0 mM, for NADPH 2.9 μM, for NADH 11 μM. (b) NAD-NR; pH 6.5; K_m values for NO_3^- 0.19 mM, for NADPH 2.60 μM, for NADH 2.9 μM. The inducible NAD-NR from nr_1 plants grown on NO_3^- had a pH optimum of 7.5 and K_m values for NO_3^- 0.13 mM, for NADPH 100 μM, and for NADH 2.5 μM. Thus the inducible NR of soybean is similar to the inducible NR most commonly found in other crop plants (Harper 1987).

Harper also found that leaves of wild-type soybean plants grown on urea (devoid of inducible NR) would evolve NO from NO_3^- during the anaerobic *in vivo* assay. In contrast, leaves of nr_1 plants grown on NO_3^- did not. Preliminary field trials indicated that nr_1 and wild-type plants were similar in appearance, growth and grain yield.

These observations lead me to speculate that the constitutive NR is a dissimilatory type enzyme that utilizes NO_3^- as electron acceptor. The metabolic function of the constitutive NR is not clear as it does not appear to be critical for assimilation of NO_3^-.

Component parts and functions. The component parts, electron donors and acceptors, inhibitors, path of electron flow and the multiple functions of NR of higher plants have been diagrammed and discussed by Beevers and Hageman (1980).

Nitrite Reductase

Early history. By 1894, it was proposed that the assimilation of NO_2^- proceeded by three successive two-electron-reductive-steps that produced in sequence hyponitrous acid, hydroxylamine and ammonia. Later studies found nitrite and hydroxylamine reductases in cell free extracts of *Bacillus, Neurospora* and soybean. Neither, hyponitrous acid or its reductase were identified and the possibility that hyponitrous acid was never released from nitrite reductase (NIR) was considered (McElroy and Glass 1956). For higher plants, the early reports of reduction of NO_2^- and NH_2OH involving pyridine nucleotides as the electron donor were not conclusive and not verified by subsequent work (Beevers and Hageman 1969).

Developments. New insight into NO_2^- reduction by cell-free extracts of higher plants was gained in 1962 (Hageman et al. 1962). In initial work, benzyl viologen, reduced chemically or enzymatically by NADPH, was used as the electron donor and NO_2^- was reduced stoichiometrically to ammonia. Crude extracts from plants were able to reduce both NO_2^- and NH_2OH to NH_3, however the latter was converted more slowly. These findings and later studies (Hewitt et al. 1968) indicated that NH_2OH was not a free intermediate. The view that a single enzyme catalyzed the reduction of NO_2^- to NH_3 without release of free intermediates was consistent with the conclusions reached with NiR extracted from *E. coli* (Lazzarini and Atkinson 1961). However, the *E. coli* enzyme utilized NADPH rather than the single electron donor, benzylviologen.

The identification of the natural electron donor for the plant NiR followed in logical sequence. First, it was shown that NO_2^- was reduced by illuminated spinach chloroplasts supplemented with extra plant extract. Second, it was suggested that the electron carrier of the photochemical system might be the electron acceptor of the newly identified photosynthetic pyridine nucleotide reductase. Third, ferredoxin (Fd) was shown to be this carrier. By 1963, several groups had demonstrated that reduced Fd was the natural electron donor for NiR (Beevers and Hageman 1969). Reduced

NADPH did not serve as a direct electron donor, however it could reduce Fd in the presence of an appropriate reductase.

With *in vitro* illuminated chloroplast systems, 1.0 mol NH_3, 1.5 mol 0_2 and 3 mol ATP were produced for every mol of NO_2^- reduced (Losada et al. 1963). In 1966, Ritenour et al. used non-aqueous extraction techniques to demonstrate that NiR was localized in the chloroplast while NR was primarily a cytoplasmic enzyme. Collectively these studies provide an explanation for a positive role of light in NO_3^- and NO_2^- reduction. Competition between NiR and Fd-NADP reductase for light energy has been demonstrated (Hewitt et al. 1968). Although NADP reduction was favored over NO_2^- reduction, the effects of competition for light energy by sugar and amino acid production by crop plants under field conditions is not clear.

By 1976, NiR, from photosynthetic tissues of diverse species, was purified and characterized (Hewitt et al. 1968; Beevers and Hageman 1980). The identification of a siroheme molecule (Murphy et al. 1974) and an iron-sulfur [Fe_xS_x] molecule (Aparicio et al. 1975; Hucklesby et al. 1976) as constituent parts of NiR, helped clarify its mode of action. Ida and Mikami (1986) contend that the iron-sulfur is a tetranuclear cluster. In roots, NiR has been localized in plastids. The root also contains a Fd (similar to the leaf Fd) which is reduced by NADPH and an appropriate root reductase (Oaks and Hirel 1985).

Component parts and functions. The component parts, electron donors and acceptors and proposed path of electron flow have been detailed (Beevers and Hageman 1980). The siroheme operates in conjunction with the Fe_x-S_x center. A suggested sequence in the reduction of NO_2^- is as follows: The siroheme, reduced by an electron from Fd binds NO_2^- to form a NO_2-Fe complex. Formation of this complex facilitates the reduction of the Fe_x-S_x center. Electrons may then be transferred from the Fe_x-S_x group to the NO_2-siroheme complex producing a NO-siroheme complex. The sequence of addition of the additional electrons and hydrogen, required to produce NH_3 and the structure of the bound intermediates is not known.

Glutamate Dehydrogenase

About 80 years ago, Knoop postulated the reductive amination of keto acids by NH_3 for the pathway of amino acid synthesis. Subsequent work has shown that glutamate is the major if not the only amino acid so synthesized. The occurrence of glutamate dehydrogenase (GDH), that catalyzes this reaction has been demonstrated in many plant species (Stewart et al. 1980). A GDH isolated from the chloroplasts has a preference but not an absolute requirement for NADPH as the electron donor. In contrast the mitochondrial isoform of the enzyme prefers NADH over NADPH. By the

1950's, the view was that GDH in conjunction with an array of transaminases was the mode of entry of NH_3 into the amino acids. Because of the high K_m of GDH for ammonia and other reasons (Miflin and Lea 1980) the current view holds that the major metabolic role for this enzyme is deamination.

Glutamine Synthetase

Enzymatic formation of glutamine from glutamate and NH_3 was first observed in 1935 with animal tissue. Subsequent studies with extracts of bacteria, algae and higher plants established the need for ATP and metallic cofactors (Stewart et al. 1980). The enzyme can also utilize hydroxylamine to form γ-glutamylhydroxamate and catalyze γ-glutamyl transfer reactions.

Despite its widespread distribution, great activity and low K_m for ammonia, glutamine synthesis was not accorded a major role in the assimilation of ammonia prior to 1970 (Miflin and Lea 1980). This was true because no enzyme was known that would transfer the amide N to the α-amino position to form amino acids. In 1970, Tempest et al. identified such an enzyme, glutamate synthase. This enzyme used the energy of NADPH or NADH to transfer the amide-N of glutamine to 2-oxoglutarate to yield two glutamate molecules. Later a NAD(P)H-glutamate synthase was found in non-green plant tissue and a Fd dependent enzyme in green leaves (Miflin and Lea 1980). The coupling of these two enzyme systems is now considered to be the primary pathway for the incorporation of NH_3 into organic N.

Nitrogen Metabolism in Maize: Basis for Biotechnology

The importance of N supply for achieving larger yield of grain in the US Corn Belt can be illustrated as follows: (a) Based upon N content of shoot, root and grain, the production of 9434 kg grain ha^{-1} requires the accumulation of 221 kg N ha^{-1}. (b) The omission of the annual application of fertilizer N causes drastic reductions in yield. The annual application of N is estimated by a "rule of thumb" as 1.4 kg N ha^{-1} for every unit (63 kg grain ha^{-1}) of yield expected. (c) By the 1950's the US farmers had recognized the economic importance of N in the production of maize. For example, in Illinois, grain yields over the past 35 years are closely associated with increased application of fertilizer N. The percentage of grain protein (N) did not change during this period. These observations and the fact that under field conditions suitable for growth, the maize plant uses NO_3^- as its major source of N, indicates that knowledge about NO_3^- metabolism could be important to improvement of productivity.

By the 1960's my students had established that: NR was substrate inducible; NR had a short half-life *in situ* and NR activity usually disappeared upon depletion of NO_3^- from the tissue or organism; NR activity varied widely among genotypes; NR was heritable; 10 inbred-45 hybrid half diallels showed the specific and general combining abilities were significant and the selections for NR would be effective in a synthetic cultivar made from these 10 inbreds; NR was subject to manipulation by classical breeding techniques and in most but not all instances, NR activity of maize hybrids was the mid-parent mean; based on the absence of intermediate products, NR was considered the rate limiting step between NO_3^- and amino acids; and suitable methods, although more complex and time consuming than desired, had been developed for assay.

Thus NR met the requirements for a physiological trait useful for selection of superior genotypes with the exception that a close association between NR activity and yield had not been established. The first attempt to establish an association between specific seasonal NR activity and yield was made in 1960 with four hybrids. Based on field yield trials, one hybrid consistently produced more grain than the other two. The fourth hybrid was selected because it had been specifically bred for high yield. When tested under field conditions in 1960 and 1961 both of the "high" yielding hybrids had significantly higher specific *in vitro* NR activity than the two "low" yielding hybrids. However, for the four hybrids there was no significant difference in yields. Numerous other experiments with maize and wheat made with variations in sampling, assay and mode of expression of activity also failed to establish an association between NR activity and yield among diverse genotypes.

Subsequently, it was shown that for a given genotype, NR activity of the whole plant, integrated over time, was closely associated with the actual amount of N accumulated by the plant. However, the view that such measurements of NR activity would provide an estimate of the input of reduced N to the plant was not valid among diverse genotypes. Among genotypes, the *in situ* accumulation of organic-N could also be affected by the NO_3^- uptake and transport characteristics that in turn could alter the amount of NO_3^- at the induction and assimilation sites. The availability of reductant and metabolites for regeneration of reductant, the half-life of NR and messenger RNA, etc. could also affect the relationship.

An attempt was made to minimize this genetic diversity in N characteristics by selecting for divergence in the level of NR in plant material derived from a common parent. By the early 1970's an *in vivo* NR assay had been developed that was well adapted for large scale selection of maize plants having divergent levels of NR activity (NRA). In 1974, a divergent phenotypic recurrent selection program was initiated with *in vivo* NR assay of leaf tissue at 10 and 20 days after anthesis as the only selection trait. The objectives were to select plants from a common genetic background that differed in level of NRA and to evaluate the effect of selection on dry matter and N accumulation and grain yield of these plants. A super stiff-stalk synthetic maize was used as the parental material. This synthetic was selected because it had been the source of several superior inbreds. For each individual selection class, 200 selfed plants were sampled. Leaf samples were taken at 10 and 20 days after anthesis for NRA. The selection intensity was 20%.

Selection cycles 0, 2, 4 and 6 for both high and low NRA strains were grown under field conditions for evaluation in 1982. The results are summarized as follows: (a) selection for NRA was significant and continuous; cycle 6 high had 59% more and cycle 6 low had 38% less NRA than cycle 0; (b) with the exception of grain of cycle 6 low, selection had no marked effect on stover and grain dry weight and organic-N content; (c) the grain yield and grain N content of cycle 6 low was significantly lower than for cycle 0 or cycle 6 high; (d) selection for high NRA was associated with a more rapid loss of chlorophyll from the leaves (increased rate of senescence); and (e) selection for high NRA was associated with an increase in specific leaf weight relative to the low selection on cycle 0.

The program was continued through eight cycles of selection which were evaluated for direct and correlative responses. The divergence in NRA between cycle eight high and low continued but at a slower rate than observed for the earlier cycles of selection. For these experiments the plants were grown at three N fertility rates, even though the selections were all made at the lowest N rate (112 kg N ha^{-1}). Although, in some instances, these data indicated an effect of N fertility on selection, for all experiments the NRA response was largely unaffected by N environment. Relative to cycle 0 the high NRA strains showed a small but consistent decrease in grain yields (average of three years of testing) at all fertilizer rates. These yield decreases could be accounted for by inbreeding effects. The low NRA strains had a large correlated yield decrease that could not be accounted for by inbreeding. NRA levels at or lower than those of cycle 6 low were associated with significant decreases in grain yield. Increased N fertility increased grain yield of the low NRA strain (especially cycle 6), more than of the high NRA strain and control. The increase in yield of the low NRA strain with increasing N fertility resulted from an increase in kernel number per plant. The results for grain N content were similar to that of grain yield. Little change was observed in either the high or low NRA strain in dry weight or organic-N content of stover at anthesis, regardless of N fertility level.

In summary, the plants selected for high *in vivo* NRA (divergence in *in vivo* NRA between the high and low NRA strains was confirmed by *in vitro* assays) accumulated slightly less organic-N and dry matter in stover and grain with each successive cycle of selection. Thus, the increase in NRA did not result in an effective increase in input of N to the plant relative to plants with common genetic background having lower NRA. The reason for the lack of relationship between NRA and organic-N accumulation is not known. However, the availability of NO_3^- at the induction and assimilation sites, availability of electron donor(s), presence of regulators of enzyme activity or half-life of NR and the appropriate m-RNA are some factors that could affect the input of N to the plant. Full details in Sherrard et al. (1986) and Hageman and Lambert (1988).

In general, knowledge about the effect of individual enzymes and metabolic systems on growth and yield are lacking. It is most difficult to identify one component or one system that is closely correlated to yield. These same restrictions apply to the effective use of gene insertion that directly or indirectly increases the amount of a given enzyme, unless this increased amount of enzyme increases a desired end product.

It is highly probable that genetic engineering can be used to insert an isozyme that differs in electron donor requirement, pH optimum, K_m for substrate or cofactor or response to various regulators. For example, the insertion into maize of the gene that codes for the NADP-NR would seem desirable as this could possibly link NO_3^- reduction to light generated energy. However, as mentioned in the chapter on NR the existence of both NADP- and NAD-NR in wild type soybean did not appear to be beneficial to growth or productivity, relative to the NAD-NR type plant. Because NADP-NR has been shown to occur in corn scutella of many cultivars but not in the leaves, a problem of expression of the inserted gene could be incountered. In soybean, the NADP-NR activity decreases with tissue age. If this instability also occurs in corn, it is possible that NADP-NR would be absent during grain fill. Even though the gene insertion was successful, NO_3^- supply, metabolic regulators, etc. could limit actual input of N to the plant. Assuming the input of N is actually increased, the effect on grain yield is not clear as the plant may regulate N input in accordance with availability of carbon skeletons and energy.

Expansion of the basic knowledge of the interaction of the various metabolic systems with growth and productivity appears to be the most pragmatic approach to these complex problems. The plant breeders, using end products as the selection indices, have been most successful in producing better genotypes. While this system is successful because it integrates genetically affected metabolic interactions as modified by environment, it provides little basic knowledge of cause and effect. Cooperative studies by breeders, genetic engineers and biochemists of these complex problems affecting crop yield provides a logical approach to improvement of productivity of crop plants.

References *

Apparicio PJ, Knaff DB, Malkin R (1975) The role of an iron-sulfur center and siroheme in spinach nitrite reductase. Arch Biochem Biophys 169: 102-107

Beevers L, Hageman RH (1969) Nitrate reduction in higher plants. Annu Rev Plant Physiol 20: 495-522

Beevers L, Hageman RH (1980) Nitrate and nitrite reduction. In: Miflin BJ (ed) The Biochemistry of Plants, Vol 5. Amino Acids and Derivities. Academic Press, New York, pp 115-168

* Editorial constraints limited the number of references

Evans HJ, Nason A (1953) Pyridine nucleotide nitrate reductase from extracts of higher plants. Plant Physiol 28: 233-254

Hageman RH, Waygood ER (1959) Methods for the extraction of enzymes from cereal leaves with especial reference to the triosephosphate dehydrogenases. Plant Physiol 34: 396-400

Hageman RH, Cresswell CF, Hewitt EJ (1962) Reduction of nitrate, nitrite and hydroxylamine to ammonia by enzymes extracted from higher plants. Nature 193: 241-50

Hageman RH, Lambert RJ (1988) The use of physiological traits for corn improvement. In: Sprague GF, Dudley J (eds) Corn and Corn Improvement. Agronomy Monograph No. 18, Madison WI, pp 438-461

Harper J (1987) Nitrogen metabolism. In : Wilcox JR (ed) Soybeans: Improvement, Production and Uses. ASA Monograph, Madison WI, Amer Soc Agron Monograph No. 16, Madison WI, pp 497-533

Hewitt EJ, Hucklesby DP, Betts GF (1968) Nitrogen assimilation in plants. In: Hewitt EJ, Cutting CV (eds) Recent Aspects of Nitrogen Metabolism in Plants. Academic Press New York, pp 20-40

Hucklesby DP, James DM, Bannell MR, Hewitt ET (1976) Properties of nitrite reductase from *Cucurbita pepo*. Phytochemistry 15: 599-603

Ida S, Mikami B (1986) Spinach ferredoxin-nitrite reductase: A purification procedure and characterization of chemical properties.Biochim Biophys Acta 871: 167-176

Jolly SO, Campbell W, Tolbert NE (1976) NADPH- and NADH-nitrate reductase from soybean leaves. Arch Biochem Biophys 173: 431-439

Lazzarini RA, Atkinson DE (1961) A triphosphopyridine nucleotide specific nitrite reductase from *Escherichia coli*. J Biol Chem 236: 3330-3335

Losada M, Paneque A, Ramírez JM, Del Campo FF (1963) Mechanism of nitrite reduction in chloroplasts. Biochem Biophys Res Commun 10: 298-303

McElroy WD, Glass B (1956) Inorganic Nitrogen Metabolism. John Hopkins Press, Baltimore, pp 1-728

Miflin BJ, Lea PJ (1980) Ammonia assimilation. In: Miflin BJ (ed) The Biochemistry of Plants, vol 5, Academic Press New York, pp 169-202

Murphy JF, Siegel LM, Tove SR, Kamin H (1974) Siroheme: A new prosthetic group participating in six-electron reduction reactions catalyzed by both sulfite and nitrite reductases. Proc Natl Acad Sci USA 70: 2386-2390

Oaks A, Hirel B (1985) Nitrogen metabolism in roots. Annu Rev Plant Physiol 36: 345-365

Ritenour GL, Joy KW, Bunning J, Hageman RH (1966) Intracellular localization of nitrate reductase, nitrite reductase, and glutamic acid dehydrogenase in green leaf tissue. Plant Physiol 42: 233-237

Sherrard JH, Lambert RJ, Below F, Dunand RT, Messmer MJ, Willman MR, Winkles CS, Hageman RH (1986) Use of physiological traits, especially those of nitrogen metabolism for selection in maize. In: Neyra CA (ed) Biochemical Basis of Plant Breeding. Nitrogen Metabolism. CRC Press, Boca Raton FL, pp 110-130

Stewart GR, Mann AF, Fentem PA (1980) Enzymes of glutamate formation: Glutamate dehydrogenase, glutamine synthetase and glutamate synthase. In: Miflin BJ (ed)The Biochemistry of Plants, vol 5, Academic Press New York, pp 271-327

Streeter JG, Basler ME (1972) Comparison of *in vitro* and *in vivo* assays for nitrate reductase in soybean leaves. Plant Physiol 49: 448-450

Tempest DW, Meers JL, Brown CM (1970) Synthesis of glutamate in *Aerobacter aerogenes* by a hitherto unknown route. Biochem J 117: 405-407

LIGHT ENERGY UTILIZATION IN OXYGENIC PHOTOSYNTHESIS

G. Forti [1]

Introduction

In oxygenic photosynthesis of green plants and cyanobacteria, the photochemical process produces a reductant of E_m = -440 mV, ferredoxin. Reduced ferredoxin is used to reduce NADP which is utilized in CO_2 assimilation to the level of carbohydrate, or to reduce nitrite directly within the chloroplast, or to reduce dinitrogen to the level of NH_2- in nitrogen-fixing photosynthetic procaryotes.

The object of this discussion is the photochemical process of oxygenic photosynthesis, i.e. the sequence of events and the mechanism of electron transport from water to NADP, and the synthesis of ATP coupled to it through the catalysis of the thylakoid bound ATP synthase. ATP is known to be a necessary reagent for the assimilation of CO_2 and for N_2 fixation, whilst it is not required for nitrite reduction.

The complete photosynthetic process occurs within the chloroplast: light is absorbed by the photosynthetic pigments which are bound to proteins in the thylakoids which also contain the electron carriers involved in the process and the ATP synthase. All the enzymes involved in carbon metabolism are present as a dense protein solution in the stroma of the chloroplast, as is also nitrite reductase. In cyanobacteria the photosynthetic membrane system occupies a large part of the cell volume.

Photochemical Reactions and Electron Transport in Oxygenic Photosynthesis

The electron transport system from H_2O to NADP requires two photochemical reactions in series connected by an array of carriers which includes quinones, cytochromes, iron-sulfur centres and a copper protein, plastocyanin.

The scheme was originally proposed by Hill and Bendall (Hill and Bendall 1960), and has received since then an enormous amount of experimental evidence by a large variety of methods and approaches, which it would be impossible to review here (see recent reviews by Ort 1986, and by Forti 1987).

The theory of Hill on photosynthetic electron transport gives a satisfactory explanation of the earlier observations of Emerson and his associates on the drop of photosynthetic efficiency at the red end of the chlorophyll absorption spectrum and the

[1] Centro CNR Biologia Cellulare e Molecolare delle Piante, Dipartimento di Biologia, Università di Milano, Via Celoria 26, 20133 Milano, Italy

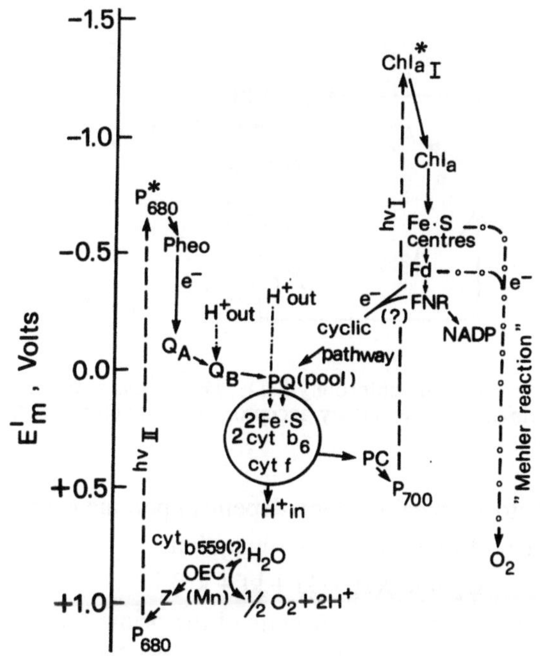

Fig 1. Scheme of electron transport in oxygenic photosynthesis. The arrows indicate the direction of electron transport, —>; proton transport, --->; electron transport to O_2, -o-o->(Mehler reaction)

superadditive effect of lower wavelenght light added to far-red light. A recent version of the Hill scheme is presented in Figure 1. The overall gain in electrochemical potential, in the transfer of two electrons from H_2O to ferredoxin is of 1.25 electronvolts. In addition, 1 or maybe 1.2 moles of ATP are formed in the process per two electrons transferred. The overall reaction may be represented by the equation:

$$H_2O + NADP + P_i + nh\nu \rightarrow NADPH + 1/2O_2 + ATP \qquad (1)$$

The free energy change of the chemical components of the system is approximately 60 kcal/mol, if the relevant thermodynamic parameters are taken at their standard value.

The two photochemical reactions are catalysed by two distinct photosystems, photosystem 1 (PS1) and photosystem 2 (PS2), each endowed with a reaction centre chlorophyll *a* molecule. This initiates the electron transport sequence when its excited state donates an electron to a primary acceptor, and then the oxidized reaction centre is reduced by a primary donor, according to the scheme presented in Figure 2.

Each reaction centre can collect the excitation energy from an array of "light harvesting" pigment molecules (approximately 300, but the actual size of this "antenna" may vary according to the growth conditions of the plant, and is certainly different in different plants). The photochemical oxidation of water is catalysed by PS2 through the action of an enzyme complex containing manganese, while ferredoxin is reduced by PS1.

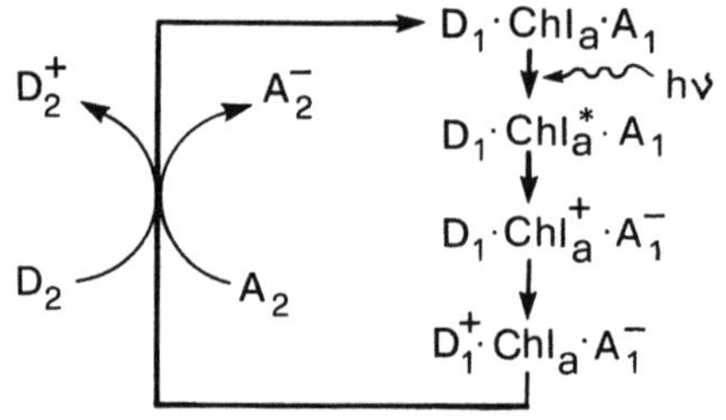

Fig. 2. Photochemical reactions of chlorophyll. D_1, D_2: primary donor and secondary donor, respectively; A_1, A_2: primary and secondary acceptor, respectively; Chla*, reaction centre chlorophyll in the excited state

The rate of the photochemical reaction depends upon the probability, ψp, that the excited reaction centre performs the charge separation

$$\psi p = kp/(kp + kt + kf + kd) \qquad (2)$$

where kp, kt, kf and kd are the rate constants of the primary photochemical reaction, the back transfer of excitation energy to the antenna, the fluorescence emission and the non-radiative dissipation (thermal decay), respectively. In other words, the probability of performing photochemistry depends on the efficiency of the photochemical reaction in competing with the other processes that use the same excitons. Kp is very large, of the order of 10^{12} s^{-1}, while the constants of the competing processes are one to two orders of magnitude smaller. This is why primary photochemistry is more than 95% efficient. The "quantum yield" of the photochemical reactions of photosynthesis, defined as the ratio electrons transferred/quanta absorbed, is obviously dependent on the time needed to regenerate the reaction centres in the "open" state, i.e. in the state where the acceptor is oxidized and the donor is reduced: this is a requirement for the forthcoming photon to be utilized. So, the quantum yield of photochemistry can be expressed as

$$\varphi p = kp\, P/(kp + kf + kd) \qquad (3)$$

where P is the fraction of centres open. In the same manner one can define the fluorescence yield as

$$\varphi f = kf/(kf + kd + kp\, P). \qquad (4)$$

The value of f is defined as F_o when P = 1, and increases to F_m when P = 0. As defined above, kp includes the probability of energy transfer from the antenna to the reaction centre,

$$\psi T = kT/(kT + kd + kf) \qquad (5)$$

where kT is the rate constant for energy transfer to the reaction centre.

The photochemical reaction catalysed by P_{680} (the reaction centre of PS2) reduces a molecule of plastoquinone (Q_a) strongly bound to the reaction centre protein, through the intermediation of a molecule of pheophytin also bound to this protein. Electron transfer from the excited P_{680} to pheophytin occurs in the picosecond time scale (Klimov et al. 1977), and Q_a is reduced in less than a nanosecond (see reviews by Ort 1986; Forti 1987; Mathis and Rutherford 1987). At the oxidizing side of PS2 a tyrosine residue of the reaction centre protein D1 reduces P_{680}^+ in 20 - 40 nanoseconds. The oxidized tyrosine (Z in many published schemes) oxidizes the oxygen evolving enzyme complex (OEC) containing manganese. When the latter has accumulated 4 oxidizing equivalents, O_2 is evolved from two molecules of H_2O and 4 H^+ are released into the lumen of the thylakoids (see reviews by Witt et al. 1986; Renger 1987). OEC is therefore the "gate" of the transition from the one quantum-one electron primary photochemical process to the 4 electron-H_2O oxidation reaction as shown by Joliot et al. (1969). Q_a^- is reoxidized by another molecule of PQ bound in the oxidized form (Q_b) to the protein D2 (previously indicated as "B"). Q_b is reduced in two steps by Q_a to the quinol state, and becomes protonated on the outer surface of the membrane, taking up protons from the external phase. Protonation leads to a large decrease of the binding constant, so that PQH_2 is released in diffusible form in the lipid phase of the membrane. PQH_2 is reoxidized by the cytochrome f-b_6-FeS complex (see reviews by Ort 1986, Forti 1987), and protons are released in the lumen of the thylakoids owing to the topology of the complex in the membrane. Increasing evidence indicates that the reoxidation of the plastoquinol may be accomplished through a Q_b cycle (Ort 1986, Rich 1988). Reoxidation of cyt *f* by PS1 is mediated by the copper protein plastocyanin, which is in solution in the internal water phase and binds reversibly to a PS1 reaction centre protein. P_{700} (the reaction centre chlorophyll *a* of P S1) is reduced by plastocyanin in ca. 12 - 18 microseconds. On the oxidizing side of PS1 a series of three acceptors including iron-sulfur centres and a phylloquinone are reduced in a very fast reaction (see Mathis and Rutherford 1987). Ferredoxin, present in solution in the outer phase, is the next electron acceptor. It binds then reversibly to a flavoprotein (ferredoxin-NADP reductase, FNR) located on the outer surface of the membrane (Forti 1987). The flavoprotein acts as the "one to two electron gate" on the reducing side of PS1: NADP and ferredoxin are bound to it in a ternary complex and NADP is reduced in a two step process, though the detailed mechanism of the reaction is not known.

A cyclic electron transport system has been shown to operate around PS1 and to be coupled to the synthesis of ATP both *in vivo* and *in vitro* (see review by Forti 1987). Cyclic electron transport has been shown to involve FNR, ferredoxin and cyt b_{563}. It seems therefore that the cyclic pathway utilizes the cyt f-b_6 complex in competition with the Q_b cycle.

Another pathway for the reoxidation of the primary acceptor of PS1 is the transfer of electrons to O_2 ("Mehler reaction") with the superoxide anion and H_2O_2 as intermediates. This process is coupled to ATP synthesis and this is its physiological role (see review by Forti 1987).

Photophosphorylation

The primary photochemical reactions of PS1 and PS2 give rise to the separation of charges across the thylakoid membrane, owing to the fact that the primary electron acceptors are located close to the outer surface of the membrane, whilst the positive charge of the oxidized reaction centres and the primary donors are close to the internal (lumenal) surface (see review by Witt 1979). A large membrane potential (monitored by the electrochromic shift of the pigments) is generated in less than one nanosecond, negative outside, which decreases to a low steady state value of 10 - 50 mV (depending on the ionic composition of the medium) because Cl^- diffuses into the lumen through a specific channel (Schönknecht et al. 1988). The protons released inside the lumen by the oxidation of water and of PQH_2 generate a pH difference across the thylakoid membrane (acidic inside) of up to 3 - 3.5 units, which is maintained in steady-state electron transport due to the extremely low proton conductance of the membrane. Thus, ΔpH and $\Delta\psi$ are additive in generating an electrochemical proton potential, $\Delta\mu H^+$, which is the source of free energy driving the synthesis of ATP according to Mitchell's chemiosmotic theory. The synthesis of ATP is performed by the CF_0-CF_1 ATP synthase enzyme complex, which spans the membrane with the CF_0 component and binds ADP and P_i on the CF_1 component. ATP synthesis occurs according to the equation

$$ADP + P_i + nH^+_{in} \rightarrow ATP + H_2O + nH^+_{out}\,.$$

The synthesis of ATP occurs when the sum of the free energy changes for ATP synthesis and the coupled proton extrusion has a negative value; ATP is hydrolysed when the reverse is true, provided the enzyme is in the active state. However, it has been demonstrated that the free energy of activation of the synthase is larger than the free energy for ATP synthesis: as a consequence, the enzyme is active only when ATP synthesis occurs and not its hydrolysis (see review by Schlodder et al. 1982). The active state of the enzyme can be stabilized by -SH groups, and under this condition ATP synthase can hydrolyse ATP.

The stoichiometry of ATP synthesis and proton extrusion is still under debate: obviously the precise assessment of the ratio H^+/ATP is important, together with the ratio of protons transferred into the lumen to electrons transported, to assess the quantum requirement of photosynthesis on the basis of the well-known ratio of 3 molecules of ATP and 2 of NADPH per CO_2 assimilated. Other minor ATP requirements are possible, to carry on metabolic processes necessary to maintain the chloroplast in the condition of activity. As a consequence, if the ratio ATP/NADPH produced by the linear electron transport is lower than 1.5, the activity of cyclic electron transport and/or the Mehler reaction are needed to provide the extra ATP stoichiometrically required for CO_2 assimilation. The highest ratios ATP/NADPH (or ATP/2 electrons) reported in the literature are 1.2 - 1.3 for linear electron transport, the value of 1 being more frequently observed. This means that extra photons have to be used to operate cyclic photophosphorylation or the Mehler reaction. *In vivo* the quantum requirement for photosynthesis seems to be not less than 9.5 - 10 quanta/O_2 evolved, and in isolated intact chloroplasts 11 - 12 quanta/O_2 have been observed (see review by Forti 1987).

Regulation of Light Energy Flow to the Two Photosystems

Optimal utilization of light quanta in the linear electron transport system through the two photoreactions in series requires that the two photosystems perform photochemistry at the same rate. In principle, they should absorb the same flux of photons. On the other hand, analysis of the chlorophyll-protein complexes and their distribution in granal chloroplasts reveal that about 60% of the pigments (chlorophylls *a+b* and the carotenoids) are connected with PS2 in the appressed regions of the membranes, whilst only 40% belong to PS1 (Anderson 1980) which is considered to be concentrated in the stroma-exposed membranes: the margins and end membranes of grana and the stromatic thylakoids connecting the grana to each other. The presence of relevant amounts of PS1 in the grana partitions is still under debate (see review by Forti 1987).

The unequal optical cross section of the two photosystems should be compensated by some kind of regulatory mechanism for the distribution of excitation energy between the two photosystems in order to account for some observations reported on quantum yield of linear electron transport well over 0.8 obtained when plots of the rate of NADP reduction/hv absorbed were extrapolated to zero photon flux (Sun and Sauer 1971). More recently however, measurements of quantum yields of PS1 in granal thylakoid preparations under conditions of linear dependence of the electron transport rate on light intensity indicated lower values of the quantum yield (Forti and Fusi, unpublished results), in agreement with previous results (Schwartz 1967). Two main mechanisms have been envisaged for the regulation of excitation energy distribution between the two photosystems: (a) The transfer to PS1 of energy from the LHCII-PS2 matrix (considered as a homogeneous antenna in the grana because all evidence indicates that all photons absorbed by LHCII can be transferred to P_{680} with the same efficiency as those absorbed by the PS2 core antenna, provided that the Mg^{2+} ion concentration is not below ca. 0.5 mM, a situation encountered *in vivo*). (b) The regulation of the size of the antennae of the photosystems, in response to the prevailing illumination conditions. The two processes can be distinguished experimentally by measuring the changes of fluorescence and of the electron transport rate (*in vivo*, O_2 evolution is usually measured), because PS1 is not fluorescent at room temperature (referring to equation 4, kf is very small compared to kd). Therefore, when excitation energy is transferred to PS1 the fluorescence is quenched. If the rate constant for this energy transfer is indicated as k_{21}, the fluorescence yield can be expressed as

$$f = kf/(kf + kd + k_{21} + kp\ P) \qquad (7)$$

and the probability of energy transfer from LHCII-PS2 to PS1 (the so-called "spillover") is

$$\psi_{21} = k_{21}/(k_{21} + kf + kd + kp\ P) \qquad (8)$$

Obviously, k_{21} will be much larger when the traps of PS2 are closed (P = 0), because the most efficient competition from the photochemistry will be suppressed in this condition. So, the spillover will affect mostly F_m while F_o will be substantially unaffected. The spillover can be regulated by the redox state of Q_a (the value of P in equation 4) and balance the excitation of PS2 and PS1. On the other hand, any change in the antenna size of PS2 will affect equally F_o and F_m; therefore the ratio F_v/F_o will not change (F_v is the difference F_m - F_o) upon any change in the size of the antennae.

The spillover requires that PS1 and PS2-LHCII come into contact, a requirement for excitation energy transfer: in isolated thylakoids large increases of spillover (reversible) are observed when the concentration of cations in the medium is drastically reduced, and the grana are disrupted by the unstacking of the membranes (Barber et al. 1980). Under these conditions the PS1 and PS2-LHCII chlorophyll-protein complexes diffuse laterally in the plane of the membranes, and the second order kinetics of fluorescence decrease monitor their collision. The opposite changes are observed when the cations are added. Spillover, however, is unlikely to change substantially *in vivo*, because the concentration of cations in the stroma of chloroplasts *in situ* is adequate to maintain the granal structure and the segregation of PS2-LHCII into the grana partitions away from PS1 which is more concentrated in the stroma-exposed thylakoids. The granal structure has indeed always been observed *in vivo*. The regulation by mechanism (b), i.e. the reversible changes in the relative antenna size of PS2 and PS1, is indicated by the classical experiments of Bonaventura and Myers (1969). These authors observed that when PS2 was over-excited with respect to PS1 *Chlorella* cells adapted in several minutes from a state of high fluorescence and lower rate of O_2 evolution ("state 1") to a state of lower fluorescence and higher photosynthetic activity ("state 2"). The opposite changes were observed when only PS1 was excited (under far-red illumination). This slow adaptation could be clearly distinguished from the fast changes in fluorescence and O_2 evolution rate due to changes in the redox state of Q_a, the primary acceptor of PS2 (Bonaventura and Myers 1969).

The mechanism of the state 1- state 2 transition is now understood, after the discovery of a thylakoid-bound protein kinase (Bennett 1983) which phosphorylates a fraction of LHCII (0.15 - 0.20). Part of the phosphorylated LHCII migrates out of the grana partitions to the stroma-exposed membranes where it can associate with PS1 (see reviews by Bennet 1983, Staehelin 1986). The result of this process is the decrease of the antenna of PS2 and the increase of the one of PS1. The kinase is activated when the electron carriers between the two photosystems are reduced: PQH_2 and cytochrome b_6 seem to be involved. When these electron carriers are oxidized by PS1 (or chemically by any suitable oxidant), the kinase is inactivated and a phosphatase, also present in the membranes, dephosphorylates LHCII which migrates back to the grana partitions where it associates again with PS2. The initial phosphorylation of LHCII is indeed correlated with the decrease of PS2 fluorescence, which is complete in 5 to 10 min, whilst the phosphorylation of LHCII goes on long thereafter, no more correlated with fluorescence changes (Islam and Jennings 1985). This change is reversible upon dephosphorylation (see review by Bennet 1983) and affects F_m and F_o to the same extent. Thus, it indicates a decrease in the optical cross section of PS2. The increase of PS1 activity under rate-limiting light intensity has also been shown, indicating an increase of its antenna (Forti and Vianelli 1988).

The time course of the phosphorylation and dephosphorylation of LHCII as observed in intact chloroplasts and in isolated thylakoids is in fair agreement with the one of state transitions observed *in vivo* by Bonaventura and Myers (1969), and a number of experiments with higher plants are also in agreement with the proposal that phosphorylation of LHCII is the mechanism of state 1 - state 2 transitions. However, no change in the rate of linear electron transport or of the associated ATP synthesis can be observed upon phosphorylation of LHCII in isolated thylakoids (Forti and Grubas 1986). This has been explained quantitatively on the basis of a model which evaluates the effects of the relative rates of PS2 and PS1 turnover on the overall electron transport rate from water to NADP and on PS2 fluorescence (Jennings and Zucchelli 1986).

It is possible that the increase of PS1 antenna caused by LHCII phosphorylation serves the function of providing the excitation on PS1 required to run cyclic electron transport necessary for the synthesis of ATP in the stoichiometric ratio of 3 ATP/2 NADPH required for CO_2 assimilation in the Calvin cycle, and maybe in other functions. In agreement with this, Tsujimoto et al. (1969) reported that Emerson enhancement in the red region of the spectrum was observed in isolated intact chloroplasts reducing CO_2, but not in the same preparations reducing NADP after osmotic shock.

References

Anderson JM (1980) P_{700} content and polypeptide profile of chlorophyll-protein complexes of spinach and barley thylakoids. Biochim Biophys Acta 591: 113-126

Barber J, Mills J, Love A (1977) Electrical diffuse layers and their influence on photosynthetic processes. FEBS Lett 74: 174-181

Bennett J (1983) Regulation of photosynthesis by reversible phosphorylation of the light-harvesting chlorophyll *a* /*b* protein. Biochem J 212: 1-13

Bonaventura C, Myers J (1969) Fluorescence and oxygen evolution from *Chlorella pyrenoidosa*. Biochim Biophys Acta 189: 366-383

Forti G (1987) Energy conversion in higher plants and algae. In: Amesz J (ed) Photosynthesis, New Comprehensive Biochemistry, vol 15, Elsevier, Amsterdam New York Oxford, pp 1-20

Forti G, Grubas PMG (1986) Influence of thylakoid protein phosphorylation on photosynthetic electron transport and photophosphorylation. Photosynth Res 10: 277-282
Forti G,Vianelli A (1988) Influence of thylakoid protein phosphorylation on photosystem I photochemistry. FEBS Lett 231: 95-98
Hill R, Bendall F (1960) Function of the two cytochrome components in chloroplasts: A working hypothesis. Nature 186: 136-137
Islam K, Jennings RC (1985) Relative kinetics of quenching of photosystem II fluorescence and phosphorylation of the two light-harvesting chlorophyll *a/b* polypeptides in isolated spinach chloroplasts. Biochim Biophys Acta 810: 158-163
Jennings RC, Zucchelli G (1986) Studies on thylakoid phosphorylation and non-cyclic electron transport. Arch Biochem Biophys 246: 108-113
Joliot P, Barbieri G, Chabaud R (1969) Un nouveau modèle des centres photochimiques du système II. Photochem Photobiol 10: 309-329
Klimov VV, Klevanik AV, Shuvalov VA, Krasnovsky AA (1977) Reduction of pheophytin in the primary reaction of photosystem II. FEBS Lett 82: 183-186
Mathis P, Rutherford AW (1987) The primary reactions of photosystems I and II of algae and higher plants. In: Amesz J (ed) Photosynthesis, New Comprehensive Biochemistry, vol 15, Elsevier, Amsterdam New York Oxford, pp 63-96
Ort D (1986) Energy transduction in oxygenic photosynthesis: An overview of structure and mechanism. In: Staehelin LA, Arntzen CJ (eds) Encyclopedia of Plant Physiology, NS vol 19, Springer, Berlin Heidelberg New York, pp 143-196
Renger G (1987) Biological exploitation of solar energy by photosynthetic water splitting. Angew Chem, Int Ed Engl 26: 643-660
Rich PR (1988) A critical examination of the supposed variable proton stoichiometry of the chloroplast cytochrome *bf* complex. Biochim Biophys Acta 932 : 33-42
Schlodder E, Graber P, Witt HT (1982) Mechanism of phosphorylation. In: Barber J (ed) Electron Transport and Photophosphorylation. Elsevier, Amsterdam New York Oxford, pp 65-176
Schönknecht G, Hedrich R, Junge W, Raschke K (1988) A voltage-dependent chloride channel in the photosynthetic membrane of a higher plant. Nature 336: 589-592
Schwartz M (1967) Wavelenght-dependent quantum yields of ATP synthesis and NADP reduction in normal and dichlorodimethylphenylurea-poisoned chloroplasts. Biochim Biophys Acta 131: 559-570
Staehelin LA(1986) Chloroplast structure and supramolecular organization of photosynthetic membranes. In: Staehelin LA, Arntzen CJ (eds) Encyclopedia of Plant Physiology, NS vol 19, Springer, Berlin Heidelberg New York, pp 1-84
Sun ASK, Sauer K (1971) Quantum requirements for the two light reactions in spinach chloroplasts. Biochim Biophys Acta 234: 399-414
Tsujimoto HY, McSwain BD, Chain RK, Arnon DI (1969) Ferricyanide and ferredoxin-NADP as alternate electron acceptors in a single photochemical reaction by chloroplasts. In: Metzner H (ed) Photosynthesis Research, Metzner, Tübingen, pp 1241-1261
Witt HT (1979) Energy conversion in the functional membrane of photosynthesis. Analysis by light pulses and electric pulse methods. The central role of the electric field. Biochim Biophys Acta 505: 355-427
Witt HT, Schlodder E, Brettel K, Saygin O (1986) Reaction sequences from light absorption to the cleavage of water in photosynthesis. Routes, rates and intermediates. Ber Bunsenges Phys Chem 90: 1015-1024

SEQUENTIAL ENERGY TRANSDUCTION IN PHOTOSYNTHESIS

M. Losada, M.G. Guerrero, M.A. de la Rosa, A. Serrano, M. Hervás and J.M. Ortega [1]

Photosynthesis consists essentially in the biological conversion at the expense of sunlight energy of simple oxidized inorganic substrates completely devoid of chemical potential - water, carbon dioxide, nitrate or dinitrogen, sulfate and phosphate - into complex reduced organic products - carbohydrates, lipids, proteins, nucleic acids and molecular oxygen (Losada et al. 1987 a,b).

From a physicochemical point of view photosynthesis is intrinsically a light-driven oxidation-reduction/acid-base process. In the course of the photosynthetic process there is a sequential transduction of light energy through electronic energy - that is to say, either photoexcitation of electrons from one orbital to another of higher energy or localization of electrons within any given orbital by restricting its boundaries into redox energy, acid-base energy, orthophosphate-metaphosphate energy and other types of chemical-bond energy. As a consequence, light energy is transformed and eventually stored as chemical energy (cell material and molecular oxygen). Oxygen from water (oxidation state -2) is oxidized to molecular oxygen (0), whereas the oxidized primordial bioelements are stoichiometrically reduced: carbon from carbon dioxide (+4) to carbohydrate (0); nitrogen from nitrate (+5) or dinitrogen (0) to ammonia (-3), and sulfur from sulfate (+6) to sulfide (-2). In contrast, phosphorus from orthophosphate ($H_2PO_4^-$) does not change its oxidation state (+5) but becomes energized to metaphosphate ($\sim PO_3^-$) in a peculiar acid-base process. Metaphosphate is in turn mostly used for the reduction and further assimilation of the bioelements as well as for the polymerization of the various building blocks - sugars, lipid components, amino acids, nucleotides -, thus deenergizing itself back to orthophosphate, again in an acid-base process (Losada et al. 1987a, b).

One of the least understood aspects of photosynthesis is the transfer of electrons and translocation of protons driven by two light reactions and coupled to noncyclic photophosphorylation. A *b*-type cytochrome, namely, cytochrome b_3 or b_{559}, was identified in algal and higher plant chloroplasts during the second half of the 1960s. There is, however, much uncertainty and controversy at present regarding the location and function of cytochrome b_{559} in photosynthetic electron transport, if such were definitely the role of this elusive and enigmatic cytochrome (Arnon and Tang 1988; Losada et al. 1988).

The chemiosmotic hypothesis (Mitchell 1961) has contributed a penetrating insight

[1] Instituto de Bioquímica Vegetal y Fotosíntesis, Consejo Superior de Investigaciones Científicas y Universidad de Sevilla, Facultad de Biología, Apartado 1113, 41080-Sevilla, Spain

into the mechanism of membrane-linked ATP formation in both oxidative and photosynthetic phosphorylation. Mitchell proposed that energy released by respiratory or photosynthetic electron transport generates an electrochemical proton gradient that is utilized by membrane-spanning ATP synthetase to produce ATP. It must, however, be emphasized that the mechanism of these two membrane-bound energy-coupling

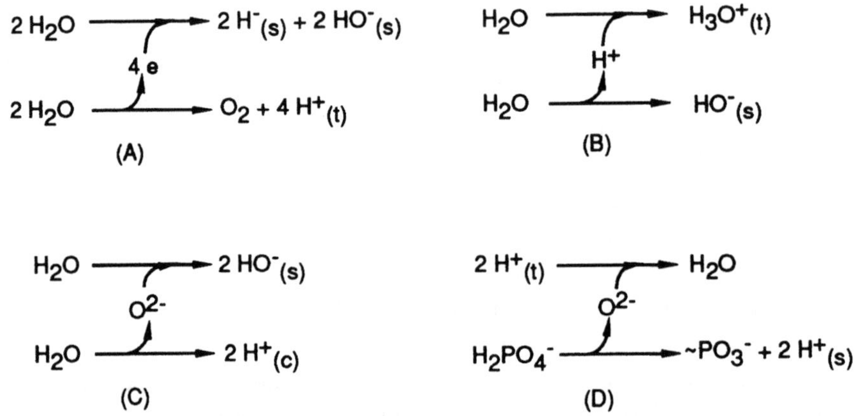

Fig. 1. Intermediate energy-transducing processes in photosynthesis: (A), water autelectrolysis; (B) water autoprotolysis; (C), water autoxidolysis and (D), phosphorylation

Symbols used: (t), (s) and (c) stand for thylakoidal lumen, chloroplast stroma and cytosol, respectively; **P** and **P*** for reaction center pigments in their ground and photoexcited states; **ferricyt** and **ferricyt*** for cytochrome b_{559} in its basal and electronically energized states; **e** and **e*** for electrons at low and high pressure, i.e., electrons of high and low redox potential pairs, respectively; $\mathbf{H^+}$ and $\mathbf{H^{+*}}$ for protons at low and high concentration, i.e., protons of high and low pK_a pairs, respectively

processes remains one of the most challenging and as yet unsolved problems of modern biology (Boyer et al. 1977).

Water is a weak reductant when it is oxidized to molecular oxygen (E_o', pH 7, +0.82 V) and a weak oxidant when it is reduced either to molecular hydrogen (E_o', pH 7, -0.42 V) or to hydride ion (E_o', pH 7, -0.32 V). Notwithstanding, water acts as an amphielectronic molecule in photosynthesis and undergoes a redox-driven autelectrolysis (oxidation-reduction). On the other hand, water is also, according to the Lowry-Brönsted concept, a weak acid (pK_a, H_2O/HO^-, 14) and a weak base (pK_a, H_3O^+/H_2O, 0). However, water acts as an amphiprotic compound in photosynthesis, as well as in respiration, and undergoes a redox-linked autoprotolysis (ionization). Water is thus an ideal substrate for the conversion of visible light energy into redox energy and acid-base energy (Fig. 1A, B).

Fig. 2 represents schematically the sequential transduction of light energy into redox energy and acid-base energy in photosynthetic noncyclic electron transport. The two key energy-transducing steps in this uphill electron transport chain are performed by reaction center chlorophyll *a* of photosystem II (P_{680}) and photosystem I (P_{700}), which, connected in series, transduce light energy through electronic energy - resulting from electron excitation - into redox energy by operating at two alternative midpoint redox potentials, i.e., high potential (P/P^+) and low potential (P^*/P^+) because of photoexcitation of the reduced form of the pair:

$$e \xrightarrow{2\,h\nu} e^*$$

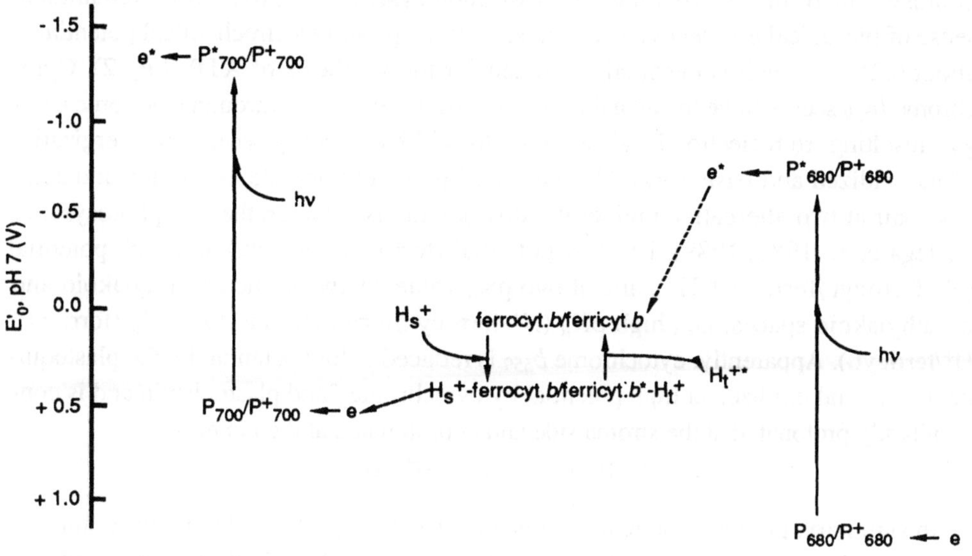

Fig. 2. Sequential transduction of light energy into redox energy and acid-base energy in noncyclic electron transport in thylakoids

This light-driven redox energy (redox potential difference = -2.2 V) is then used for the oxidation-reduction of water:

$$2\,H_2O \longrightarrow O_2 + 4\,H^+(t) + 4\,e$$

$$4\,e \xrightarrow{8\,h\nu} 4\,e^*$$

$$2\,H_2O + 4\,e^* \longrightarrow 2\,H^-(s) + 2\,HO^-(s)$$

$$4\,H_2O \xrightarrow{8\,h\nu} O_2 + 4\,H^+(t) + 2\,H^-(s) + 2\,HO^-(s)$$

According to these equations water is oxidized at high-potential inside the thylakoid vesicles by P_{680}, via a manganese enzyme, and stoichiometric amounts of molecular oxygen and protons are then released. Water is concomitantly reduced at low potential outside the thylakoids by high-energy electrons supplied by P_{700}, via ferredoxin, and stoichiometric amounts of hydride and hydroxide ions are then produced. Water autelectrolysis is thus accompanied by water autoprotolysis, and redox energy is converted not only into redox energy (O_2 and H^-) but also into acid-base energy (H^+, H^- and

HO^-). The primordial oxidized bioelements are subsequently reduced - also outside the thylakoids - by electrons or hydride ions, and stoichiometric amounts of protons from water are compulsorily used for removal of oxide anions from the oxidized bioelements as well as for further protonation of the reduced bioelements themselves (see below).

The electron potential gap of about 0.4 V generated between the reducing side of photosystem II and the oxidizing side of photosystem I allows, with a remarkable sense of biological efficiency, its conversion into a proton electrochemical potential of about 0.25 V, which is eventually utilized for the synthesis of ATP (Fig. 2). Cytochrome b_{559} seems to be the actual transducer of redox energy through electronic energy - resulting from electron localization - into acid-base energy because of energization of its oxidized and basic form. Cytochrome b_{559} would operate as a redox and acid-base pair at two alternative midpoint redox potentials between the two photosystems (Ortega et al. 1988, 1989), i.e., low potential (ferrocyt./ferricyt.) and high potential (H^+-ferrocyt./ferricyt.*-H^+), and at two pK_a values between the extrathylakoid and intrathylakoid spaces, i.e., high pK_a (H^+-ferrocyt./ferrocyt.) and low pK_a (ferricyt* H^+/ferricyt.). Apparently, cytochrome b_{559} is reduced at low potential by the plastoquinone pool and oxidized at high potential by cytochrome *f* and plastocyanin and is concomitantly protonated at the stroma side and deprotonated at the lumen side:

$$e^* + H^+(s) \longrightarrow e + H^{+*}(t)$$

In summary, photosynthetic noncyclic electron transport coupled to water autelectrolysis and autoprotolysis involves the transduction of about half of the trapped light energy (8 hν = 14 eV) not only into redox energy (5 eV) but also into acid-base energy (2 eV):

$$8\ H_2O \xrightarrow{8\ h\nu} O_2 + 8\ H^+(t) + 2\ H^-(s) + 6\ HO^-(s)$$

For energetic reasons, the reduction of nitrate to ammonia does not require metaphosphate (Losada et al. 1987 a,b). At the stroma and cytosol pHs, protons are obviously energetic enough to remove by themselves oxide anions from nitrate, yielding water. This is not, however, the case for both the reduction of carbon dioxide to carbohydrate and the reduction of sulfate to sulfide, which require high-energy phosphate bonds:

$$NO_3^- + 8\ e + 9\ H^+ \longrightarrow NH_3 + 3\ H_2O$$

$$CO_2 + 4\ e + 4\ H^+ \xrightarrow{3\ \sim P} (CH_2O) + H_2O$$

$$SO_4^{2-} + 8\ e + 10\ H^+ \xrightarrow{3\ \sim P} H_2S + 4\ H_2O$$

In the last two instances, metaphosphate energy is required, first of all, as a previous step to reduction, for the removal by protons of oxide anions from the corresponding oxidized bioelements, either from carbon dioxide (actually two carboxylate anions) or from sulfate. As an example, the ATP-dependent reduction of 3-phosphoglycerate (PGA) to glyceraldehyde-3-phosphate (G3P) by NADPH occurs, via the corresponding acylium cation, in the following two steps:

$$R\text{-}COO^- + 2\,H^+ + {\sim}PO_3^- \longrightarrow (R\text{-}C^+{=}O) + H_2PO_4^-$$
$$(R\text{-}C^+{=}O) + H^- \longrightarrow R\text{-}CHO$$

$$R\text{-}COO^- + H^- + 2\,H^+ \xrightarrow{\sim P} R\text{-}CHO + H_2O$$

Removal of an oxide anion by metaphosphate (or active adenylate) is also essential for the activation of amino acids to form amide (or peptide) bonds, as exemplified by the synthesis of glutamine from glutamate and ammonia:

$$R\text{-}COO^- + {\sim}PO_3^- + 2\,H^+ \longrightarrow (R\text{-}C^+{=}O) + H_2PO_4^-$$
$$(R\text{-}C^+{=}O) + NH_4^+ \longrightarrow R\text{-}CONH_2 + 2\,H^+$$

$$R\text{-}COO^- + NH_4^+ \xrightarrow{\sim P} R\text{-}CONH_2 + H_2O$$

Metaphosphate ($\sim PO_3^-$) is thus (Losada et al. 1987 a,b) a strong Lewis acid (phospho cation), which fixes an oxide anion (O_2^-), then becoming a strong base (PO_4^{3-}), which binds two protons ($H_2PO_4^-$). In this way, although metaphosphate deenergizes itself to orthophosphate by hydration, its acid-base energy (1/3 eV molecule^{-1}) is chemically transduced rather than dissipated as heat. Orthophosphate is energized back to metaphosphate (phosphorylation) at the expense of the transmembrane acid-base energy of the electrochemical proton potential (Fig. 1D):

$$2\,HO^-(s) + H_2PO_4^- + 2\,H^+(t) \longrightarrow {\sim}PO_3^- + 3\,H_2O$$

On a global scale it can be estimated that about 1 Gmol (10^{33}) of electrons per second are transferred from water to the oxidized bioelements during photosynthesis. This huge redox process occurs concomitantly with the ionization of water and the subsequent dehydration of orthophosphate and hydration of metaphosphate in similar astronomical figures. From the above considerations, it becomes evident that, although water is the ultimate electron donor for the reduction of the bioelements, it is also continually ionized and regenerated by a variety of strategies of acid-base neutralization.

As stated above, while oxide anions are removed only in part when carbon dioxide is reduced, nitrate and sulfate reduction proceed with the concomitant release of all the oxygen atoms present in their respective molecules. The following sequence of reactions shows that assimilation of carbon dioxide (the anhydride of carbonic acid) starts actually with the fixation of one molecule of CO_2 by ribulose-bisphosphate (R-CO-R) along with one molecule of H_2O, thus forming two identical molecules of PGA (R-COO$^-$) that are subsequently reduced to G3P (R-CHO) with the concourse of two molecules of ATP; carbon appears finally reduced at the carbohydrate level (CH_2O) and ribulose-bisphosphate is regenerated with the concourse of one additional molecule of ATP:

$$\text{R-CO-R} + CO_2 + H_2O \longrightarrow 2\ \text{R-COO}^- + 2\ H^+ \quad \text{(a)}$$

$$2\ H_2O \xrightarrow{\text{light}} O_2 + 4\ e + 4\ H^+ \quad \text{(b)}$$

$$2\ \text{R-COO}^- + 4\ e + 6\ H^+ \xrightarrow{2\ \sim P} 2\ \text{R-CHO} + 2\ H_2O \quad \text{(c)}$$

$$2\ \text{R-CHO} \xrightarrow{\sim P} \text{R-CO-R} + (CH_2O) \quad \text{(d)}$$

$$CO_2 + H_2O \xrightarrow{\text{light}} (CH_2O) + O_2$$

From this overall reaction, it can be inferred that 2/3 of the oxygen in cell material comes from CO_2 and 1/3 from H_2O.

As a corollary, up to 2/3 of the O_2 evolved in carbon photosynthesis might have its origin in CO_2 itself and 1/3 in H_2O, provided that water generated in reaction (c) is used in reaction (b). These considerations might help to clarify controversy around the origin of O_2 evolved in photosynthesis (Warburg 1964).

Photosynthetic eukaryotic cells contain two NADP-dependent glyceraldehyde-3-phosphate dehydrogenases (Iglesias et al. 1987), one in the chloroplast stroma (phosphorylating) and the other in the cytosol (non-phosphorylating). The triose phosphate-3-phosphoglycerate-phosphate translocator (phosphate translocator) of the chloroplast inner membrane (Flügge and Heldt 1984) allows the exit of G3P from the chloroplast stroma to the cytosol and the entrance of PGA in the opposite direction. We have suggested (Losada et al. 1987 b) that both NADP-dependent G3P-DHases may work in

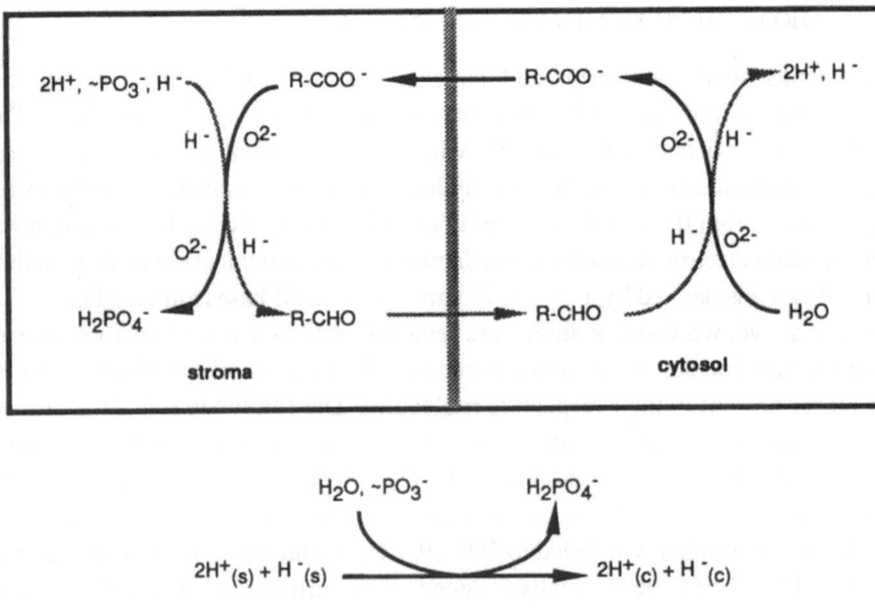

Fig. 3. ATP-dependent "translocation" of hydride ions and protons from the chloroplast stroma to the cytosol by the PGA-G3P shuttle system in photosynthesis

cooperation according to the scheme of Fig. 3, thus allowing the ATP-dependent "translocation" not only of reducing power (H^-) but also of acid (H^+) by water autoxidolysis (Fig. 1D) from the stroma to the cytosol.

The PGA-G3P shuttle might eventually cooperate with water photolysis and proton translocation by cytochrome b_{559} in allowing translocation of reducing power and acidification of the intrathylakoid and cytosol regions with regard to the chloroplast stroma, as summarized by the overall equation:

$$2\,H_2O + 6\,H^+(s) \xrightarrow{8\,h\nu} O_2 + 4\,H^+(t) + 2\,H^-(c) + 4\,H^+(c)$$

Acknowledgments. Research supported by grant SEUI-PB87/401 from Dirección General de Investigación Científica y Técnica and by Junta de Andalucía, Spain. J.M. Ortega is the recipient of a post-doctoral fellowship from CSIC, Spain. We wish to thank Dr J. Florencio for the computerized design of figures.

References

Arnon DI, Tang MS (1988) Cytochrome b_{559} and proton conductance in oxygenic photosynthesis. Proc Natl Acad Sci USA 85: 9524-9528

Boyer PD, Chance B, Ernster L, Mitchell P, Racker E, Slater EC (1977) Oxidative phosphorylation and photophosphorylation. Annu Rev Biochem 46: 955-1026

Flügge UI, Heldt HW (1984) The phosphate-triose phosphate-phosphoglycerate translocator of chloroplast. Trends Biochem Sci 9: 530-533

Iglesias A, Serrano A, Guerrero MG, Losada M (1987) Purification and properties of NADP-dependent non-phosphorylating glyceraldehyde-3-phosphate dehydrogenase from the green alga *Chlamydomonas reinhardtii*. Biochim Biophys Acta 925: 1-10

Losada M, Hervás M, Ortega JM (1987 a) Photosynthetic assimilation of the primordial bioelements. In: Ullrich WR, Aparicio PJ, Syrett PJ, Castillo F (eds) Inorganic Nitrogen Metabolism. Springer, Berlin Heidelberg New York, pp 3-15

Losada M, Guerrero MG, Hervás M, Ortega JM, Serrano A (1987 b) La asimilación fotosintética de los bioelementos primordiales. In: Química agrícola y alimentaria-2 , vol I. European Federation of Chemical Engineering, Publ. Series no 68, pp 19-42

Losada M, Hervás M, Ortega JM (1988) Cytochrome *b*-559 as a transducer of redox energy into acid-base energy in photosynthesis. In: Grassi G, Hall DO (eds) Photocatalytic Production of Energy-Rich Compounds. Elsevier Appl Sci, London, pp 169-179

Mitchell P (1961) Coupling of phosphorylation to electron and hydrogen transfer by a chemiosmotic type of mechanism. Nature 191: 144-148

Ortega JM, Hervás M, Losada M (1988) Redox and acid-base characterization of cytochrome *b*-559 in photosystem II particles. Eur J Biochem 171: 449-455

Ortega JM, Hervás M, Losada M (1989) Location of cytochrome *b*-559 between photosystem II and photosystem I in noncyclic electron transport. Biochim Biophys Acta 975: 244-251

Warburg O (1964) Prefatory chapter. Annu Rev Biochem 33: 1-14

THE EFFECT OF NITROGEN ON THE QUALITY OF MAIN CROPS

A. Bozzini [1]

Introduction

In recent years world food and non-food agricultural production has reached levels nevver before attained in the history of humanity. Table 1 summarizes, by main commodities, the 1988 harvests. Taking cereals - the main food and feed crops - into consideration, we have witnessed an ever-increasing rise in production over the last 30 years. Only in 1987 and 1988, and perhaps also in 1989, has there been a drop in production from the peak of 1,866 million tons harvested in 1986.

Table 1. World agricultural production, by commodity, harvested in 1988 (million metric tons)

Total cereal production	1,730.8
Wheat	509.9
Rice	478.4
Coarse grains	742.5
Roots and tubers	587.1
Grain legumes	57.4
Annual oil crops	255.8
Sugar	102.8
Cocoa	2.1
Coffee	5.5
Tea	2.5
Cotton	17.3
Tobacco	6.3
Meat	163.3
Milk	523.4
Eggs	34.5

Considering a population of 6 billion (6×10^9) people, theoretically each person should have available nearly 800 g of cereal per day. Given an average protein content of 8%, some 64 g of protein of vegetable origin could represent the basis for individual nutrition. To this we should add the protein input of grain legumes, of vegetables and of products of animal origin (meat and milk).

At the moment, although food is theoretically available on a world scale, the situation is not so brilliant, since physical and economical availability of food is not always

[1] Regional Representative for Europe, Food and Agriculture Organization of the United Nations, Via delle Terme di Caracalla, I-00100 Roma, Italy

guaranteed to at least 500 million people, more than 8% of the entire world population.

Nitrogen represents the basic element of the most noble part of food: the proteins. The amount of nitrogen (and proteins) present in the main crops often determines the quality of the product, not only from the nutritional point of view, but also with respect to the technological value of the derived products.

Nitrogen as a Fertilizer

Even if the total amount of nitrogen in a given product is relatively modest, nitrogen is certainly the main element, on a quantitative basis, among fertilizers. In fact, world nitrogen fertilizer consumption is now some 76 million tons and represents 54% of the total fertilizers used (FAO 1989). Unfortunately, as is well known, the rate of absorption of nitrogen from the soil is rather low. Generally not more than 20 - 25 % of the fertilizers added to the soil are taken up and used by the plant. Also in energy terms nitrogen assimilation has a high cost. It has been calculated that 5.9 and 6.2 g of C (as carbohydrates) are needed for each g of N_2 fixed or nitric nitrogen assimilated, respectively (Minchin and Pate 1973).

The availability of nitrogen in the soil, plus water and good sunshine, normally permit the maximum exploitation of nitrogen uptake and assimilation for a given crop genotype. However, the difficulty in increasing the protein content in cereals and the generally low grain productivity in legumes have resulted in a series of studies on factors limiting nitrogen accumulation, in particular in cereal seeds.

It seems that in the seed endosperm, during ripening, protein acccumulation is strictly related to the availability of free amino acids, which also control RNA (Brunori et al. 1977). When the free amino acids are not available, protein synthesis and accumulation stop and the kernel accumulates only the last fraction of starch. In legumes, in fact, protein synthesis continues up to the end of ripening (Smith 1973) since there is a large availability of free amino acids in the plant.

Nitrogen and Crop Quality

In determining the influence of nitrogen on crop quality we should distinguish between nitrogen content of seeds used directly or representing the basis for a transformation, and nitrogen content in vegetative parts (such as leafy vegetables, forages, roots and tubers, etc.). In the first group most of the nitrogen is present as protein, while in the second group of commodities a relatively good share can be present in an inorganic or simple organic form (as nitrate, nitrite, free amino acids, etc.) with quite different effect at the nutritional and technological level.

Cereals

Considering cereal seeds (caryopses), high protein content is normally highly valuable, particularly if in a form of high technological value. It is the case of bread wheat, where protein content can vary from 8% to 22% (of dry matter) in the cultivated varieties, but with a wide variation as to protein quality. A similar range of protein content is also found in *durum* wheat, with some wild forms (*Triticum dicoccoides*) reaching 20% protein content (see Table 2).

Table 2. Protein content of the most important food crops

Species	Protein content (g / 100 g seed dry wt) Mean	Range
Cereals		
Barley	13.1	8.5 - 21.2
Maize	10.4	7.5 - 16.9
Oat	13.3	7.4 - 23.2
Rice	7.6	4.0 - 18.8
Rye	13.4	9.0 - 18.2
Sorghum	12.5	4.7 - 18.2
Wheat	13.0	7.0 - 22.0
Legumes		
Broad bean	24.0	22.0 - 38.2
Chick pea	22.0	19.1 - 31.3
Common bean	23.9	15.2 - 36.0
Common pea	23.1	14.2 - 36.1
Cowpea	24.0	21.1 - 34.2
Groundnut	26.2	17.1 - 31.0
Pigeon pea	21.0	17.9 - 25.8
Soybean	40.3	28.7 - 50.1
Oilseeds		
Cotton	20.2	18.2 - 23.4
Rape	24.5	24.0 - 25.0
Safflower	14.0	12.2 - 16.9
Sesame	20.1	18.1 - 26.2
Sunflower	17.0	12.6 - 33.0

Generally speaking, the economic value of wheat increases with the increase of protein content, particularly if the protein quality is the one devised for technological transformation into bread or other oven products. While protein quality is under a stricter genetic control, total protein amount is more influenced by environmental factors, particularly in relation to the nitrogen and water availability in the soil.

Also in rice, forage barley, maize, sorghum, millets, oat, rye and triticale high protein content is considered of high value. The main exception is malting barley, where the lowest protein content in the seed is considered necessary for making good quality beer. Selection for low protein content in barley has been so strong in Europe that it is now difficult to find there material with a high protein content. Only in certain areas of the world (Morocco, India, Nepal, Tibet, China) where barley flour (mixed with wheat flour) is normally used for making bread or bread-like products (e.g. Chapati in India and in the Himalayan region) are still spread varieties with naked grains and with high protein content up to 18% of dry matter.

Of all the cereal species, rice has the lowest average protein content (7.6% at 14% moisture in milled rice). Biological studies with man and rats have shown that the utilization of milled rice protein is basically related to its protein content (Juliano et al. 1973; Hegsted and Juliano 1974); protein content is the main nutritional goal for this crop. In fact, rice has a fairly well-balanced amino acid composition, and with an increase in protein content by as much as 40 - 72 % only a slight drop in the first limiting amino acid (lysine) was noted (Juliano 1977). Grain yield and protein appear to have a quadratic relationship, so that both characteristics can be increased simultaneously only up to a point beyond which an increase in protein implies a decrease in yields. Physiological studies indicate that high protein content in rice grain is due more to the more efficient mobilization of foliar nitrogen into the grain than to an enhanced nitrogen uptake from the soil (Perez et al. 1973). Rice cultivars with a very high protein content (as high as 18%) have been obtained through selection or mutation breeding, but their impact on rice production has been low so far, mainly because of their low yielding capacity.

High protein rices have been obtained also by improved cultural practices. A favourable effect of top dressing with nitrogen fertilizer on the protein content of rice grains has been reported as well.

Among all cereals, only oat shows a behaviour similar to the one known for rice. In fact, as the negative correlation coefficient between lysine and protein content in oat is low (-0.183), increasing protein in oat need not be accompanied by a decreased lysine concentration in this protein (Pomeranz 1975). This response to nitrogen application of the amino acid composition of oat is likely to be due to the fact that oat, like rice, has a low prolamine content.

The effect of increasing nitrogen applications, and consequently nitrogen concentrations, on the nutritional value of oat grain has been estimated in nitrogen balance studies with rats. Oat protein, especially at high levels, has a higher nutritional value than that of other cereals; the studies suggest that it might be useful to concentrate breeding efforts on increasing the protein content of oat.

The selection of higher-protein cultivars, although desirable, cannot by itself compensate for the nutritional inadequacy of cereal species other than rice and oat. Furthermore, not only higher protein content but also a more balanced amino acid pattern is needed for barley, maize, rye, sorghum, wheat and millet. Negative relationships between lysine and the protein content have usually been found in cereal species other than rice and oat.

Starchy Roots and Tubers

Starchy roots and tubers have a very low nitrogen content (Table 3), with the important exceptions of tuberous roots produced by legume species and potatoes, which show protein contents in the ranges of 3.9 to 14.1 % and 4 to 5 %, respectively. In these foodstuffs, non-protein nitrogen (NPN) represents more than half of the total, and about 50% of it consists of free amino acids, the remainder being amide nitrogen (glutamine and asparagine), nitrate, ammonia and other compounds.

Table 3. Protein and non-protein nitrogen (NPN) in starchy roots and tubers

	Total N (g/ 100 g dry matter) Mean	Range	Protein N (% of total N)	NPN (% of total N)
Cassava	0.3	0.1 - 1.1	20 - 40	60 - 80
Potato	1.6	–	50	50
Sweet potato	0.7	1.4	–	–
Tuberous roots from legume species *	2.5	2.1 - 3.1	20 - 78	22 - 80
Yam	1.0	0.7 - 1.4	–	–

* African yam bean, Mexican yam bean, winged bean, cowpea

The presence of such a large amount of free amino acids is relatively unfavourable from the nutritional standpoint, because of the significant losses of this nitrogen occurring during processing, or even during cooking in water. Thus, the average protein content of most important roots and tubers is extremely low; e.g. cassava roots contain only 0.9 g/ 100 g dry matter of roots. Chronic protein shortages are found in regions where starchy roots and tubers, and particularly cassava, constitute the basic diet. In such regions the so-called "cassava problem", related to kwashiorkor, is present. It appears that persons with low incomes tend to consume cassava in excessive quantities because it supplies sufficient calories and gives a feeling of satiation but, as a result, many suffer from protein and/or vitamin deficiencies. It is not possible to eat as much cassava, in the form of gaplek, sago, etc., as would be needed to satisfy the protein requirements (about 1 g/day and kg of body weight for adults and even more for children); for example, when 400 g gaplek, giving a feeling of satiation, are eaten per day, only 6 g protein are consumed. Even if the deficit of 45 g protein is available in the form of about 200 g meat, fish or pulses, the supply of proteins usually remains below the required level, when cassava is the principal food. It is well known, in fact, that the main food determines the protein supply, and that supplementary food has no predominant influence on it. Therefore, the availability of higher-protein cassava cultivars for human consumption would be very helpful in relieving the protein problem where starchy roots and tubers provide the main food. In this case, the improvement of the protein quality in cassava is only of second priority. Although there are considerable differences in the crude protein contents of different cultivars of cassava, yam and sweet potato, there is little evidence that breeding or improved culture practices can produce a satisfying balance of protein in these starchy foodstuffs. In particular for cassava, there are indications that cultivars with higher protein contents in the roots, obtained through breeding or culture methods, have significantly higher HCN contents or other unfavourable features, including lower yields (HCN can, however, be eliminated during cooking).

Legumes and Oilseeds

Grain legumes are characterized among cultivated plants by a relatively high protein content ranging, on average, from 24 to 40 % (Table 2). Within each species, there is a large variability in protein content, due to both genotypical and environmental factors, including symbiotic nitrogen-fixing rhizobia living in the legume root nodules. *Rhizobium* symbiosis is an important source of nitrogen, but the protein content of legume seeds is relatively high even when the plants are grown in the absence of nitrogen-fixing bacteria, and this demonstrates the importance of the genetic component. Rhizobia, however, are a very important component for grain legume yield and, if proper inoculation is provided, are often responsible for high economic returns in production.

Grain legumes are of primary importance in complementing and supplementing the protein-deficient cereal diets of populations in the developing countries. However, breeding for high-protein legume varieties, in general, cannot be regarded as important, since in many species the increase in limiting essential amino acids does not parallel the increase in protein content. Negative relationships have usually been found between legume protein percentage and methionine content per unit of protein. This is due to the fact that in legumes higher protein contents are usually associated with higher amounts of globulin fractions that are low in sulphur-containing amino acids. The negative association between protein content and crop yield is another major handicap in breeding legume crop plants.

Oilseeds are rich in both calories (oil) and proteins. However, a negative correlation between percentages of oil and protein has been found in most oilseed crops. The oils are largely used for human nutrition, the residues after oil extraction being used mostly as animal feed; only a very small portion of oilseed meals is used directly for human nutrition. Hence, for the present, most efforts should be directed towards eliminating, by breeding, culture or technological methods, the thermostable toxic factors not inactivated during heating and preparation. Owing to the presence of certain anti-nutritional factors, some unprocessed oilseed proteins in the dietary pattern can, in fact, create problems such as agglutination of red blood cells, goitrous symptoms and haemolytic effects.

Vegetables

Three main groups are to be considered: leafy vegetables (such as spinach, beets, salads, etc.); vegetables where the fruit is the edible part (tomatoes, peppers, eggplants, squashes, etc.) and finally bulbs and underground edible products (onion, garlic, carrots, turnips, etc.).

It is well known that nitrogen promotes the development of vegetative parts and therefore is a key element in leafy vegetable and bulb production. As in roots and tubers, in these vegetative organs often the non-protein nitrogen is about half the total nitrogen content. However, when nitrogen is in excess and enough water is available, a nutritional imbalance can result as revealed in high amounts of free nitrate and nitrite which are normally present in these products.

Excess nitrogen also influences the flavour and texture of such products. Excessive vegetative development can lead to an increase in water content and to less flavour and taste. This is particularly true of fruits. Tomatos, zucchini, eggplants fed too much nitrogen have an insipid flavour and their consistency as well as their shelf life are reduced. It should also be stressed that products from plants treated with too much nitrogen are more susceptible to a number of fungal and bacterial diseases which also reduce shelf life - a characteristic of paramount importance in modern horticulture. On the other hand, balanced nitrogen nutrition is essential for high production, a requirement in modern, advanced agriculture.

References

Brunori A, Mannino P. Ancora G, Bozzini A (1977) Protein accumulation, RNA and soluble amino-nitrogen content in developing endosperm of two varieties of *Triticum aestivum* with high and low protein seed. Theoret Appl Genet 50: 73-77

Hegsted DM, Juliano BO, (1974) Difficulties in assessing the nutritional qualities of rice proteins. J Nutr 104: 772-781

Juliano BO, Antonio RA, Eshama BV (1973) Effects of protein content on the distribution and properties of rice proteins. J Sci Food Agric 24: 295-306

Minchin FR, Pate JS (1973) The carbon balance of a legume and the functional economy of its root nodules. J Exp Bot 24: 259-271

Pomeranz Y (1975) Proteins and amino acids of barley, oats and buckwheat. In: Protein Nutritional Quality of Foods and Feeds, vol 1, IM Dekker Inc, New York, pp 13-78

Smith DL (1973) Nucleic acid, protein, and starch synthesis in developing cotyledons of *Pisum arvense* L. Ann Bot 37: 795-804

Part 2 Transport and Assimilation of Inorganic Nitrogen

GROWTH AND NITRATE UPTAKE PROPERTIES OF HIGHER PLANTS: RELATIONS TO EXTERNAL AND INTERNAL NITROGEN.

C.-M. Larsson [1] and P. Oscarson [2]

Introduction

Nitrate-N supplied by fertilization or produced via microbial activity becomes available to plants by mass flow and/or diffusion. Under well-irrigated and -fertilized conditions in the field, it does not appear that movement of solutes to the root surface limits yield, even at very high levels of productivity (Barraclough 1986). However, in a natural environment or in less intensive agricultural practice, nutrient flow to the plants may become limiting for growth. Under these conditions, the overall limitation of nutrient availability is that set by integration of nutrient flux and root growth, not by nutrient concentration (for reviews on these matters, see e.g. Nye and Tinker 1977; Ingestad 1982; Haynes 1986; Robinson 1986). Physiological effects of nutrient flux limitation has, however, only received minor attention in experimental studies on plant nutrition, where the concept of concentration control has been favoured. As alternatives to concentration control, nutrients can be supplied to the plants in a desired relation to either a previously established growth curve (programmed nutrient addition; Asher and Cowie 1970, reviewed by Asher and Edwards 1983), or the amount of nutrient already bound in biomass (the relative addition rate concept: Ingestad and Lund 1979; reviewed by Ingestad 1982; Ingestad and Lund 1986). In the sense that nutrients are added in a fixed relation to biomass and/or nutrients bound in the culture, these concepts have principal similarities to that used for culture of microorganisms in chemostats. In the present communication growth and nitrate uptake kinetics will be described for plants maintained under flux limitation, using the relative addition rate concept and nitrate-N as the limiting nutrient.

Principle

Growth of a plant culture in the exponential phase can be described according to:

$$W_t = W_0 \, e^{RGR\, t} \qquad (1)$$

where W_0 and W_t are the weights at the start and end of the time interval t, respectively, and RGR, the relative growth rate. The increment in nutrients over the same time and at stable internal nutrient levels can be described in analogy to (1), i.e:

[1] Department of Botany, University of Stockholm, S-106 91 Stockholm, Sweden.

[2] Department of Crop Genetics and Breeding, The Swedish University of Agricultural Sciences, S-268 00 Svalöv, Sweden

$$N_t = N_o\, e^{RGR\, t} \qquad (2)$$

where N_o and N_t are the initial and final nutrient contents, respectively. This mathematical representation of nutrient acquisition can be regarded as representative for a nutrient-sufficient condition when the growth rate determines the long-term rate of nutrient uptake. However, nutrients can be made available to the plants at a desired rate relative to the amount of nutrients bound in biomass (RA):

$$N_t = N_o\, e^{RA\, t} \qquad (3)$$

RA is the treatment variable and will determine the relative rate of increase in plant nutrient content (RN). Under conditions of exponential growth with stable nutrient concentrations in the plant tissue, the situation will be:

$$RGR = RN = RA \qquad (4)$$

The amount of nutrients that will have to be introduced over the time interval t will thus be given by $N_t - N_o$. In practice, addition intervals from 24 h down to a few minutes in computer-controlled systems have been used. The ultimate perfection of the concept would be a completely continuous addition system which would account for a completely smooth exponential increase in nutrient supply with time. A system of this kind has apparently so far only been used by Freijsen and Otten (1987) and Hirose et al. (1988). Addition intervals of 24 h has, however, in the authors experience no specific effect on either growth or uptake properties of the plants, although there will be certain fluctuations in internal N pools and rates of N-assimilatory processes (Ingemarsson 1984; Oscarson and Larsson 1986).

Growth under N-Limitation

We have adopted the relative addition rate concept for the culture of a variety of annual plants. This is accompanied with some problems in controlling growth rates due to the short generation time of the plant as a whole and the short life-span of each individual leaf. Data will here be given for *Lemna gibba* L. and *Pisum sativum* L. (cv. Marma). *Lemna* (duckweed) is in a sense a useful "reference" since growth in this species defines expansion of a population, where age-related effects are randomized. Furthermore, expansion takes place solely in the horizontal plane which eliminates risks for self-shading.

Table 1 shows growth data for *Pisum* (non-nodulated) and *Lemna* under conditions where nitrate-N was fed to an otherwise sufficient nutrient solution at a range of RA. A good agreement was obtained between RA and RGR. Since only trace residual N levels were found in the nutrient solution, these data clearly indicate that the relationship described in eqn. (4) is valid. Maximum growth rates under the prevailing conditions were approximately 0.22 day^{-1} for *Pisum* and 0.35 day^{-1} for *Lemna*. Although the lower limit was not investigated, these data show that growth rates corresponding to 15 % of maximum can be obtained and controlled, thus allowing for exploration of physiological performances of plants over a wide range of N-limited growth rates. Table 1 also shows that the root contribution to total plant weight decreased, and the average tissue N concentrations increased, with increasing RA.

Table 1. Growth data for *Pisum* and *Lemna* adapted to different RA

RA day^{-1}	RGR day^{-1}		Root DW, % of total plant DW		Tissue N concentration $mg\ N\ g^{-1}\ DW$	
	Pisum	*Lemna*	*Pisum*	*Lemna*	*Pisum*	*Lemna*
0.03	0.036		56.5		14.4	
0.05		0.055		45.0		5.4
0.06	0.066		51.9		18.0	
0.09	0.086		49.4		18.2	
0.10		0.102		35.2		8.9
0.12	0.124		46.5		20.8	
0.15	0.140	0.152	44.8	31.4	24.4	11.9
0.18	0.189		42.2		26.1	
0.20		0.202		27.5		16.4
0.21	0.218		39.6		31.2	
0.30		0.291		23.4		24.8

Nitrate Uptake

The performance of the nitrate uptake system at different RA was investigated in experiments on net uptake kinetics. However, a problem with kinetic interpretations of net uptake data is the possibly large discrepancy between net uptake and influx, caused by efflux of nitrate. The question of the relative importance of efflux during uptake was addressed in experiments where ^{13}N-labelled nitrate was introduced to plants during steady rates of uptake. For the following few minutes, net uptake and uptake of radioactivity were recorded simultaneously and continuously (methods described by Oscarson et al. 1987; Ingemarsson et al. 1987). The short duration of these experiments, where the specific activity of the medium presumably remains considerably above the specific activity of cytoplasmic nitrate, allows for experimental separation of net uptake (v) and influx (i) rates. The results in Table 2 show that, at an external concentration of approximately 50 μM, net uptake corresponded to about 90 % of influx in *Pisum,* whereas no efflux at all could be detected in *Lemna.* At 5 μM external nitrate, however, the discrepancy between net uptake and influx increased greatly in both species, whereas influx was little affected.

These observations indicate that: 1) At concentrations approaching saturation, the difference between v and i decreases and thus, V_{max} values obtained in net uptake experiments are probably valid approximations of maximum influx rates, I_{max}. 2) Apparent K_m values obtained in net uptake experiments may be highly unrepresentative of

Table 2. Influx rates and ratios of net uptake (v) over influx (i) in *Pisum* and *Lemna* at different external nitrate concentrations.

RA	$[NO_3^-]$	i		v/i	
		Pisum	*Lemna*	*Pisum*	*Lemna*
day^{-1}	µM	µmol g^{-1} weight h^{-1} (dry weight)	(fresh weight)		
0.06	50	11.4		0.96	
0.06	5	10.2		≤0.25	
0.12	50	10.5		0.83	
0.12	5	9.9		≤0.14	
0.20	50		1.96		1.03
0.20	5		1.92		≤0.67

the actual uptake (= influx) system. 3) The application of Michaelis-Menten kinetics to net uptake experiments can only be justified from an empirical point of view; speculations as to the biochemical characteristics of the presumed nitrate carrier may be highly ambiguous. 4) The data indicate minor interference of unstirred layers with the uptake measurements. The unrestricted uptake of isotope also when both external concentration and net fluxes are low indicates that unstirred layers are virtually absent, or that the resistance to diffusion through the unstirred layer is insignificant relative to other resistances involved in nitrate uptake.

Kinetic constants for net nitrate uptake in the concentration range 10 to 100 µM (i.e. system I of uptake; Epstein 1976) are given in Table 3. V_{max} increased with RA which, bearing the above considerations in mind, can be attributed to increased I_{max}. A similar relationship was seen in *Lemna* fronds, although the uptake rates were considerably lower (Oscarson et al. 1989 b). When calculated on a plant N basis (relative V_{max}), values increased with RA for *Pisum* but decreased for *Lemna* (Table 3). The relative V_{max} must at least be equal to RGR in order to meet the N demand set by plant growth. The data show that the relative V_{max} over the entire RA range investigated exceeds the minimum requirement, at the lowest RA used by factors 5 and 44 for *Pisum* and *Lemna*, respectively.

A further demonstration of the uptake capacity can be made by computing, from the kinetic data, the balancing nitrate concentration required to sustain growth after adaptation to different RA. Although possibly subject to errors caused by inaccurate determinations of K_m (values in the order of 15 µM for *Pisum* and 10 µM for *Lemna*; Oscarson et al. 1989 b) and the increased contribution by efflux at lower external concentrations (Oscarson et al. 1987; Ingemarsson et al. 1987), this operation may nevertheless identify the order of concentrations relevant in terms of growth limitation. These data are included in Table 2, and range from 3.5 to 22 µM for *Pisum* growing in the RA range 0.03 to 0.21 day^{-1}, and 0.2 to 5.4 µM for *Lemna* in the range 0.05 to 0.30 day^{-1}.

Table 3. V_{max} and steady state concentrations required to maintain growth (i.e. balancing nitrate concentrations) for *Pisum* and *Lemna* grown at different RA. Kinetic data for fronds (not shown) are included in calculations of relative V_{max} and balancing nitrate concentrations.

RA day^{-1}	V_{max}, µmol NO_3^- g^{-1}DW(root) h^{-1}		Relative V_{max} gN g^{-1}N day^{-1}		Balancing $[NO_3^-]$, µM	
	Pisum	*Lemna*	*Pisum*	*Lemna*	*Pisum*	*Lemna*
0.03	11.8		0.16		4.4	
0.05		68.8		2.22		0.2
0.06	19.2		0.19		12.0	
0.09	27.5		0.25		8.3	
0.10		95.2		1.69		0.5
0.12	38.6		0.29		9.2	
0.15	52.2	123	0.32	1.49	9.9	0.9
0.18	68.6		0.37		17.4	
0.20		164		1.16		2.2
0.21	78.2		0.33		45.0	
0.30		199		1.00		5.1

In Figure 1, V_{max} values for root nitrate uptake are plotted against the root N-concentrations. Contribution by soluble N was less than 5% of total N for both species. Thus, V_{max} in both *Pisum* and *Lemna* during N-limited growth increases linearly with root protein concentration. The curve for *Pisum* intercepts at zero activity at a tissue N concentration of approximately 14 mg N g^{-1} dry weight. Interestingly, if growth itself (Oscarson et al. 1989 a) or growth respiration of the roots (P. Duarte et al., in preparation) are plotted against tissue N concentrations, linear relationships with very

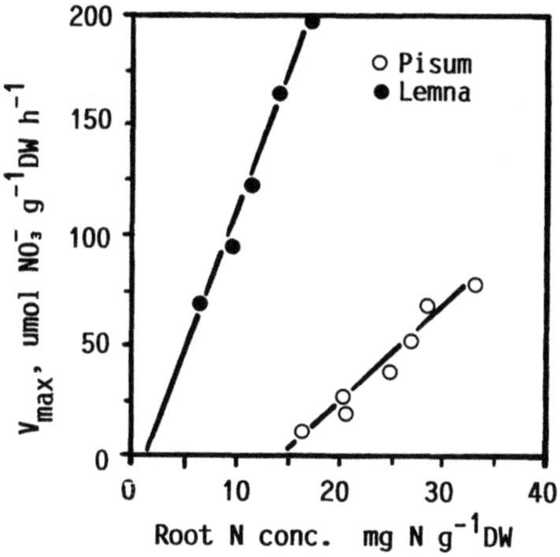

Fig. 1. V_{max} for nitrate uptake in roots of *Pisum* and *Lemna* against root N concentration

similar intercepts are obtained. In a formal manner this identifies two pools of protein N, one associated with growth and the other engaged in maintenance work or simply not accessible to retrieval for metabolic use. This remains, however, a formal identification, and the physiological identity of the two N-fractions is, at present, uncertain. It appears a tentative explanation to the results that nitrate-carriers under these conditions are formed in a fixed proportion to the proteins engaged in growth processes. Further elucidation of this problem requires tools to identify expression of the uptake system at the molecular level, which we presently lack.

Conclusions

Data presented in this communication indicate that adaptation of the nitrate uptake system to long-term limitations in N availability is primarily a growth response. Uptake capacity on a unit root weight basis decreases with decreasing RA, in parallel with the root N concentration. However, this decrease in tissue N levels, together with growth retardation and increased root contribution to total biomass, all contributes to increased overcapacity to absorb nitrate (when put in relation to the demand set by growth) the lower RA becomes. This is also manifested in the balancing nitrate concentrations, which can possibly be used as measures of affinity of the uptake system, thus integrating nitrate fluxes, external nitrate concentrations, and N demand set by growth.

Acknowledgment. Financial support from the Swedish Natural Science Research Council is acknowledged.

References

Asher CJ, Cowie AM (1970) Programmed nutrient addition - A simple method for controlling the nutrient status of plants. Proc Aust Plant Nutr Conf, Mt Gambier, S Aust, Sect 1(b), pp 28-32

Asher CJ, Edwards DG (1983) Modern solution culture techniques. In: Läuchli A, Bieleski RL (eds) Inorganic Plant Nutrition. Encyclopedia of Plant Physiology, Vol 15A, Springer, Berlin Heidelberg New York, pp 94-119

Barraclough PB (1986) The growth and activity of winter wheat roots in the field: nutrient inflows of high-yielding crops. J Agric Sci 106: 53-59

Freijsen AHJ, Otten H (1987) A comparison of the response of two *Plantago* species to nitrate availability in culture experiments with exponential nutrient addition. Oecologia 74: 389-395

Haynes RJ (1986) Mineral nitrogen in the plant-soil system. Academic Press, London

Hirose T, Freijsen AHJ, Lambers H (1988) Modelling of the responses to nitrogen availability of two *Plantago* species grown at range of exponential nutrient addition rates. Plant Cell Environ 11: 827-834

Ingemarsson B, Johansson L, Larsson C-M (1984) Photosynthesis and nitrogen utilization in exponentially growing nitrogen-limited cultures of *Lemna gibba*. Physiol Plant 62: 363-369

Ingemarsson B, Oscarson P, af Ugglas M, Larsson C-M (1987) Nitrogen utilization in *Lemna*. II. Studies of nitrate uptake using $^{13}NO_3^-$. Plant Physiol 85: 860-864

Ingestad T (1982) Relative addition rates and external concentration: Driving variables used in plant nutrition research. Plant Cell Environ 5: 443-453

Ingestad T, Lund A-B. (1979) Nitrogen stress in birch seedlings. I. Growth technique and growth. Physiol Plant 45: 137-148

Ingestad T, Lund A-B (1986) Theory and techniques for steady state mineral nutrition and growth of plants. Scand J Forest Res 1: 439-453

Nye PH, Tinker PB (1977) Solute movement in the soil-root system. Blackwell Scientific Publ, London

Oscarson P, Larsson C-M (1986) Relations between uptake and utilization of NO_3^- in *Pisum* growing exponentially under nitrogen limitation. Physiol Plant 67: 109-117

Oscarson P, Ingemarsson B, af Ugglas M, Larsson C-M. (1987) Short-term studies of NO_3^- uptake in *Pisum* using $^{13}NO_3^-$. Planta 170: 550-555

Oscarson P, Ingemarsson B, Larsson C-M (1989 a) Growth and nitrate uptake properties of plants grown at different relative rates of nitrogen supply. I. Growth of *Pisum* and *Lemna* in relation to nitrogen. Plant Cell Environ 12: 779-785

Oscarson P, Ingemarsson B, Larsson C-M (1989 b) Growth and nitrate uptake properties of plants grown at different relative rates of nitrogen supply. II. Activity and affinity of the nitrate uptake system in *Pisum* and *Lemna* in relation to nitrogen availability and nitrogen demand. Plant Cell Environ 12: 787-794

Robinson D (1986) Limits to nutrient inflow rates in roots and root systems. Physiol Plant 68: 551-559

TRANSPORT OF INORGANIC NITROGEN COMPOUNDS: PHYSIOLOGICAL STUDIES ON UPTAKE AND ASSIMILATION

W.R. Ullrich, S. Lesch, L. Jarczyk, M. Herterich and G.D. Trogisch [1]

Introduction

In recent years nitrate uptake, in addition to nitrate reduction, has become subject to manifold investigations with respect to its regulatory role in nitrate assimilation. Some special situations have been described for algae in which most probably nitrate uptake rather than nitrate reductase (NR) has been limiting for nitrate utilization (for reviews, Syrett 1981; Ullrich 1983). Data have been presented that show regulation of nitrate uptake as a process at least partly independent of NR activity (Ullrich 1987). Butz and Jackson (1977) brought up a hypothesis that NR, whose localization in the plant cell was not fully clarified, could also be responsible for nitrate transport across the plasmalemma, a hypothesis supported at that time only by data from different organisms. This concept was renewed recently together with a possible role of NR in extracellular reduction of, e.g. ferricyanide (Jones and Morel 1988). Ward et al. (1988) postulated that the (apparently constitutive) part of NR localized in the plasmalemma might be responsible for nitrate transport. Both claims are supported by immunological evidence showing specific anti-NR reactions in the plasmalemma and corresponding transport inhibition by antibodies (see Tischner, in this volume), but a role in transport has been claimed to be unlikely by Warner and Huffaker (1989).

This article presents physiological data in favour of an independence of nitrate uptake and NR activity or nitrate assimilation and a comparison with ammonium uptake in a green alga and in *Lemna* under different conditions without and with artificial inhibitors.

Materials and Methods

Monoraphidium braunii Legnerová, strain 202-7d from the Göttingen algal collection was grown in synchronous culture at 27°C with a 14:10 h light/dark rhythm (Eisele and Ullrich 1977). Nitrogen starvation was achieved by N-free medium within 2 to 3 hours. *Lemna gibba* G 1, obtained from Prof. R. Kandeler, Vienna, cultivated autotrophically for 5 days prior to the experiments, was either used directly from this culture supplied with 4 mM NO_3^- (unstarved *Lemna*), or nitrogen starved for 5 days and, unless stated otherwise, "induced" with 40 µM $Ca(NO_3)_2$ 18 hours prior to the

[1] Institut für Botanik, Technische Hochschule, Schnittspahnstrasse 3, D-6100 Darmstadt, FRG

experiments in order to have nitrate uptake, rather than NR, in a partly induced state. (Ullrich et al. 1984; Ullrich 1987). Treatment of *Lemna* with tungstate (WO_4^{2-}) was also done in an autotrophic culture for 5 days prior to the experiments with the concentrations indicated in Figure 6.

Results and Discussion

Nitrate uptake follows saturation kinetics in most organisms tested, but often a second, low-affinity ("linear") system is superimposed which may become saturated at high external nitrate concentrations. The saturable system, probably a one-step system, is inducible and has a half-saturation constant (K_s) between less than 1 μM in some marine algae and about 25 or 30 μM NO_3^- in some freshwater algae and in many higher plants (cf. Ullrich 1983). By contrast, NR has a K_m in the order of 100 to 200 μM in most algae and higher plants (cf. also Menacho and Vega in this volume).

In Figure 1 nitrate uptake kinetics in *Monoraphidium* is compared at different CO_2 concentrations and in the absence of CO_2 (in N_2). Apparently the saturating nitrate uptake system is almost completely CO_2-dependent, has a half-saturation of 23 μM nitrate and seems to depend on energy and metabolism. From the concentration dependence it can be concluded that the saturating nitrate transport system accumulates nitrate against its concentration gradient within the cytosol, thus increasing the overall affinity of nitrate assimilation compared with NR.

The "linear" system continues to function at low levels of carbon metabolism, i.e. in N_2, is pH dependent and ammonium insensitive (cf. Serra et al. 1978) and shows considerable uptake rates with higher nitrate concentrations at neutral or alkaline pH, but only relatively poor rates in the acidic pH range (Eisele and Ullrich 1977; Di Martino Rigano et al. 1985). The nitrate taken up is also reduced to ammonium but instead of being metabolized, the ammonium is almost quantitatively released to the medium. The "linear" system may reflect the functioning of anion channels which have been shown to display their highest activities at low membrane potentials (E_m = -40 to -50 mV) in guard cell protoplasts and which are highly permeable for nitrate (Keller et al. 1989). At low nitrate concentrations within the cytosol (50 μM or less) and at low E_m (-50 mV inside) which is likely at the high K^+ concentrations of 10 mM used in our experiments (Köhler et al. 1985, for *Eremosphaera*) the equilibrium nitrate concentration in the medium would be 0.336 mM. Anion channels might transport nitrate into the cytosol as long as there is a concentration gradient and nitrate is consumed. This may proceed even far below the K_m of NR.

Two nitrate uptake components have been shown to function also in *Lemna*, in this case by inhibiting the saturable component with external ammonium (Ullrich et al. 1984). Electrophysiological experiments supported the view of anion/H^+ cotransport for at least one component in *Lemna* (Ullrich and Novacky 1981; Ullrich et al. 1984)

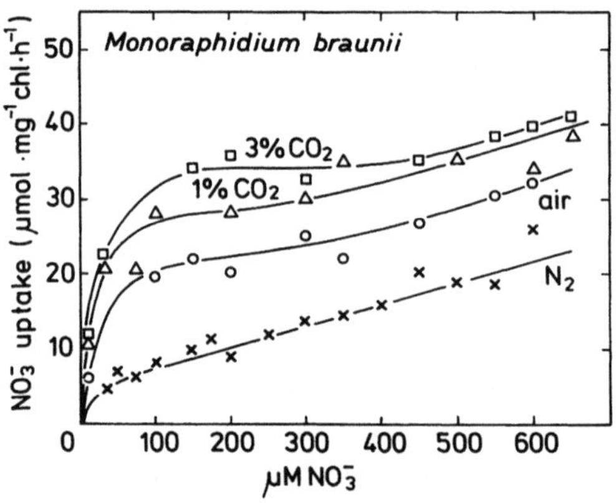

Fig. 1. Concentration dependence of nitrate uptake in *Monoraphidium braunii* in N_2 and at different CO_2 concentrations. Light, pH 6.2, 10 mM K^+ (K_2SO_4), 15 mM Na^+ (Na_2SO_4 and phosphate)

and more recently also in roots of terrestrial plants (Ullrich 1987; Ullrich et al. 1989), where the often observed hyperpolarization could also reflect transport by anion channels. *Lemna* cells have large vacuoles and can store nitrate and, to some extent, also ammonium, hence a differentiation between nitrate uptake and storage on one hand, and assimilation by enzymes on the other hand was possible by following the time course or by using different physiological states but without artificial inhibitors.

In *nitrate uptake* experiments with *Lemna* (Fig. 2 A) whose NR activity was high enough to consume all nitrate taken up, neither nitrate nor ammonium was accumulated measurably within the experimental period of time. However, in an *ammonium uptake* experiment under otherwise the same conditions (Fig. 2 B), ammonium was taken up at a clearly higher rate during the initial phase of the experiment, and an accumulation of ammonium was visible in the pool size whose initial rate of increase accounts for most of the overall rate of ammonium uptake. Only after 15 to 20 min becomes the accumulation much slower than total uptake and a considerable part of the ammonium taken up is metabolized. Nitrate is apparently almost completely metabolized from the beginning (Fig. 2 A). There is no change in the pool size of ammonium in figure 2 A, showing that once nitrate is reduced it is fully assimilated into amino acids, nor is ammonium converted to nitrate (Fig. 2 B).

Similar experiments were carried out with *Lemna* of *different physiological states*, i.e. with unstarved, N-starved, and N-starved but nitrate-induced plants. In unstarved *Lemna*, nitrate uptake showed a moderate rate, the activity of NR would allow a higher overall assimilation rate, hence only little or even no nitrate is stored as such during the experiment (Fig. 3 A). But this applies only to plants with saturating CO_2 supply, whereas under N_2 NR activity is reduced and, probably, the decrease in membrane potential and some K^+ efflux forces the plants to release nitrate upon its addition to the

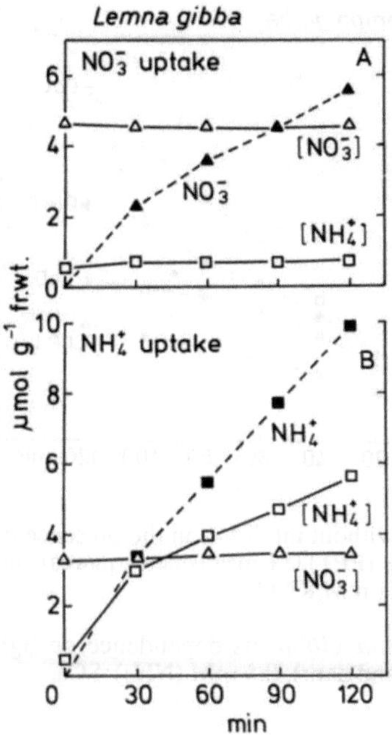

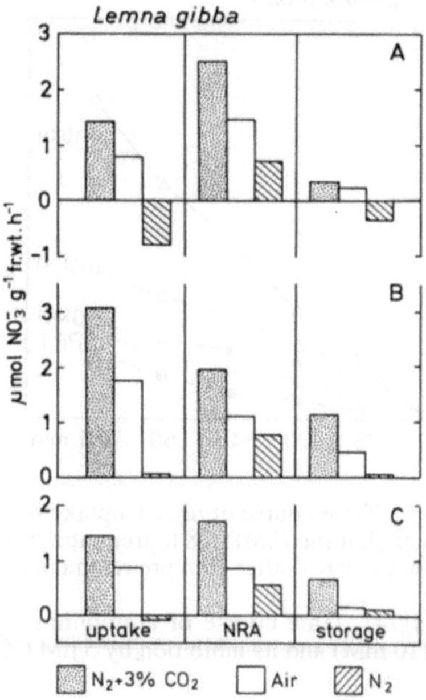

Fig. 2 *(left)***.** Uptake experiments with N-starved and induced *Lemna gibba*. Uptake (closed symbols) calculated from disappearance in the medium; pool sizes (open symbols) measured in the tissue. **A**, NO_3^- uptake experiment; **B**, NH_4^+ uptake experiment. KNO_3 or $(NH_4)_2SO_4$, 0.5mM

Fig. 3 *(right)***.** Comparison of nitrate uptake (30 min, 0.1 mM KNO_3), nitrate reductase activity (NRA) and pool size changes (storage) in *Lemna gibba* of different activation states. **A**, unstarved, i.e. from the normal culture with 4 mM NO_3^-; **B**, N-starved for 5 d and "induced" with 40 µM NO_3^- 18 h prior to the experiment; **C**, starved for 5 d but without "induction". Experimental gassing started 15 min before addition of nitrate

medium. This nitrate must come from the vacuole. A release of anions has been found at low ATP concentrations in tonoplast vesicle studies (Roos 1989). Figure 3 B shows the same type of experiment with N-starved but nitrate-induced *Lemna* plants. In these plants nitrate uptake is still relatively active while NR with a shorter half-life has already lost a greater part of its activity. Here uptake shows the highest rates, and nitrate not assimilated is apparently stored. In N_2 atmosphere no nitrate is released, probably due to the lower pool sizes after N-starvation. Without induction the rates of both nitrate uptake and reduction are low, but as in unstarved plants, NR activity is higher than the available nitrate uptake capacity (Fig. 3 C). Correspondingly there is also little nitrate left for storage. Under N_2 very little nitrate is released to the medium from these N-starved plants.

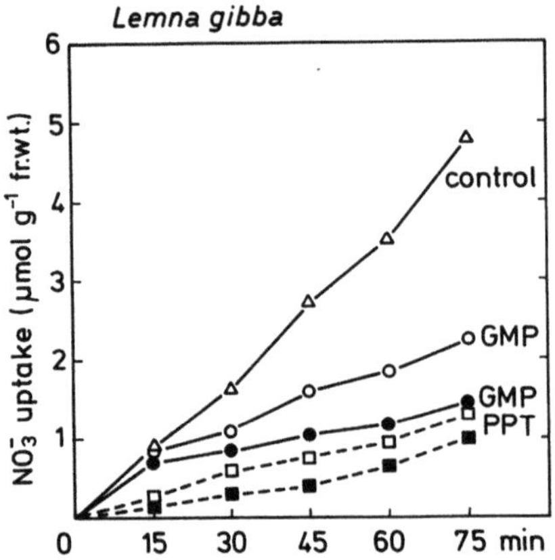

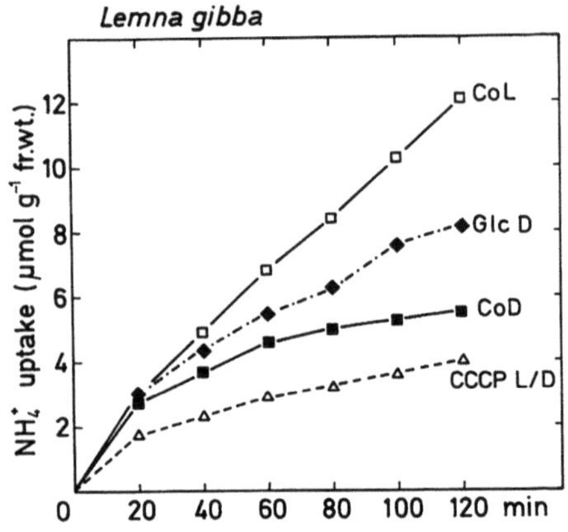

Fig. 4 *(left)*. Time course of nitrate uptake in *Lemna gibba* without inhibitor, in the presence of 0.1 mM 6-methylpurine (6MP, 18 h pretreatment) or glufosinate (PPT) 0.1 mM (open squares) and 1.0 mM (closed squares) after 19 h pretreatment in both cases. 0.1 mM KNO_3

Fig. 5 *(right)*. Time course of ammonium uptake in *Lemna gibba*, its dependence on light or glucose (10 mM) and its inhibition by 5 µM CCCP (in light and dark). 0.5 mM $(NH_4)_2SO_4$

Another approach was tried with *inhibitor experiments*. 6-Methylpurine (6MP), used to prevent protein synthesis by inhibiting transcription, causes a more severe decrease in NR activity than in nitrate uptake. Glufosinate (PPT), on the other hand, could be shown to interfere with glutamine synthetase (Wild et al. 1985) and directly with anion uptake (Trogisch et al.1989). In nitrate uptake and assimilation experiments the different action of the two inhibitors appears especially in the first minutes showing almost unaffected nitrate uptake in the presence of 6-methylpurine, but strong inhibition in the presence of glufosinate (Fig. 4). Under normal conditions in *Lemna* (Fig. 2 A), initial nitrate accumulation by mere uptake is completely prevented by high assimilation rates. Thus inhibitor experiments allow a separation in the time course.

This is even more so with *ammonium uptake* which can be regarded as energy-independent influx of a cation into the negatively charged cells. Hence the uptake rates normally do not differ between light and dark as demonstrated for the initial uptake phase after N-starvation (Fig. 5). Later (cf. Fig. 2 B), ammonium assimilation rates determine the uptake rate. Accordingly much more ammonium is consumed in the light than in the dark after the initial phase, and addition of glucose may establish intermediate conditions in the dark. Only if the entire energy supply of the cell is cut off and E_m of the plasmalemma is low, as in the presence of the uncoupler CCCP, is the initial ammonium uptake rate partly inhibited, in the light as much as in the dark.

Tungstate can replace molybdate in the iron-molybdenum complex of NR and thus leads to an inactive form. In spite of side effects as expected with many heavy metals,

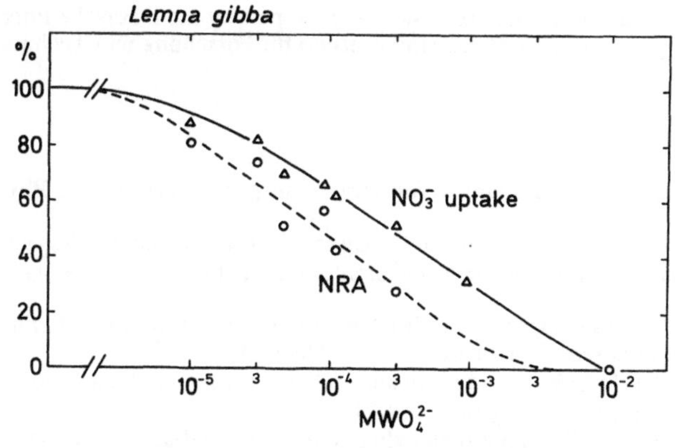

Fig. 6. Relative nitrate uptake rates and nitrate reductase activity (NRA) in *Lemna gibba* after 5 d treatment with tungstate (in the absence of Mo). 0.1 mM KNO_3, pH 6.5

tungstate inactivation can serve as a criterion for common or separate prosthetic groups in NR and the nitrate carrier of the plasmalemma. In *Lemna* plants pretreated for 5 days with various concentrations of WO_4^{2-} (Fig. 6), nitrate uptake was also sensitive to tungstate, but considerably less than NR, especially visible as relative inhibition at higher concentrations. This result adds again to the evidence in favour of different systems for nitrate uptake and reduction.

Conclusion

The data presented in this article, resulting from several independent *in-vivo* experiments, show that nitrate uptake and nitrate reduction can be at least partly separated by physiological conditions as well as by using inhibitors. It is unlikely, therefore, that the inducible nitrate uptake system uses the protein of NR, but this does not exclude that the nitrate carrier at the plasmalemma and nitrate reductase in cytosol or plasmalemma have some peptide chains in common, which respond to antibodies. There may be a special membrane NR with low enzyme activity and a carrier function for nitrate. Independence of the saturable nitrate uptake system from reduction has been shown e.g. with mutants of the cyanobacterium *Synechococcus* (Omata et al. 1989, and this volume; Guerrero et al., this volume) and with NR-deficient mutants of *Arabidopsis* (Doddema and Telkamp 1979). The "linear", constitutive, system may reflect the action of anion channels. Because of its insensitivity to tungstate (Serra et al. 1978) an identity of this system with NR is also doubtful.

Acknowledgments. The work was financially supported in part by the Deutsche Forschungsgemeinschaft, and in part by a grant from the Bundesministerium für Forschung und Technologie (Bonn).

References

Butz RG, Jackson WA (1977) A mechanism for nitrate transport and reduction. Phytochemistry 16: 409-417

Di Martino Rigano V, Martello A, Di Martino C, Rigano C (1985) Effect of CO_2 and phosphate deprivation on the control of nitrate, nitrite and ammonium metabolism in *Chlorella*. Physiol Plant 63: 241-246

Doddema H, Telkamp GP (1979) Uptake of nitrate by mutants of *Arabidopsis thaliana*, disturbed in uptake or reduction of nitrate. II. Kinetics. Physiol Plant 45: 332-338

Eisele R, Ullrich WR (1977) Effect of glucose and CO_2 on nitrate uptake and coupled OH^- flux in *Ankistrodesmus braunii*. Plant Physiol 59: 18-21

Jones GJ, Morel FMM (1988) Plasmalemma redox activity in the diatom *Thalassiosira*. A possible role for nitrate reductase. Plant Physiol 87: 143-147

Keller BU, Hedrich R, Raschke K (1989) Voltage-dependent anion channels in the plasma membrane of guard cells. Nature 341: 450-453

Köhler K, Steigner W, Simonis W, Urbach W (1985) Potassium channels in *Eremosphaera viridis*. I. Influence of cations and pH on resting membrane potential and on action-potential-like response. Planta 166: 490-499

Omata T, Ohmori M, Arai N, Ogawa T (1989) Genetically engineered mutant of the cyanobacterium *Synechococcus* PCC7942 defective in nitrate transport. Proc Natl Acad Sci USA 86: 6612-6616

Roos W (1989) Cytosolic control of vacuolar amino acid transport in *Penicillium cyclopium*. In: Dainty J, De Michelis MI, Marrè E, Rasi-Caldogno F (eds) Plant Membrane Transport: The Current Position. Elsevier, Amsterdam New York Oxford, pp 393-396

Serra JL, Llama MJ, Cadenas E (1978) Nitrate utilization by the diatom *Skeletonema costatum*. II. Regulation of nitrate uptake. Plant Physiol 62: 991-994

Syrett PJ (1981) Nitrogen metabolism of microalgae. Can Bull Fish Aquat Sci 210:182-210

Syrett PJ (1987) Nitrogen assimilation by eucaryotic algae. In: Ullrich WR, Aparicio PJ, Syrett PJ, Castillo F (eds) Inorganic Nitrogen Metabolism. Springer, Berlin Heidelberg New York, pp 25-31

Trogisch GD, Köcher H, Ullrich WR (1989) Effects of glufosinate on anion uptake in *Lemna gibba* G1. Z Naturforsch 44c: 33-38

Ullrich WR (1983) Uptake and reduction of nitrate: Algae and fungi. In: Läuchli A, Bieleski RL (eds) Inorganic Plant Nutrition. Encyclopedia of Plant Physiology, NS vol 15A. Springer, Berlin Heidelberg New York, pp 376-397

Ullrich WR (1987) Nitrate and ammonium uptake in green algae and higher plants: Mechanism and relationship with nitrate metabolism. In: Ullrich WR, Aparicio PJ, Syrett PJ, Castillo F (eds) Inorganic Nitrogen Metabolism. Springer, Berlin Heidelberg New York, pp 32-38

Ullrich WR, Larsson M, Larsson C-M, Lesch S, Novacky A (1984) Ammonium uptake in *Lemna gibba* G1, related membrane potential changes, and inhibition of anion uptake. Physiol Plant 61: 369-376

Ullrich WR, Jaenicke H, Brandl G (1989) Nitrate uptake in roots: Induction and charge balance. In: Dainty J, De Michelis MI, Marrè E, Rasi-Caldogno F (eds) Plant Membrane Transport: The Current Position. Elsevier, Amsterdam New York Oxford, pp 335-338

Ullrich WR, Novacky A (1981) Nitrate-dependent membrane potential changes and their induction in *Lemna gibba* G1. Plant Sci Lett 22: 211-217

Ward MR, Tischner R, Huffaker RC (1988) Inhibition of nitrate transport by anti-nitrate reductase IgG fragments and the identification of plasmamembrane associated nitrate reductase in roots of barley seedlings. Plant Physiol 88: 1141-1145

Warner RL, Huffaker RC (1989) Nitrate transport is independent of NADH and NAD(P)H nitrate reductases in barley seedlings. Plant Physiol 91: 947-953

Wild A, Sauer H, Rühle W (1987) The effect of phosphinothricin (glufosinate) on photosynthesis. I. Inhibition of photosynthesis and accumulation of ammonia. Z Naturforsch 42c: 263-269

NEW REGULATORY STEPS IN NITRATE ASSIMILATION OF LOWER AND HIGHER PLANTS

R. Tischner [1]

Introduction

Nitrate is the major nitrogen source for most plants in the environment. However its availibility is low due to strong competition and to the low binding capacity to soil particles. Ammonium, the alternative nitrogen source, is bound by the cation exchange capacity of soils. However, major pH problems often arise from ammonium utilization by plants.

In nitrate assimilation nitrate reductase has been considered as the main rate-limiting enzyme (Beevers and Hageman 1983). Only recently has the respective role of nitrate uptake been considered. In this paper new data on nitrate uptake are reported and evidence for a nitrate carrier function of membrane-bound nitrate reductase is presented along with data on a possible regulatory step at xylem loading.

Material and Methods

The experiments presented in this paper were performed using the unicellular alga *Chlorella sorokiniana* and barley (*Hordeum vulgare* L. var. *distichon* Alef cv. Aura). The methods used have been published recently (Behl et al. 1988; Ward et al. 1988; Tischner et al. 1989).

Results and Discussion

Uptake of nitrate and potassium

After a transfer of nitrogen-starved excised barley roots into 0.2 mM potassium nitrate, uptake of potassium started immediately with a high rate, close to the highest rate observed during the whole experiment. The uptake of nitrate began with a rate of 1 µmol g^{-1} fwt h^{-1}, increased very slowly during the first hour, as shown by high time-resolution experiments, then continuously and more rapidly and reached the K^+-uptake rate after 4 hours. This indicates that during the induction period a nitrate uptake system is continuously synthesized as soon as an induction signal has reached the genetic code.

---

[1] Pflanzenphysiologisches Institut der Universität, Untere Karspüle 2, 3400 Göttingen, FRG

During the initial lag the nitrate uptake rate was never zero, a low-capacity uptake system was always present. This system seemed to act as a sensing mechanism which indicated the presence of even low amounts of nitrate accessible to the roots. Lee and Drew (1986) presented similar results. For the induction process protein synthesis was required as shown by nitrate uptake inhibition with p-fluorophenylalanine (pFPA). However, this protein synthesis inhibitor had no effect on the low capacity uptake mechanism. The decline in K^+-uptake rate due to the application of pFPA was caused by an inhibitory effect on K^+ translocation, i.e. the xylem loading mechanism (Behl et al. 1988). This mechanism was also affected in the case of nitrate-induced plants. In such plants the uptake rate was around 10 µmol g^{-1} fwt h^{-1} and pFPA inhibited also translocation to the shoot. The rate of accumulation accounted for the uptake rate in this case. We conclude from these results that also the xylem-loading mechanism might be a regulatory step in nitrate assimilation for the whole plant. It should be especially important in plants which reduce most of the nitrate in the shoot. One can expect a feed-back regulation of nitrate uptake when the xylem loading mechanism is inhibited. When long-distance translocation starts, the accumulation of nitrate in the roots is already completed. Thus, under steady-state conditions translocation seems to be the most important sink for the nitrate taken up, and xylem loading seems to be a controlling step.

From the data mentioned afore one can conclude the existence of two nitrate uptake mechanisms, a basic constitutive system and an "inducible" system. From recent results on sugarbeets (Mäck 1988) it seems that both systems differ only in the number of identical carriers rather than in their quality (K_s is the same, but V_{max} is higher after induction).

Plasma membrane-bound nitrate reductase

In both *Chlorella* cells and barley roots a nitrate reductase activity bound at the plasmalemma has been found (Ward et al. 1988; Tischner et al. 1989). This protein was also detected after gel electrophoresis and Western blotting of purified plasma membranes using an antibody raised against soluble *Chlorella* nitrate reductase. The subunit molecular weight after SDS-gel electrophoresis was 60 kDa (data not shown). It is not clear yet what the molecular weight of the native protein may be, but it is, at the moment, difficult to explain nitrate reduction with NADH by a protein with 60 kDa subunit size. The subunit size of nitrate reductase reported so far was in the range of 90-105 kDa. Recent results from our laboratory showed that all partial reactions typical for the soluble nitrate reductase are also catalyzed by the plasmalemma-bound form. Hence the difference in subunit molecular weight must be based rather on smaller interdomain regions than on changed sizes of the catalytically active sites. The plasmalemma-bound nitrate reductase is a glycosylated protein and can be activated by phospholipids after being purified by affinity chromatography (unpublished results).

In *Chlorella* the two forms of nitrate reductase seemed to be regulated in a different way with respect to the effect of NH_4^+. In cells grown in NH_4^+-medium the plasmalemma-bound nitrate reductase was present, while the soluble form of this enzyme was totally lacking. It seems as if the repression of the nitrate reductase gene by NH_4^+ affected only the gene coding for the soluble form. But the membrane-bound enzyme may also be more stable due to protection against proteases.

The membrane-bound form of nitrate reductase could be involved in the uptake of nitrate. Fragments of the antibody raised against soluble nitrate reductase inhibited nitrate uptake (but not nitrite uptake) in both *Chlorella* cells and barley roots. The intact antibody showed the same activity as the fragments in inhibiting nitrate uptake of protoplasts, indicating that its access to the plasmalemma of intact cells was limited by size.

Butz and Jackson (1977) proposed that nitrate reductase may be a part of the enzyme complex in the plasmalemma responsible for nitrate uptake and partial reduction. Our results support this idea. The uptake of nitrate seems to be mediated by a 60 kDa protein (subunit) which is located within the plasmalemma. Some characteristics of this protein are reported in this paper. Based on reconstitution experiments we will demonstrate the carrier function of this protein. It is necessary to postulate a mechanism which might involve reduction of part of the nitrate during transport. However, according to the widely known nitrate accumulation in many plant tissues the stoichiometry between transport and reduction must be different from one.

References

Beevers L, Hageman RH (1983) Uptake and reduction of nitrate: Bacteria and higher plants. In: Läuchli A, Bieleski RL (eds) Encyclopedia of Plant Physiology, NS vol 15A, Springer, Berlin Heidelberg New York, pp 351-375

Behl R, Tischner R, Raschke K (1988) Introduction of a high-capacity nitrate-uptake mechanism in barley roots prompted by nitrate uptake through a constitutive low-capacity mechanism. Planta 176: 235-240

Butz RG, Jackson WA (1977) A mechanism for nitrate transport and reduction. Phytochemistry 16: 409-417

Lee RB, Drew MC (1986) Nitrogen-13 studies of nitrate fluxes in barley roots. J Exp Bot 37: 1768-1779

Mäck G (1988) Untersuchungen zum Stickstoffmetabolismus in Zuckerrübenpflanzen unterschiedlicher Entwicklungsstadien. Dissertation 130 p, Göttingen

Tischner R, Ward MR, Huffaker RC (1989) Evidence for a plasmamembrane-bound nitrate reductase involved in nitrate uptake of *Chlorella sorokiniana*. Planta 178: 19-24

Ward MR, Tischner R, Huffaker RC (1988) Inhibition of nitrate transport by anti-nitrate reductase IgG fragments and the identification of plasma membrane associated nitrate reductase in roots of barley seedlings. Plant Physiol 88: 1141-1145

ACCUMULATION AND VACUOLAR REMOBILIZATION OF NITRATE IN *CATHARANTHUS ROSEUS* CELLS

G. Marigo [1], H. Bouyssou [1], A.M. Pennarun [2] and J.P. Rona [2]

Introduction

Most of the higher plants growing in soils obtain their nitrogen in the form of nitrate. Control of the availability of intracellular nitrate has generally been considered to function via regulation of enzymes involved in nitrate assimilation. However, nitrate which can be accumulated in large quantities in certain crop plants, was shown to be mainly located in the vacuoles, and this vacuolar pool is probably a nitrogen reserve. The transfer processes for nitrate into the different compartments of the cells can thus be considered as limiting steps in nitrate assimilation.

In cell suspensions of *Catharanthus roseus* we demonstrated that nitrate is rapidly absorbed from the culture medium and accumulated reversibly (Fig. 1). Moreover, an inverse correlation was observed in the time course between changes in the nitrate concentrations and the cellular levels of another anion present in large amounts, namely malate. Direct analysis of isolated protoplasts and vacuoles showed that the bulk of malate and nitrate in *Catharanthus roseus* cells is located inside the vacuole. We took advantage of this particular situation to investigate the mechanism involved in nitrate absorption (plasmalemma transport), nitrate transfer into the vacuolar compartment and the subsequent remobilization from this storage site. Special attention was paid to the relationships between the displacements of nitrate and malate.

Nitrate Absorption by *Catharanthus roseus* Cells: Role of the Vacuolar Efflux of Malate

Transfer of *C. roseus* cells into a fresh medium was marked during the first 48 hours by a decrease in the malate level (Fig. 1) which can be considered as a net vacuolar efflux since malate is mainly located in the vacuoles (Marigo et al. 1985). After its transfer out of the vacuole, malic acid is not detectable in the culture medium. Moreover, the decrease in malate concentration is accompanied by an alkalinization of the medium and by CO_2 release from the cells. Malate coming from the vacuole must therefore be essentially metabolized to CO_2 in the cytoplasm and the oxidation products (CO_2 + HCO_3^-) are released by the cells.

[1] Université Paul Sabatier, route de Narbonne, F-31062 Toulouse Cédex, France;
[2] Université Paris VII, 2 place Jussieu, F-75251 Paris Cédex, France

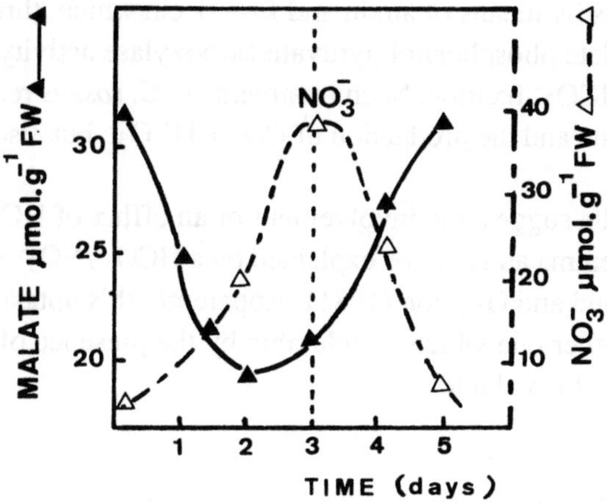

Fig. 1. Time course of changes in malate (▲) and nitrate (△) concentrations in cells of *Catharanthus roseus* after transfer to a fresh medium

The decrease in malate concentration occurs concurrently with an intracellular accumulation of nitrate (Fig. 1). Different observations indicate that the consumption of malate can be considered as a source of HCO_3^- for nitrate uptake:

- First, malate is not exported into the medium, which excludes the possibility of a malate/nitrate antiport at the plasmalemma. Another possible mechanism explaining the inverse relationships could involve malate/nitrate exchange at the tonoplast. Nevertheless, the decrease in malate is also observed with cells incubated in a medium without NO_3^-. These data do not exclude the existence of anionic exchange at the tonoplastic level (see data in the next paragraph), but they show that the initial step of the modifica-tions observed is directly linked to vacuolar remobilization and to subsequent metabo-lization of malate.

- Second, the products resulting from nitrate metabolism are not involved in nitrate uptake. In this way, the remobilization of malate and its oxidation could produce NADH for NO_3^- reduction. Complete reduction of NO_3^- to NH_4^+ produces OH^- which has been suggested to take part in antiport mechanisms involved in nitrate uptake (Thibaud and Grignon 1981). To test this hypothesis, NO_3^- reduction was inhibited by tungstate which suppresses nitrate reductase activity (Srivastava 1980), but this treatment did not decrease either NO_3^- uptake or malate consumption. These results indicate a relative independence between NO_3^- reductase activity and NO_3^- uptake in *C. roseus* cells.

- Finally, the last possibility to explain the increase in NO_3^- uptake could imply HCO_3^- as an oxidation product of malate. With this idea in mind, we modulated the level of HCO_3^- in the cells by means of auxin and fusicoccin which, through changes in cytoplasmic pH, stimulate phosphoenol pyruvate carboxylase activity (Guern et al. 1982) and consequently HCO_3^- fixation. Such treatments in *C. roseus* reduce not only the remobilization of malate and the production of $CO_2 + HCO_3^-$, but also the accumulation of nitrate.

Altogether, these results suggest the involvement of an efflux of HCO_3^- in nitrate uptake across the plasmalemma and may be explained by a HCO_3^-/NO_3^- antiport as already suggested by Thibaud and Grignon (1981). Apparently this uptake system has a basal component and another one which is inducible by the presence of nitrate in the medium (cf. Rona et al. in this volume).

Vacuolar Uptake of Nitrate

The dissipation of the membrane potential, which is generated by proton pumps, has been used to obtain qualitative estimates of anion permeability and information about the mechanism involved in trans-tonoplastic ion transfer (Kaestner and Sze 1987). In *Catharanthus roseus* we applied this procedure to study vacuolar NO_3^- uptake. Vacuoles were isolated by an osmotic lysis of protoplasts; they hydrolysed ATP with the generation of a proton-electrochemical gradient, positive and acidic inside (Marigo et al. 1988). The membrane potential (E_m) was determined from equilibrium distribution of the highly permeant labeled anion ^{14}C-thiocyanate. After development of E_m, the ATPase was inhibited by chelating Mg^{2+} with EDTA to exclude side effects of anions on the enzyme activity. In this experimental system various anion species were compared for their ability to break down E_m (Table 1). The results show that NO_3^- was by far the most effective, Cl^- had 60% less effect, malate only little, HPO_4^{2-} and SO_4^{2-} no effect at all. As indicated by other results the decreased potential exhibited saturation kinetics for Cl^-, whereas that produced by NO_3^- consisted of a saturable and a linear component (data not shown). The saturable component had a K_m of about 2 mM for NO_3^- and 10 mM for Cl^-. As already suggested by others (Kaestner and Sze 1987), these results demonstrate that NO_3^- and Cl^- are transported into the vacuole by carrier proteins. According to the uptake kinetics (K_m, V_{max}), however, NO_3^- appears to be the major anionic species moving into the vacuolar compartment under physiological conditions. This observation is consistent with the high concentrations of NO_3^- found in vacuoles of *C. roseus* and with the role of the vacuole as a storage compartment for NO_3^-.

In intact cells, the vacuolar uptake of nitrate is accompanied by an efflux of malate (Fig. 1). To examine a possible effect of nitrate on malate remobilization, vacuoles were preloaded with ^{14}C-malate in the presence of Mg-ATP, a condition which favours its uptake (Marigo et al.1988). When, after 2 hours, the distribution equilibrium was attained, the ATPase was inhibited by EDTA and the relative rate of malate efflux estimated from the decrease in malate concentration. Among the inorganic anions studied only NO_3^- caused a significant efflux of ^{14}C-malate (Table 1). This exchange NO_3^- /mal^{2-} can not be depicted as a strict antiport mechanism since SCN^-, a highly permeant anion, showed a similar effect. Based on this observation, it may be concluded rather that the release of malate is due to dissipation of E_m by nitrate transport.

Table 1. Effect of different anions on the vacuolar remobilization of ^{14}C-SCN^- (as evidence of membrane potential dissipation) and of ^{14}C-malate

Salts added (50 mM)	^{14}C-SCN- mobilized	^{14}C-Malate mobilized
	[pmol h^{-1} (10^6 vacuoles)$^{-1}$]	
Tris-NO_3	220	88
Tris-Cl	94	12
Tris-PO_4	8	0
Tris-SO_4	6	0
Tris-Malate	34	112
NaSCN	186	72

Vacuolar Nitrate Remobilization: a Coupling Mechanism with Malate Uptake

As a physiologically important ion, nitrate must be not only stored in the vacuoles but also released from them for metabolic use. The reversibility of accumulation could be clearly distinguished in *C. roseus* cells from the third day of growth onwards (Fig. 1). From this time NO_3^- concentrations decreased gradually during the next 2 days and this was related to an increase in intracellular malate concentration. Both anions, which may accumulate up to high concentrations (about 35 mM) are, as expected, essentially located in the vacuole (Marigo et al. 1985). As their changes in total cell concentration can be roughly attributed to changes in vacuolar concentrations, this system was adequate to study, in intact cells, nitrate remobilization from the vacuoles, particularly in relation to malate accumulation.

Previous work showed that malate uptake into the vacuoles occurred together with uptake of 1 H^+ (1 mal^{2-} + 1 H^+) (Marigo et al. 1983). To answer the question of ionic balance, we tried to show an exchange mechanism between mal^{2-} - H^+ and NO_3^- by modulating the level of malate and investigating the effects on the whole system. From these experiments it followed that each modification which increased (addition of sucrose) or decreased (treatment with metabolic inhibitors) the level of malate also induced, in an opposite way, changes in nitrate concentration. In any case, NO_3^- and mal^{2-} - H^+ transfers were not stoichiometrically equilibrated (about 3 NO_3^- for 1H mal^-) suggesting that other ions were involved in the exchange mechanism.

It is largely accepted that the vacuolar accumulation of many solutes is reversible, but little is known about the effluxes and their regulation, more especially in relation to a control by cytoplasmic events. In *C. roseus* cells we studied the extent of vacuolar nitrate remobilization in response to modifications of nitrate reduction, a step which consumes a great amount of reducing power in the cytoplasm. We therefore applied to the cells various treatments adequate to alter cytosolic proton concentration, and we checked their possible effects on the mobilization of vacuolar nitrate. Thus 5,5-dimethyloxazolidine-2,4-dione (DMO) was applied at different concentrations. This weak acid crosses the plasma membrane in its non-ionized form and releases protons within the cytosol. In the presence of DMO, nitrate mobilization occurs more rapidly than in the control, and the effect is directly correlated with the DMO concentrations used (Boudet et al. 1987). These results strongly suggest that:

- reducing power in the cytosol could in part control the rate of nitrate reduction and secondarily nitrate efflux from the vacuole,
- mobilization of nitrate in the vacuole could play a part in the maintenance of cytoplasmic homeostasis.

It is widely accepted that nitrate inhibits the tonoplast H^+-ATPase and this property is quite consistent with a possible autoregulation by cytoplasmic nitrate concentration. A reduction of proton influx at the tonoplast would indeed increase nitrate assimilation.

Conclusion

Interactions between malate and nitrate metabolism have been described several times in the literature (Beevers and Hageman 1980), malate oxidation providing a major source of NADH for nitrate reduction. The evidence summarized in this paper demonstrates another level of interactions in which malate plays a central role in controlling NO_3^- transfer across the cell membranes. First, it will provide a source of HCO_3^- for nitrate uptake across the plasmalemma according to a proposed HCO_3^-/NO_3^- antiport mechanism, which may be acting especially in cells with relatively low membrane potentials. Second, it is involved as a counter-ion in trans-tonoplast influx and efflux of NO_3^-. By this and in addition to their trophic significance, the reciprocal malate and nitrate accumulation may also play a role in the ionic and osmotic balance of the cell.

References

Beevers L, Hageman RH (1980) Nitrate and nitrite reduction. In: Stumpf PR, Conn EE (eds) The Biochemistry of Plants, vol 5. Acad Press, New York London, pp 115-168

Boudet AM, Alibert G, Marigo G, Ranjeva R (1987) The vacuole: Possible role in signal transduction versus cytoplasmic homeostasis. In: Marin B (ed) Plant Vacuoles. Life Sciences, Plenum, New York, pp 427-437

Guern J, Kurkjian A, Mathieu Y (1982) Hormonal regulation of intracellular pH: Hypotheses versus facts. In: Wareing PF (ed) Plant Growth Substances. Aberystwyth, pp 427-437

Kaestner KH, Sze H (1987) Potential-dependent anion transport in tonoplast vesicles from oat roots. Plant Physiol 83: 483-489

Marigo G, Delorme YM, Lüttge U, Boudet AM (1983) Rôle de l'acide malique dans la régulation du pH vacuolaire dans les cellules de *Catharanthus roseus* cultivées *in vitro*. Physiol Vég 21: 1135-1144

Marigo G, Bouyssou H, Belkoura M (1985) Vacuolar efflux of malate and its influence on nitrate accumulation in *Catharanthus roseus* cells. Plant Sci 39: 97-103

Marigo G, Bouyssou H, Laborie D (1988) Evidence for malate transport into vacuoles isolated from *Catharanthus roseus* cells. Botanica Acta 101: 187-191

Srivastava HS (1980) Regulation of nitrate reductase activity in higher plants. Phytochemistry 19: 725-728

Thibaud JB, Grignon C (1981) Mechanism of nitrate uptake in corn roots. Plant Sci Lett 22: 279-289

NITRATE UPTAKE IN *CATHARANTHUS ROSEUS* CELLS: ELECTROPHYSIOLOGICAL EFFECTS

J.P. Rona [1], M. Monestiez [1], A.M. Pennarun [1], M. Convert [1], D. Cornel [1], U. Bousquet [1], R. Kiolle [1] and G. Marigo [2]

Introduction

Data on electrophysiological parameters of nitrate uptake are rather rare in plant cells. The distinction between the electrical effect of NO_3^- at the plasmalemma and at the tonoplast turns out to be too complex for a heterogeneously polarized system, especially because the localization of the microelectrode tip in the cytosol or vacuole is often unknown. Transcellular potential changes connected to an external alkalinization upon addition of nitrate have often been considered as an indication of an unbalanced stoichiometry of a nitrate cotransport mechanism located at the plasma membrane. The wide-spread neglect of the positive trans-tonoplast potential in studies of nitrate effects on membrane potential can lead to errors in interpreting hyperpolarization phenomena (Rona et al. 1980 a,b; Barbier-Brygoo et al. 1985; Chedhomme and Rona 1988). For cell hyperpolarization, it has been suggested that NO_3^-/OH^- antiport functions as the main transport mechanism for nitrate across the plasmalemma (Thibaud and Grignon 1981; Monestiez et al. 1987), in the case of cell depolarization upon addition of nitrate a $NO_3^-/2H^+$ symport (Ullrich and Novacky 1981; Ullrich 1987) has been proposed. Nevertheless, it has been reported in the literature that the electropositive gradient across the tonoplast is also affected by nitrate uptake into the vacuole, causing a dissipation of the passive trans-tonoplast potential (Poole and Blumwald 1987; Leigh and Pope 1987) and a decrease of the positive electrogenic component in connection with the partial inhibition of the tonoplast ATPase (Bennett and Spanswick 1984; Sze 1984; Jochem et al. 1984; Griffith et al. 1986; Chedhomme and Rona 1986). In both cases, depolarization of the vacuoles (inside positive), whether isolated or not (Monestiez et al. 1987), and tonoplast vesicles (Blumwald and Poole 1985; Kaestner and Sze 1987) was observed. The classic technique for measuring cell polarization and membrane potential changes upon addition of nitrate was using microelectrodes inserted into the cells, often into the vacuoles. What is the meaning of nitrate-induced hyperpolarization? Does it result (i) from a hyperpolarizing nitrate transport system on the plasmalemma, or (ii) from a depolarizing system on the tonoplast, or both, or (iii) from a very rapid over-reaction of repolarization? In order to clarify the possible role of both membranes and to explain the hyperpolarization of plant cells during nitrate up-

[1] Université Paris VII, 2 Place Jussieu, 75251 Paris Cedex 05.
[2] Université Paul Sabatier, Route de Narbonne, 31062 Toulouse, France

take, we present some data on the electrical characteristics of isolated *Catharanthus roseus* cells.

Materials and Methods

Culture techniques. The cells of *Catharanthus roseus* G. Don used in this study were kindly provided by Prof. Jean Guern (Laboratoire de Physiologie Cellulaire, CNRS, Gif sur Yvette, France). The free-cell suspension cultures (cell line C 20) were grown at 27°C in 1 liter flasks, containing 200 ml of Gamborg medium (Gamborg et al. 1968). The culture medium was modified as previously described (Marigo et al. 1985). The flasks, inclined by 45° and rotated at 60 r.p.m. were inoculated every week with 50 ml of cell suspension (density about 10^6 cells ml^{-1}). The multiplication rate of our cultures with a rotating shaker (doubling in 7 days) was about one half of that obtained with an orbital shaker at 120 r.p.m. (Courtois and Guern 1980; Marigo et al. 1985) and thus comparable with the rates achieved with the orbital shaker at 20°C (Courtois and Guern 1980). For this work we used 1 to 7 day old cultures, when the cells were in the exponential growth phase.
Electrical measurements. For the measurement of transcellular electrical profiles of *C. roseus* cells in Gamborg liquid medium, we used the method previously described for *Acer pseudoplatanus* cells (Rona et al. 1980 a; Tamponnet et al. 1988). The isolated cells were immobilized by means of a microfunnel (40 to 70 μm outer diameter) and controlled by a micromanipulator (De Fonbrune). They were impaled with high-impedance microelectrodes (Chedhomme and Rona 1988); the glass micropipettes were filled with 3 M KCl. The methods for controlling the location of the microelectrode tip within the cell (cytoplasm or vacuole) have been described previously (Rona et al. 1980 b).
Analysis. Potassium concentrations were determined by flame photometry (Technicon Mark V) and those of nitrate by colorimetry (Technicon Auto Analyser method).

Results

Nitrate uptake. During the initial seven days of growth, nitrate concentration in the external medium decreased from 28 to 5 mM, while the cellular dry weight increased (Fig. 1). At the same time, as shown previously by Marigo et al. (1985), a strong alkalinization (around 2 pH units) was observed in the bathing medium.

From a lower initial rate net nitrate uptake increased up to a steady state rate (Fig. 2). When this rate became constant, NO_3^- net accumulation was about nil. The time course of changes in nitrate content and nitrate concentrations in *Catharanthus roseus*

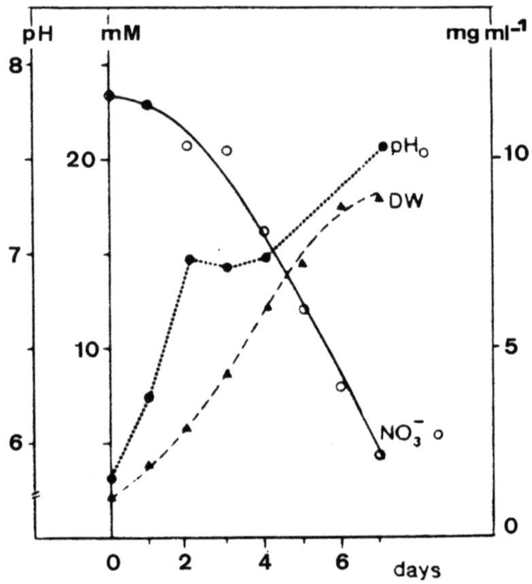

Fig. 1. Time course of external nitrate concentration, the external pH and dry weight during the cell growth in Gamborg medium. Dry weight is given for 1 ml cell suspension

cells is illustrated by Figure 3. The intracellular NO_3^- concentration decreased due to the vacuolar dilution by cell extension. Thus, during NO_3^- net uptake at steady state, the net transport of nitrate mainly occurred at the plasmalemma, and part of the NO_3^- ions absorbed were reduced directly in the cytosol. In our culture conditions a vacuolar accumulation of nitrate occurred during the acidification phase of the vacuolar compartment, as previously described (Marigo et al. 1986; see also Marigo et al., this volume). This phenomenon appeared after 6 days, when the vacuolar extension had ceased.

Table 1. Mean values of membrane potential (E_m) in *Catharanthus roseus* cells. Data given as means ±SD (n = number of measurements with separate cells). Measurements were done either in stationary conditions or in flux equilibria

Cell compartment	E_m (mV)	n
Cytoplasm (E_{co})	–71 ± 14	34
Vacuole (E_{vo})	–49 ± 11	45

Electrophysiological measurements. Data of electrical cell characteristics during the growth period (7 days) are summarized in Table 1. The extreme values of transcellular

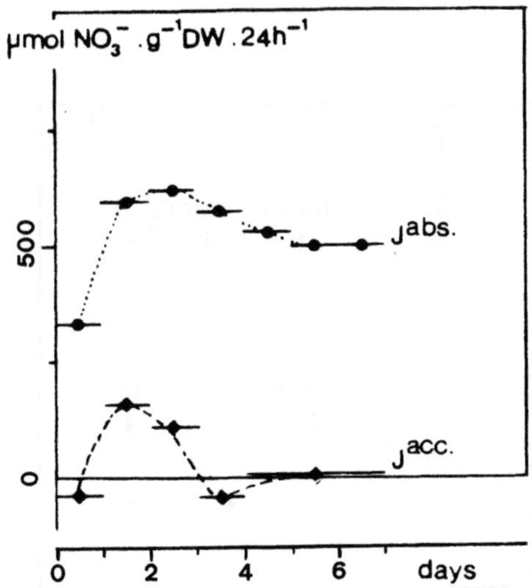

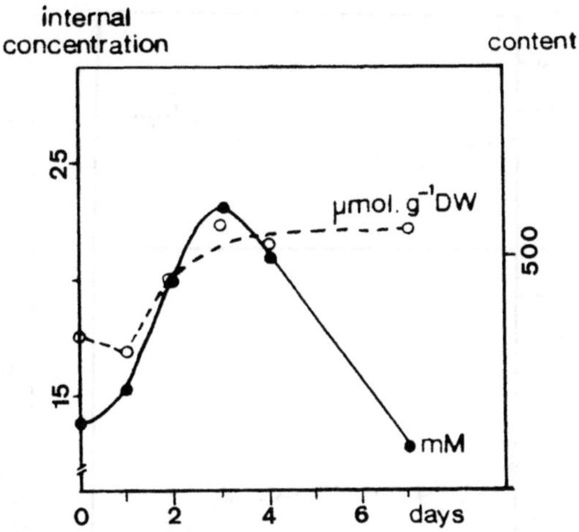

Fig. 2 *(left)*. Mean rates of net absorption (J_{abs}) and net accumulation (J_{acc}) of nitrate during *C. roseus* cell culture. The quantities absorbed were computed from the variation of the external concentration and the quantities accumulated from the cellular nitrate content. For each period of time, the given quantities were divided by the mean dry weight measured at the middle of this time

Fig. 3 *(right)*. Time course of changes in nitrate content and nitrate concentration in *C. roseus* cells. Concentrations computed from nitrate content divided by water content

electrical profiles were obtained 10 min after taking the experimental cells from the cell suspension. The electrical potential across the plasma membrane was negative (E_{co} about -71 mV, cytoplasm against outside) and the trans-tonoplast electrical potential was clearly positive (E_{vc} about +22 mV, vacuole against cytoplasm). The amplitudes of membrane potentials (plasmalemma and tonoplast) in isolated cells of *C. roseus* were similar to results obtained under the same conditions with isolated cells of *A. pseudoplatanus* (Rona et al. 1982). The high values of the standard deviation of E_{co} and E_{vo} were probably due to an increase in cell polarization with growth during the 7 days culture period (Fig. 4). Two distinct phases may be observed: (i) an initial phase (first 2 days) characterized by a strong vacuolar hyperpolarization (E_{vo} = -24 mV); (ii) a second phase (between the 2nd and the 6th day) which was characterized by stabilization of the vacuolar potentials.

More information was obtained from the time course of the trans-plasmalemma potential (E_{co}) during cell growth. While the nitrate uptake rate increased (Fig. 2), only a hyperpolarization of the plasmalemma was observed (Fig. 4) and this hyperpolarization remained stable at steady state. The slight E_m variation at the tonoplast cannot explain the time course of vacuolar E_m. In short-time experiments, 6 day old cells

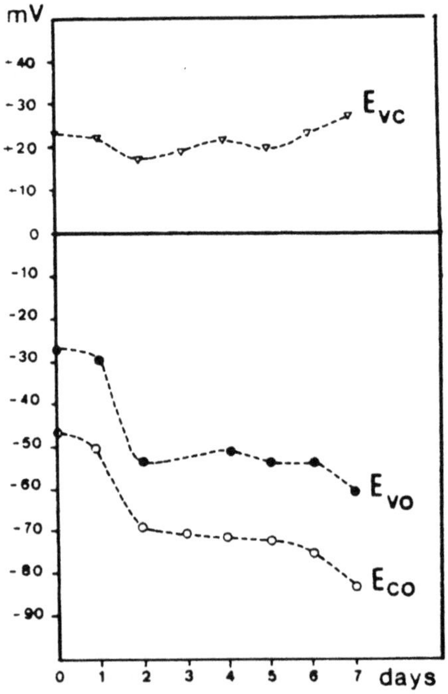

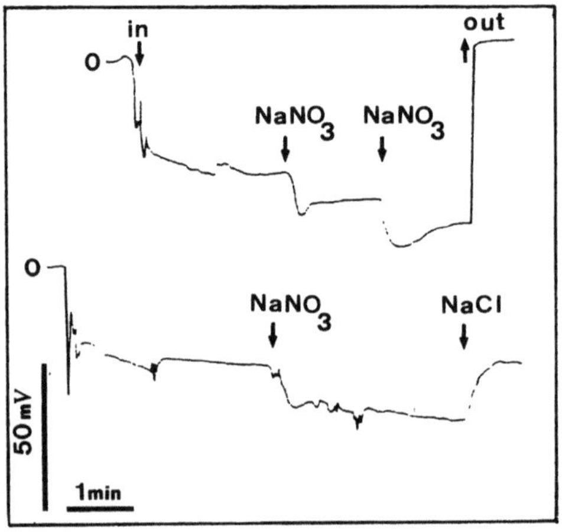

Fig. 4 *(left)*. Time course of the membrane potentials, E_{CO}, E_{VO} and E_{VC}, during the cell culture. E_{CO}: membrane potential at the plasmalemma (cytoplasm against outside); E_{VO}: vacuolar potential (vacuole against outside); E_{VC} : Membrane potential at the tonoplast (vacuole against cytoplasm)

Fig. 5 *(right)*. Membrane potential (E_{CO}) changes after addition of $NaNO_3$ (10 mM) and NaCl (10 mM, as control) to 6 day old cells in Gamborg medium

cells were sampled and variations of E_{CO} were measured after the addition of $NaNO_3$ (10 mM) to the medium. Figure 5 clearly shows that NO_3^- additions induce an immediate hyperpolarization of the plasmalemma (about -15 mV).

Conclusion

In *Catharanthus roseus* cells, the tonoplast depolarization cannot explain cell hyperpolarization linked to nitrate uptake. All results agree with the assumed functioning of an electrogenic NO_3^-/OH^- antiport at the plasmalemma, as suggested by Marigo (Marigo et al. 1985), and are in accordance with Thibaud's interpretation (Thibaud and Grignon 1981). A stimulation of the electrogenic H^+ pump due to cytoplasmic acidification by the NO_3^- influx (NO_3^- is a strong anion compared with OH^-) may also contribute to explain the hyperpolarization of the plasma membrane. To date, the stoichiometry of the antiport has not yet been determined.

Acknowledgments. This work was supported by grants from the CNRS (UA 1180) and from the MRT (85 C 1180). We thank Professor C. Grignon for his critical comments and discussion.

References

Barbier-Brygoo H, Gibrat R, Renaudin JP, Brown S, Pradier JM, Grignon C, Guern J (1985) Membrane potential difference of isolated plant vacuoles: positive or negative? II. Comparison of measurements with microelectrodes and cationic probes. Biochim Biophys Acta 819: 215-224

Bennett AB, Spanswick RM (1984) H^+-ATPase of storage tissue of *Beta vulgaris* L.: H^+/ATP stoichiometry of an anion-sensitive H^+-ATPase. Plant Physiol 74: 545-549

Blumwald E, Poole RJ (1985) Nitrate storage and retrieval in *Beta vulgaris*: Effect of nitrate and chloride on proton gradients in tonoplast vesicles. Proc Natl Acad Sci 82: 3683-3687

Chedhomme F, Rona JP (1986) Isolation and electrical characterization of tonoplast vesicles from the Kiwi fruit (*Actinidia chinensis*). Physiol Plant 67: 29-36 (1986)

Chedhomme F, Rona JP (1988) Electrical properties of tonoplast from cells and vesicles of storage Kiwi fruits. J Plant Physiol 133: 89-95

Courtois D, Guern J (1980) Temperature response of *Catharanthus roseus* cells cultivated in liquid medium. Plant Sci Lett 17: 473-482

Gamborg DL, Miller RA, Ojima M (1968) Nutrient requirements of suspension cultures of soybean root cells. Exp Cell Res 50: 151-158

Griffith CJ, Rea PA, Blumwald E, Poole RJ (1986) Mechanism of stimulation and inhibition of tonoplast H^+-ATPase by chloride and nitrate. Plant Physiol 81: 120-125

Jochem P, Rona JP, Smith AC, Lüttge U (1984) Anion-sensitive ATPase activity and proton transport in isolated vacuoles of species of the CAM genus *Kalanchoë*. Physiol Plant 62: 410-415

Kaestner HK, Sze H (1987) Potential-dependent anion transport in tonoplast vesicles from oat roots. Plant Physiol 83: 483-489

Leigh RA, Pope AJ (1987) Understanding tonoplast function: Some emerging problems. In: Marin BP (ed) Plant Vacuoles: Their Importance in Plant Cell Compartmentation and Their Applications in Biotechnology. Life Sciences Series 134, Plenum, New York, pp 101-110

Marigo G, Bouyssou H, Belkoura M (1985) Vacuolar efflux of malate and its influence on nitrate accumulation. Plant Sci Lett 39: 97-103

Marigo G, Bouyssou H, Boudet A (1986) Accumulation des ions nitrate et malate dans des cellules de *Catharanthus roseus* et incidence sur le pH vacuolaire. Physiol Vég 24: 15-23

Monestiez M, Belabed AM, Pennarun AM, Convert M, Cornel D, Rona JP (1987) Some characteristics of tonoplast NO_3^- transport processes on *Acer* cells. In: Marin BP (ed) Plant Vacuoles: Their Importance in Plant Cell Compartmentation and Their Applications in Biotechnology. Life Sciences Series 134, Plenum, New York, pp 231-237

Poole RJ, Blumwald E (1987) Transport of inorganic ions in tonoplast vesicles. In: Marin BP (ed) Plant Vacuoles: Their Importance in Plant Cell Compartmentation and Their Applications in Biotechnology. Life Sciences Series 134, Plenum, New York, pp 219-225

Rona JP, Van de Sype G, Cornel D, Grignon C, Heller R (1980 a) Plasmolysis effect on electrical characteristics of free cells and protoplasts of *Acer pseudoplatanus* L. Bioelectrochem and Bioenerg 7: 377-391

Rona JP, Pitman MG, Lüttge U, Ball E (1980 b) Electrochemical data on compartmentation into wall, cytoplasm and vacuole of leaf cells in the CAM genus *Kalanchoë*. J Membr Biol 57: 25-35

Rona JP, Cornel D, Grignon C, Heller R (1982) The electrical potential difference across the tonoplast of *Acer pseudoplatanus* cells. Physiol Vég 20: 459-463

Sze H (1984) Minireview: H^+-translocating ATPases of the plasma membrane and tonoplast of plant cells. Physiol Plant 61: 683-691

Tamponnet C, Rona JP, Barbotin NJ, Calvayrac R (1988) Effects of high external calcium concentration on etiolated *Euglena gracilis* Z cells and evidence of an internal potential. Biochim Biophys Acta 943: 87-94

Thibaud JB, Grignon C (1981) Mechanism of nitrate uptake in corn roots. Plant Sci Lett 22: 279-289

Ullrich WR, Novacky A (1981) Nitrate-dependent membrane potential changes and their induction in *Lemna gibba* G1. Plant Sci Lett 22: 211-215

Ullrich WR (1987) Nitrate and amonium uptake in green algae and higher plants: Mechanism and relationship with nitrate metabolism. In: Ullrich WR, Aparicio PJ, Syrett PJ, Castillo F (eds) Inorganic Nitrogen Metabolism. Springer, Berlin Heidelberg New York, pp 32-38

UPTAKE OF INORGANIC NITROGEN COMPOUNDS IN AN ACIDOPHILIC ALGA

A. Fuggi [1]

Introduction

Inorganic nitrogen compounds are often limiting for plant growth in natural environments. Their average concentration is at the micromolar level or below and can change largely with environmental stress (Raven 1980; Syrett 1981). Specific uptake systems, therefore, have been postulated and found at the plasmalemma to concentrate these compounds within the cell and thus sustain growth. Generally they are not constitutive and can be switched on when required to accommodate the environmental availability to the nutrition needs. Metabolic energy is eventually always required to maintain the uptake even if the primary transport processes can be directly driven by the electrochemical potential of the considered compound (Ullrich 1987). This is true for nitrate, the most widely available inorganic nitrogen compound for plant growth in terrestrial habitats as well as for ammonium or ureides more abundant than nitrate in the sea (Syrett 1981, 1987).

Here the uptake of nitrate, nitrite and ammonium in the acidophilic non-vacuolated red alga *Cyanidium caldarium* is discussed. Being strictly autotrophic and acidophilic, it exchanges basically only mineral ions with the medium and, in particular, the proton exchange dependent on uptake and metabolism of nitrogen compounds can be studied in conditions in which photosynthesis and respiration do not interfere.

Results and Discussion

Nitrate uptake

Two transport systems were shown to be operative in taking up nitrate in *Cyanidium caldarium*: a high-affinity and a low-affinity transport system (Fuggi et al. 1984). The high-affinity system was not constitutive in the alga: it was derepressed in cells transferred to low nitrate media or to nitrogen-starving conditions; nitrate was not necessary for induction, a feature found in several marine algae and cyanobacteria, but not in green algae, land plants and fungi (Syrett et al. 1981; Sivak et al. 1989; Tischner and Lorenzen 1979; Schloemer and Garrett 1974). Protein synthesis was a requirement for its expression.

[1] Dipartimento di Biologia Vegetale, Università di Napoli, Via Foria 223, I-80139 Napoli, Italy

The high-affinity nitrate uptake system was able to scavenge trace amounts of nitrate in the medium: it was operative in acidic but not in neutral conditions. Stoichiometric data of alkalinization with pulse amounts of nitrate in transitory conditions and the kinetics of nitrate uptake at various medium pHs fitted in with the mechanism of nitrate-proton symport of Figure 1 (Fuggi 1985, for details): the carrier could operate with a stoichiometry of 2 protons per nitrate and involve a tight coupling of protons and nitrate independent of the medium pH. The hypothesis that protons could be involved in taking up nitrate has been proposed in the last years, essentially from electrophysiological experiments: a stoichiometry with more than one proton per nitrate ion has been suggested to explain the electrical depolarization of the plasma membrane upon nitrate addition (Ullrich and Novacky 1981; Ullrich 1987).

The mechanism displayed in Figure 1 should be electrogenic unless other ions interact directly with the carrier. The exchange of protons and potassium ions between the cell interior and the medium upon nitrate additions (Fig. 2) suggested that this is the case (K^+ distribution in plant cells is essentially regulated by the membrane potential, Raven 1980).

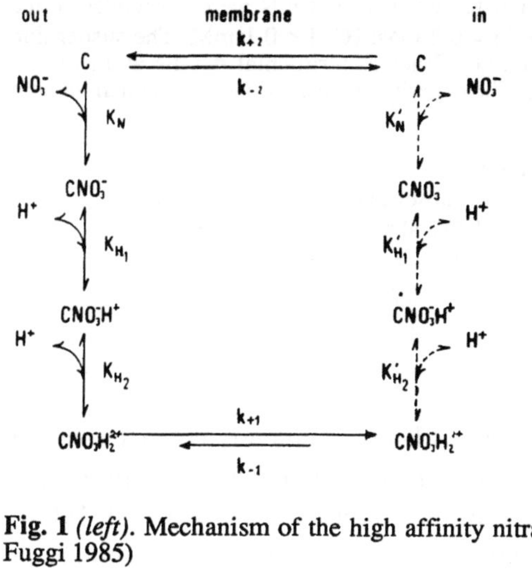

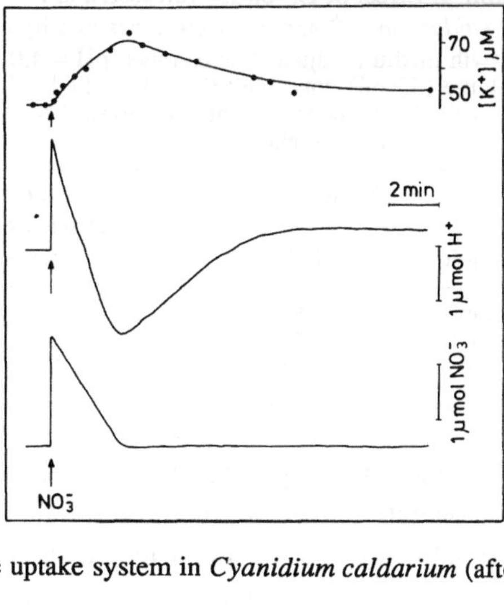

Fig. 1 *(left)*. Mechanism of the high affinity nitrate uptake system in *Cyanidium caldarium* (after Fuggi 1985)

Fig. 2 *(right)*. K^+ / H^+ exchange upon addition of a pulse of nitrate (HNO_3) to a cell suspension of *C. caldarium*. H^+ and NO_3^- were determined by specific electrodes; K^+ by atomic absorption spectrophotometry in aliquots of the suspension medium. Working suspension: volume = 60 ml; pH = 3.8; T = 37 °C; cell content = 0.38 ml packed cells; nitrate addition indicated by the arrow.

The H^+ / K^+ exchange upon addition of a pulse of HNO_3 to a cell suspension approached a stoichiometry of 2 protons taken up and 1 potassium ion released per nitrate, but only in low-chloride medium. Table 1 shows that K^+ was not tightly coupled to the nitrate carrier: K^+ release was suppressed in a medium supplemented with 40 mM Cl^-, whereas nitrate uptake suffered an inhibition of less than 10%. K^+ did not inhibit nitrate uptake significantly in acidic media even at concentrations exceeding 100 mM in the medium and, at neutrality, nitrate was not taken up even if the electrochemical potassium gradient across the plasmalemma exceeded 120 mV (not shown). These observations favour the idea of a potassium release just for charge balance: Cl^- influx or/and H^+ efflux could replace K^+ efflux (Table 1) (Hosoi et al. 1988). The acidification when nitrate was consumed (Fig. 2) indicates that the proton pump, driven by changes of the cytoplasmic pH and/or membrane potential (nitrate uptake through the carrier as a H^+ cotransport should depolarize the membrane and acidify the cytoplasm), restores the membrane potential and the charge balance for the overall assimilation of the added nitrate. A nitrate-proton symport with a stoichiometry of 2 H^+ per NO_3^-, linked to potassium release for charge balance, has been reported also in *Candida* (Eddy and Hopkins 1985).

Table 1. Effect of Cl^- on K^+ released and H^+ taken up per NO_3^- absorbed in a cell suspension of *C. caldarium.* H^+ and K^+ were measured by pH and K^+ electrodes. Cells were suspended in the growth medium adjusted as follows: pH = 4.0; $[K^+]$ = 0.2 mM; $[Cl^-]$ < 0.1 mM. The suspension volume (55 ml) contained 0.2 ml packed cell volume. Other experimental details as in Fig. 2. Basically the suspension contained 30 mM Na_2SO_4. The experiments were done at least half an hour after the additions of NaCl

Assay conditions	Additions (μmol) HNO_3	Stoichiometry K^+ released $/NO_3^-$ absorbed	Stoichiometry H^+ absorbed $/NO_3^-$ absorbed	Uptake rate μmol NO_3^- /min
Control	0.3	1.1	2.1	0.21
+ NaCl 7 mM	0.3	0.5	2.0	0.19
+ NaCl 40 mM	0.3	0.0	1.5	0.19

The low-affinity nitrate transport system has not been completely characterized. It was operative in cells grown on nitrate and in nitrogen-limited growth conditions. It was saturable and was not affected by changes of the medium pH (Fuggi et al. 1984). A NO_3^-/anion antiport mechanism as suggested in barley roots (Deane-Drummond and Glass 1983) remained questionable. However, this system has been suggested to work as a leak system in regulating the net nitrate uptake in *Cyanidium* (Fuggi 1989).

Nitrite Uptake

A general survey of the literature reveals that two mechanisms have been considered for nitrite transport within the cell: a passive diffusion of HNO_2 across the cell membranes (Ferrari and Varner 1971) and an energy-dependent process through a carrier taking up NO_2^-, as in marine diatoms (Eppley and Coatsworth 1968; Cresswell and

Syrett 1982). Both systems were operative in *Lemna* and in cyanobacteria (Ullrich 1987; Flores et al. 1987).

In *C. caldarium* both transport systems have been evidenced. Nitrous acid diffusion is normally found in acidic media. The pH gradient across the plasmalemma (the cell sap is neutral) allowed to trap NO_2^- within the cell to a high level, even in ammonium-grown cells where no significant nitrite reduction took place: the distribution of nitrite between medium and cell sap at acidic pH fitted in with the Henderson-Hasselbalch equation in the assumption that HNO_2 was the only permeating species (Fuggi, unpublished). Experiments with cells able to reduce nitrite (Table 2) allowed to discriminate between HNO_2 diffusion and a carrier-mediated NO_2^- transport by measuring the photosynthetic O_2 evolution dependent on nitrite assimilation (Fuggi 1989, for details). Nitrite uptake through diffusion of HNO_2 was largely exceeding the carrier-mediated NO_2^- uptake rate in physiological, acidic conditions: nitrite assimilation was not affected by ammonium that prevented nitrate utilization in *Cyanidium* even when $[NO_3^-]$ was much higher than $[NO_2^-]$. On the other hand, in conditions in which the uptake by nitrous acid diffusion was negligible (pH > 5 and at nitrite concentrations of few μM) (Table 2), the assimilation rate of nitrite was comparable to that of nitrate; it was not affected by ammonium but was strongly affected by nitrate added at the same concentration. No effect of nitrate was observed at higher nitrite concentration (300 μM).

Nitrite transport in the above conditions (nitrite uptake was the limiting step of the overall assimilation) shared many characteristics with the high affinity nitrate uptake system (Fuggi 1989): it was saturable and nitrate was a strong competitive inhibitor; the inhibition constant K_i of nitrate was comparable to $K_{1/2}$ of nitrate uptake; the maximal nitrite uptake rate was similar to V_{max} of nitrate uptake; at those concentrations no nitrite was taken up at neutrality, nor was nitrate.

For many organisms it has been suggested that nitrite and nitrate uptake can use the same carrier. In *Cyanidium* a comparison of K_i of nitrite uptake and $K_{1/2}$ of nitrate uptake (Fuggi 1989) suggested that the competitive inhibition by nitrate of nitrite uptake and not viceversa was due to a much higher affinity of the carrier for nitrate than for nitrite. This could be the case also in the organisms in which nitrate inhibits nitrite uptake as in green algae and fungi (Syrett and Morris 1963; Thacker and Syrett 1972; Schloemer and Garrett 1974). The reverse can occur when nitrite inhibits nitrate uptake as in brown algae and diatoms (Hanisak and Harlin 1978; Cresswell and Syrett 1982). Competitive interaction between nitrite and nitrate as in cyanobacteria and in higher plants (Madueño et al. 1987; Ullrich 1987) suggests that their affinity constants for the carrier have comparable values.

Ammonium Uptake

Ammonium transport systems have been found to be quite ubiquitous in bacteria, fungi, algae and higher plants (Kleiner 1981; Syrett 1987; Ullrich 1987). Apart from ammonia diffusion two mechanisms have been considered to drive ammonium uptake:

Table 2. Nitrite uptake and assimilation by *C. caldarium* cells grown in nitrogen-limited conditions and resuspended at two different pHs. The assimilation rates were evaluated by measuring the photosynthetic O_2 evolution dependent on nitrate and nitrite assimilation at the CO_2 compensation point (details in Fuggi 1989). Ammonium was added to the cell suspension at least 20 min before testing the activities. The initial O_2 evolution rates were determined upon addition of nitrite or nitrate to the suspension as indicated. Maximal photosynthesis is shown for comparison. Data in µmol O_2 evolved min^{-1} ml^{-1} packed cells

Suspension conditions	NO_2^- (4 µM)	NO_2^- (300 µM)	NO_3^- (4 µM)	NO_3^- (20 mM)	Photosynthesis
pH = 1.8					
control	0.99	---	0.51	0.60	4.50
+ NH_4^+ (2 mM)	0.95	---	0.01	0.02	4.40
+ NH_4^+ (2 mM)+ NO_3^- (20 mM)	0.95	---	---	---	4.45
pH = 5.5					
control	0.47	1.00	0.45	0.60	4.50
+ NH_4^+ (2 mM)	0.47	0.98	0.01	0.01	4.45
+ NH_4^+ (2 mM) + NO_3^- (4 µM)	0.03	0.98	---	---	4.50

(1) an electrophoretic uniport, taking advantage of the membrane potential (Kleiner 1981; Ullrich et al. 1984); (2) an ammonium-potassium antiport, taking advantage of the potassium gradient across the plasmalemma (Jayakumar et al. 1985).

In *C. caldarium* an ammonium transport system with a high affinity for ammonium ($K_{1/2}$ around 1 µM) has been described (Fuggi et al. 1981, 1984): it was active in acidic media and in aerobic conditions, its activity was increased when ammonium-grown cells were transferred to a nitrate- or nitrogen-limited medium, and protein synthesis was required for its expression (unpublished results).

An electrophoretic mechanism for this high-affinity ammonium transport system could be suggested from analyzing the transitory K^+/H^+ exchange dependent on ammonium uptake. NH_4^+ transport across the plasmalemma, in fact, transfers a positive charge and acidifies the cytoplasm (Bertl et al. 1984). Figure 3 shows a typical H^+/K^+ exchange pattern appearing on pulse addition of NH_4^+ to a cell suspension of *Cyanidium*. K^+ and H^+ were released just upon ammonium addition. The K^+ concentration in the medium rapidly reached a maximum that was maintained until ammonium was depleted. The H^+ release, on the other hand, exhibited a biphasical pattern: the first phase was simultaneous to potassium release, the second was delayed after ammonium addition and was dependent on further ammonium assimilation.

The ratio between released K^+ and the corresponding H^+ in the first stage for a certain suspension and a given addition of NH_4^+ was dependent on several factors. Table 3 shows the role of chloride: in cells resuspended in a low-potassium medium ($[K^+] <$ 1 mM) at low chloride ($[Cl^-] < 0.1$ mM), K^+ was the main ion released and H^+ had only a minor contribution; addition of chloride changed this figure by increasing the release of H^+ and decreasing that of K^+. Cl^- did not affect the overall initial release rate

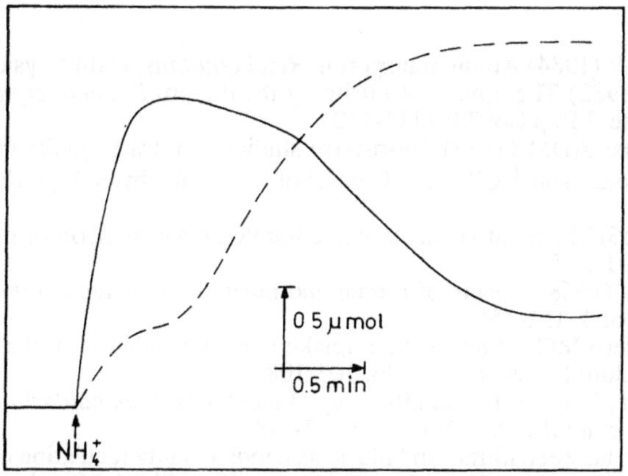

Fig. 3. H^+ (- - -) and K^+ (——) released upon addition of NH_4^+ to a cell suspension of *C. caldarium* grown in nitrogen-limited conditions. NH_4^+ (2.5 μmol) was added at the arrow. Working suspension: volume = 60 ml; pH = 3.56; $[K^+]$ = 0.6 mM; T = 37°C; cell content = 0.41 ml packed cells

Table 3. Effect of Cl^- on the K^+ and H^+ released per NH_4^+ added in a cell suspension of *C. caldarium.* Experimental details and conditions as in Table 2

Assay Conditions	NH_4^+ additions (μmol)	Stoichiometry K^+ released /NH_4^+ added	 H^+ released /NH_4^+ added	Steady uptake rate μmol NH_4^+ /min
Control	0.75	1.12	0.00	0.30
+ NaCl 7 mM	0.75	1.0	0.03	0.19
+ NaCl 50 mM	0.80	0.24	0.33	0.25

of H^+ and K^+ (not shown).

Methylammonium, the ammonium analog, taken up at a rate lower than ammonium, favoured H^+ efflux versus K^+ when added at comparable concentrations. Also in microaerobiosis, H^+ was the main ion released, the contribution of K^+ being strongly reduced. In addition, ammonium was taken up at significant rates also from a high-potassium medium ($[K^+] > 100$ mM). These results suggest that membrane potential is the driving force for ammonium uptake: K^+ efflux could be partially replaced by Cl^- influx for charge balance. The H^+ extruded (Table 3) could be related to activation of the proton pump at the plasmalemma driven by the decrease of cytoplasmic pH dependent on some H^+/Cl^- symport or equivalent mechanism.

Acknowledgments. This work has been supported by Ministero della Pubblica Istruzione and Consiglio Nazionale delle Ricerche of Italy.

References

Bertl A, Felle H, Bentrup FW (1984) Amine transport in *Riccia fluitans*. Plant Physiol 76: 75-78
Cresswell RC, Syrrett PJ (1982) The uptake of nitrite by the diatom *Phaeodactylum*: Interactions between nitrite and nitrate. J Exp Bot 33: 1111-1121
Deane-Drummond CE, Glass ADM (1983) Short-term studies of nitrate uptake into barley plants using ion-specific electrodes and $^{36}ClO_3^-$. 1. Control of net uptake by NO_3^- efflux. Plant Physiol 73: 100-104
Eddy AA, Hopkins PG (1985) The putative electrogenic nitrate-proton symport of the yeast *Candida utilis*. Biochem J 231: 291-297
Eppley RW, Coatsworth JL (1968) Uptake of nitrate and nitrite by *Ditylum brightwellii* - Kinetics and mechanisms. J Phycol 4: 151-156
Flores E, Herrero A, Guerrero MG (1987) Nitrite uptake and its regulation in the cyanobacterium *Anacystis nidulans*. Biochim Biophys Acta 896: 103-108
Fuggi A (1985) Mechanism of proton-linked nitrate uptake in *Cyanidium caldarium*, an acidophilic non-vacuolated alga. Biochim Biophys Acta 815: 392-398
Fuggi A (1989) Competition between nitrate and nitrite as a tool to study regulation of nitrate metabolism by ammonium in the microalga *Cyanidium caldarium*. Plant Physiol Biochem 27: 563-568
Fuggi A, Di Martino Rigano V, Vona V, Rigano C (1981) Nitrate and ammonium assimilation in algal cell suspensions and related pH variations in the external medium, monitored by electrodes. Plant Sci Lett 23: 129-138
Fuggi A, Di Martino C, Rigano V (1984) Ammonium transport systems in the acidophilic alga *Cyanidium caldarium*. Giorn Bot Ital 118 (Suppl 1): 60-61
Fuggi A, Vona V, Di Martino Rigano V, Di Martino C, Martello A, Rigano C (1984) Evidence of two transport systems for nitrate in the acidophilic thermophilic alga *Cyanidium caldarium*. Arch Microbiol 137: 281-285
Hanisak MD, Harlin MM (1978) Uptake of inorganic nitrogen by *Codium fragile* subsp. *tomentosoides* (Chlorophyta). J Phycol 14: 450-454
Hosoi S, Iino M, Shimazaki K (1988) Outward-rectifying K^+ channels in stomatal guard cell protoplasts. Plant Cell Physiol 29: 907-911
Jayakumar A, Epstein W, Barnes EM jr (1985) Characterization of ammonium/potassium antiport in *Escherichia coli*. J Biol Chem 260: 7528-7532
Kleiner D (1981) The transport of NH_4^+ across biological membranes. Biochim Biophys Acta 639: 41-52
Madueño F, Flores E, Guerrero GM (1987) Competition between nitrate and nitrite uptake in the cyanobacterium *Anacystis nidulans*. Biochim Biophys Acta 896: 109-112
Schloemer RH, Garrett RH (1974) Nitrate transport system in *Neurospora crassa*. J Bacteriol 118: 259-269
Sivak MN, Lara C, Romero JM, Rodríguez R, Guerrero MG (1989) Relationship between a 47 kDa cytoplasmic membrane polypeptide and nitrate transport in *Anacystis nidulans*. Biochem Biophys Res Commun 158: 257-262
Syrett PJ (1981) Nitrogen metabolism in microalgae. Can Bull Fish Aquat Sci 210: 182-210
Syrett PJ (1987) Nitrogen assimilation by eukaryotic algae. In: Ullrich WR, Aparicio PJ, Syrett PJ, Castillo F (eds) Inorganic Nitrogen Metabolism. Springer, Berlin Heidelberg New York, pp 25-31
Thacker A, Syrett PJ (1972) The assimilation of nitrate and ammonium by *Chlamydomonas reinhardii*. New Phytol 71: 423-433
Tischner R, Lorenzen H (1979) Nitrate uptake and nitrate reduction in synchronous *Chlorella*. Planta 146: 287-292
Ullrich WR (1983) Uptake and reduction of nitrate: Algae and fungi. In: Läuchli A, Bieleski RL (eds) Inorganic Plant Nutrition. Encyclopedia of Plant Physiology, NS vol 15A. Springer, Berlin Heidelberg New York, pp 376-397
Ullrich WR (1987) Nitrate and ammonium uptake in green algae and higher plants: Mechanism and relationship with nitrate metabolism. In: Ullrich WR, Aparicio PJ, Syrett PJ, Castillo F (eds) Inorganic Nitrogen Metabolism. Springer, Berlin Heidelberg New York, pp 32-38
Ullrich WR, Larsson M, Larsson C-M, Lesch S, Novacky A (1984) Ammonium uptake in *Lemna gibba* G1, related potential changes and inhibition of anion uptake. Physiol Plant 61: 369-376
Ullrich WR, Novacky A (1981) Nitrate-dependent membrane potential changes and their induction in *Lemna gibba* G1. Plant Sci Lett 22: 211-217

REGULATION OF INORGANIC NITROGEN METABOLISM IN *CHLAMYDOMONAS REINHARDTII*

J.M. Vega and A. Menacho [1]

Introduction

The primary N source for green algae and higher plants that are not associated with N-fixing bacteria is nitrate. It has been established that nitrate assimilation in the green alga *Chlamydomonas reinhardtii* involves three basic steps: (i) the uptake of nitrate, which requires metabolic energy and seems to be mediated by a specific permease, (ii) the reduction of nitrate to ammonium, which requires 8 e^- and the consecutive action of NAD(P)-nitrate reductase and Fd-nitrite reductase, and (iii) the incorporation of ammonium to carbon skeletons, as α-amino group of L-glutamate, which is catalyzed by the successive action of glutamine synthetase (GS) and glutamate synthase (GOGAT). In summary, the assimilation of 1 nitrate-N into L-glutamate requires 10 e^- and at least 2 ATP. The presence of other enzymes which may participate in the ammonium assimilatory reactions, such as glutamate dehydrogenase or alanine dehydrogenase, has been reported for *C. reinhardtii*; however, its involvement, if any, in ammonium assimilation seems to be occasional and/or restricted to specific nutritional conditions.

The synthesis of L-glutamine is catalyzed by glutamine synthetase. Two isoforms of this enzyme have been separated from *C. reinhardtii* cells, purified until electrophoretic homogeneity and characterized as plant type enzymes (Florencio and Vega 1983 a). On the other hand, two glutamate synthases with different specificity for electron donors, Fd- and NADH-GOGAT, have been distinguished in this alga (Cullimore and Sims 1981). Fd-GOGAT has been purified to electrophoretic homogeneity, and the enzyme is a single polypeptide chain containing 1 FAD, 1 FMN and 1 iron-sulfur cluster as prosthetic groups (Márquez et al. 1986 a).

Utilization of Inorganic Nitrogen

Table 1 shows the N-source consumption rate by cells under different nutritional conditions. Light is apparently required for inorganic nitrogen consumption by N-supplied cells, as it has been documented in green algae; however, cells in the presence of acetate or enriched in carbon reserves by subjecting them to N-starvation may

[1] Departamento de Bioquímica Vegetal y Biología Molecular, Facultad de Química, Apartado 553, 41080 Sevilla, Spain.

assimilate inorganic nitrogen in the dark (see Table 1), indicating that carbon skeletons are required, rather than light, as observed previously by Thacker and Syrett (1972). Consequently, under CO_2-limiting conditions the inorganic nitrogen consumption rates by the cells were significantly inhibited.

N-starvation increased (about 2 fold) the consumption of ammonium but did not modify that of nitrate or nitrite. On the other hand, ammonium consumption shown by N-starved cells was higher in the dark than in the light, whereas a different behaviour was observed with the consumption of nitrate or nitrite (Table 1). These data are consistent with previous observations in eukaryotic algae, such as *Chlorella* and *Amphidinium*, indicating that ammonium mobilizes carbohydrate reserves not available for nitrate assimilation (Syrett 1987). In addition, the high ammonium consumption rate shown by darkened cells may be related to the high intracellular NADH-GOGAT activity level observed in N-starved *C. reinhardtii* (see below), and a possible involvement of NADH-GOGAT in the assimilation of ammonium in darkness was also suggested because of the high level of this enzyme found in heterotrophically cultured cells (Márquez et al. 1986 b).

Table 1. Consumption of inorganic nitrogen under different nutritional conditions

Experimental condition	Consumption rates (%) NO_3^-	NO_2^-	NH_4^+
Normal cells			
Light	100	100	100
Dark	5	5	5
C-Limited cells			
Light	40	75	50
N-Starved cells			
Light	110	105	180
Dark	75	50	220

Normal cells were grown in the light with nitrate under air containing 5% CO_2. C-limited cells were normal cells grown for 2 h under regular air, and N-starved cells were normal cells kept for 18 h in an N-free medium. For consumption experiments, the cells were harvested and resuspended in N-free medium, and 0.25 mM of the N-source was added (zero time) to the culture. The rates shown by normal cells were 10 - 20 μmol mg^{-1} Chl h^{-1}

As a general phenomenon, ammonium is a powerful antagonist of nitrate assimilation in microalgae. In *C. reinhardtii*, the effect of ammonium can be summarized as follows: (i) it causes an immediate and complete inhibition of nitrate and nitrite uptake, (ii) NAD(P)H-nitrate reductase, but not Fd-nitrite reductase, is reversibly inactivated after treatment of the cells with ammonium, and (iii) the synthesis of these enzymes is repressed in ammonium-grown cells (Florencio and Vega 1983 b).

Under CO_2-limiting conditions (cells flushed with regular air) the ammonium-promoted inhibition of nitrate and nitrite uptake was not complete. The inhibition of nitrate uptake was still high (about 70 %), whereas that of nitrite was low (about 25 %) (Menacho and Vega, unpublished). These results support the hypothesis that consumption of nitrate may be inhibited by ammonium itself and/or a primary product of its assimilation (L-glutamine seems to be a good candidate), while that of nitrite should be decreased by one (or several) other metabolite(s) derived from the ammonium assimilation, the production of which may be low under CO_2-limiting conditions. It is interesting here that a significant consumption of nitrate was observed in the presence of ammonium by cells under air, thus suggesting that besides the compounds mentioned above, other products of ammonium assimilation may be also involved in the negative control of nitrate consumption in *C. reinhardtii* cells. A combined control system by ammonium plus ammonium derivatives has been proposed for the consumption of nitrate in *Anabaena cycadae* (Bagchi et al. 1985) and for that of nitrite in *C. reinhardtii* (Córdoba et al. 1987).

In several eukaryotic algae, nitrate uptake was inhibited under air or in the absence of CO_2, especially at low pH (≤ 6) (Ullrich 1983). When the CO_2 supply was limited in *C. reinhardtii* cells, the inhibition was stronger in nitrate than in nitrite consumption (see Table 1). Furthermore, the production of ammonium and its release to the medium was observed in cells using 0.25 mM nitrite, but not with the same concentration of nitrate, thus indicating the existence of a CO_2-dependent regulatory mechanism for nitrate consumption, which avoids its futile reduction. This CO_2-effect on nitrate utilization may be related to the ammonium-promoted inhibition of nitrate uptake, as proposed for *Anacystis nidulans* (Flores et al. 1983). Our current working hypothesis is that, under air, the low availability of carbon skeletons causes an accumulation of ammonium, and probably L-glutamine, inside the cells which inhibits nitrate utilization (but not that of nitrite). This regulatory mechanism can be overcome by increasing the concentration of nitrate, which causes accumulation of nitrite and ammonium in the medium (Azuara and Aparicio 1987).

Intracellular Activity Levels of the GS-GOGAT Enzymes under Different Nutritional Conditions

GS was inactivated by about 50% in *C. reinhardtii* cells growing with nitrate under CO_2-enriched air, either by adding ammonium to the culture or by limiting CO_2, i.e. using regular air to flush the cells. In both cases, the inactivation affected GS_1 and GS_2 to a similar extent and was also observed in cells previously treated with azaserine until the GOGAT activity was negligible. In addition, the GS inactivation produced by CO_2 limitation required the presence of a nitrogen source in the medium (Menacho and Vega, unpublished). These data suggest that ammonium itself and/or L-glutamine, accumulated inside the cells, either by adding ammonium or by limiting CO_2 in the presence of a nitrogen source, are involved in the enzyme inactivation mechanism. The decrease in GS activity may be important to prevent a wasteful formation of excess L-glutamine by decreasing the intracellular level of L-glutamate, and to preserve ATP under uncoupling conditions produced by the intracellular accumulation of ammonium (Rowell et al. 1979), but not to control the functional rate of the GS-GOGAT pathway, which is basically determined by the availability of carbon skeletons.

On the other hand, *C. reinhardtii* cells flushed with CO_2-free air in the presence of nitrite as the unique nitrogen source (GS 50% inactivated), consumed nitrite and released to the medium stoichiometric amounts of ammonium, indicating that ammonium generated from an external N-source cannot be assimilated under these conditions. However, when MSX was also present in the culture medium (GS 100% inactivated), the amount of ammonium released by cells in the presence of 0.25 mM nitrite reached up to 1 mM, indicating that in this case neither exogenous nor endogenously generated ammonium can be assimilated. Thus, GS inactivation by low-CO_2 conditions does not decrease the capacity of the cells to reassimilate the endogenously produced ammonium.

In general, the enzymes of ammonium assimilation reach the highest activities under conditions of low availability of nitrogen. Keeping *C. reinhardtii* cells in N-free medium increases GS and GOGAT activities, which may explain the stimulation of the ammonium consumption rate shown by N-starved cells, as mentioned above. The effect on GS_1 (3 fold) and NADH-GOGAT (4 fold) activities is higher than that on GS_2 (1.5 fold) and Fd-GOGAT (1.5 fold). These data suggest a functional association between GS_1 and NADH-GOGAT activities. Nitrogen starvation can be also induced in *C. reinhardtii* cells either: (i) by using methylammonium, an analog of ammonium which cannot be used by *C. reinhardtii* cells as the only nitrogen compound in the medium, or (ii) by including in the culture medium, besides any conventional nitrogen source, MSX or azaserine, inhibitors of the GS-GOGAT cycle. These metabolic situations permit to correlate the effect of any nitrogen compound accumulated inside the cells with the increase in enzyme activity produced by N-starvation. Experiments with these compounds suggest that : (i) ammonium itself and/or L-glutamine prevents maximal expression of GS activity, and (ii) ammonium decreases Fd-GOGAT activity while a metabolite resulting from its assimilation by the alga may be the negative modulator for the NADH-GOGAT activity (Menacho and Vega 1989).

Effect of Metabolites and Thiols on GS-GOGAT Activities

We found an accumulative inhibition by amino acids and nucleotides, further metabolites from ammonium assimilation, on the GS-GOGAT activities in *C. reinhardtii*. L-Alanine, L-glycine and L-arginine inhibited GS activity by 30 - 70% and, in all cases, the effect of the corresponding amino acid was similar on GS_1 and GS_2 (Florencio and Vega 1983 a). On the other hand, the sensitivity of GOGAT activities to metabolites

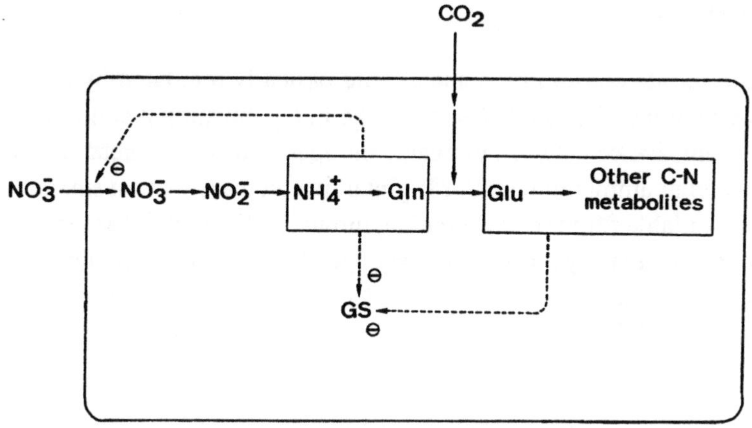

Fig. 1. Model for the regulation of inorganic nitrogen metabolism in *C. reinhardtii*. Broken lines indicate regulatory signals. For further explanation see text

showed important differences. Fd-GOGAT is significantly inhibited by amino acids, such as L-glycine and L-serine, and nucleotides, as ATP and ADP, while NADH-GOGAT is only inhibited by ATP and ADP (Gotor et al., unpublished). These results suggest a possible feedback regulation for the GS-GOGAT enzymes in *C. reinhardtii*. In addition, the photorespiratory intermediates glycolate and glyoxylate inhibited the Fd-GOGAT activity, while NADH-GOGAT activity was only affected by glycolate.

Sulfhydryl groups are essentially required for the GS and the Fd- and NADH-GOGAT activities in *C. reinhardtii*. This has been shown by two kinds of approach: (i) the activation of enzymes by thiols, and (ii) the inactivation of catalytic activity by sulfhydryl reagents. The GS isoenzymes were about 3 fold stimulated when treated with 5 mM thiol (Florencio and Vega 1983 a); NADH-GOGAT was also activated (6 fold), while Fd-GOGAT was stimulated almost 3 fold by high concentrations of DTE (Gotor et al., unpublished). In addition, the Fd-GOGAT activity was highly enhanced by thiols in a light-dependent reconstituted chloroplast system from *C. reinhardtii* (Márquez et al. 1987). These results suggest a possible regulation by thiols of the GS-GOGAT systems from *C. reinhardtii*, which may be consistent with the reported thioredoxin-mediated activation of GS and GOGAT from *Chlorella* (Tischner and Schmidt 1982). Notwithstanding, we have been unable to see any significant effect of light on the GS-GOGAT activities *in vivo* in *C. reinhardtii*, but a regulation of these enzymes by redox power not directly related to the photosynthetic electron transfer chain should be possible.

Concluding Remarks

In *C. reinhardtii*, nitrate consumption and GS are basically regulated by ammonium itself and/or L-glutamine (Fig. 1). When a change in the nutritional conditions of the cells produces an imbalance between inorganic nitrogen and carbon metabolism, in the sense that intracellular ammonium and/or L-glutamine increase when a smaller carbon skeleton pool is available for their assimilation, nitrate uptake and GS are inhibited. In addition, a feedback regulatory mechanism may operate when the intracellular levels of N-metabolites, such as aminio acids and nucleotides, increase, thus producing an inhibition of the GS-GOGAT enzymes (Fig. 1).

Acknowledgments. This work was supported by Comisión Interministerial de Ciencia y Tecnología (Research Grant no. BT87-0082-C02), and by the European Economic Community (BAP-0354-E).

References

Azuara MP, Aparicio PJ (1983) *In vivo* blue-light activation of *Chlamydomonas reinhardtii* nitrate reductase. Plant Physiol 71: 286-290

Bagchi SN, Rai UN, Rai AN, Singh HN (1985) Nitrate metabolism in the cyanobacterium *Anabaena cycadae*: Regulation of nitrate uptake and reductase. Physiol Plant 63: 322-326

Córdoba F, Cárdenas J, Fernández E (1987) Cooperative regulation by ammonium and ammonium derivatives of nitrite uptake in *Chlamydomonas reinhardtii*. Biochim Biophys Acta 902: 287-292

Cullimore J, Sims AP (1981) Occurrence of two forms of glutamate synthase in *Chlamydomonas reinhardtii*. Phytochemistry 20: 597-600

Florencio FJ, Vega JM (1983 a) Separation, purification and characterization of two isoforms of glutamine synthetase from *Chlamydomonas reinhardtii*. Z Naturforsch 38c: 531-538

Florencio FJ, Vega JM (1983 b) Utilization of nitrate, nitrite and ammonium by *Chlamydomonas reinhardtii*. Planta 158: 288-293

Flores E, Romero JM, Guerrero MG, Losada M (1983) Regulatory interaction of photosynthetic nitrate utilization and carbon dioxide fixation. Biochim Biophys Acta 722: 408-416

Márquez AJ, Gotor C, Romero LC, Galván F, Vega JM (1986 a) Ferredoxin-glutamate synthase from *Chlamydomonas reinhardtii*. Prosthetic groups and preliminary studies of mechanism. Int J Biochem 18: 531-53

Márquez AJ, Galván F, Vega JM (1986 b) Utilization of ammonium by mutants and wild types of *Chlamydomonas reinhardtii*. Studies on the glutamate synthase activities. J Plant Physiol 124: 95-102

Márquez AJ, Serra MA, Vega JM (1987) Characterization of a light-dependent glutamate synthase activity in *Chlamydomonas reinhardtii*. Photosynthesis Research 12: 73-81

Menacho A, Vega JM (1989) Effect of nitrogen starvation on ammonium assimilation by *Chlamydomonas reinhardtii*. Physiol Plant 75: 285-289

Rowell P, Sampaio MJAM, Ladha JK, Stewart WDP (1979) Alteration of cyanobacterial glutamine synthetase activity *in vivo* in response to light and ammonium. Arch Microbiol 120: 195-200

Syrett PJ (1987) Nitrogen assimilation by eukaryotic algae. In: Ullrich WR, Aparicio PJ, Syrett PJ, Castillo F (eds) Inorganic Nitrogen Metabolism, Springer, Berlin Heidelberg New York, pp 25-31

Thacker A, Syrett PJ (1972) The assimilation of nitrate and ammonium by *Chlamydomonas reinhardtii*. New Phytol 71: 423-433

Tischner R, Schmidt A (1982) A thioredoxin-mediated activation of glutamine synthetase and glutamate synthase in synchronous *Chlorella sorokiniana*. Plant Physiol 70: 113-116

Ullrich WR (1983) Uptake and reduction of nitrate: Algae and fungi. In: Läuchli A, Bieleski RL (eds) Encyclopedia of Plant Physiology, NS vol 15A, Springer, Berlin Heidelberg New York, pp 376-397

NITRATE TRANSPORT IN CYANOBACTERIA

M.G. Guerrero, J.M. Romero, R. Rodríguez and C. Lara [1]

Introduction

Prior to its reduction to ammonium and conversion of the latter into an α-amino group, nitrate has to enter the cell. Transport of nitrate into photosynthetic cells remains the least understood step in the global process of nitrate assimilation. Actually, very little is known about the biochemistry of nitrate transport in higher plants and microorganisms.

Studies on the entrance of nitrate have conventionally been conducted by following the uptake of the ion from the external medium. This is particularly true for those microalgae which lack storage organelles, including the cyanobacteria (blue-green algae). Such an approach does not allow to distinguish experimentally between the transport step and the subsequent metabolism of nitrate, although it has been valuable to indicate the participation in nitrate assimilation of transport systems with a high affinity for nitrate (K_s in the micromolar range) (Syrett 1988; Ullrich 1987). With vacuolated microalgae, such as the diatom *Phaeodactylum tricornutum*, Syrett and coworkers have addressed the study of nitrate transport by following the appearence of the substrate within the cells (see Syrett 1988, for a review). More recently, a sensitive method has been developed to measure low intracellular nitrate levels and has been applied to the study of nitrate transport in unicellular cyanobacteria (Lara et al. 1987 a; Romero et al. 1989). In parallel, studies of the polypeptide composition of cytoplasmic membranes are also providing data about proteins involved in nitrate transport (Omata 1988; Omata et al. 1989; Sivak et al. 1988, 1989). A fast development in the understanding of the process can thus be predicted.

Being the first step in nitrate assimilation, nitrate transport represents the first possible point of control of the whole process. This has been overlooked in past years, attention being especially focused on nitrate reductase as a regulatory enzyme responsible for the regulation of nitrate assimilation (Guerrero et al. 1981). More recently, however, evidence has accumulated in favor of nitrate transport as the first tightly controlled rate-limiting step in nitrate utilization (Lara et al. 1987 a; Syrett 1988), reinforcing the relevance of metabolic regulation at the level of substrate supply to the cell.

In this chapter we summarize the present state of knowledge on nitrate transport in cyanobacteria, with particular emphasis on our own results.

[1] Instituto de Bioquímica Vegetal y Fotosíntesis, Facultad de Biología, Universidad de Sevilla y CSIC, Apdo. 1113, 41080-Sevilla, Spain

Nitrate Transport

The study of nitrate transport in cyanobacteria has been an elusive task till recently, mainly because of the unavailability of a reliable method to measure transport of nitrate independently of its subsequent assimilation (Guerrero and Lara 1987). Although it is generally assumed that nitrate is actively transported into the cells, direct experimental evidence is lacking in many cases. Recently, a sensitive method has been developed to measure intracellular nitrate. It involves centrifugation of the cells through a layer of silicone oil into an acid solution and the analysis of cell lysates by ion-exchange high performance liquid chromatography followed by UV detection of nitrate at 210 nm (Lara et al. 1987 a; Romero et al. 1989).

Using this procedure, intracellular nitrate accumulation has been measured in the unicellular cyanobacterium *Anacystis nidulans* (*Synechococcus leopoliensis* 1402-1) (Lara et al. 1987 a). The intracellular nitrate level depends on the balance between the rate of net nitrate entrance into the cell and the rate of its reduction by nitrate reductase. In intact nitrate-grown cells with a fully active nitrate reductase, low but significant intracellular concentrations of nitrate can be estimated. In order to separate nitrate transport from reduction, experimental systems are most convenient in which the cells exhibit high nitrate transport activity together with negligible nitrate reduction activity. Such a system is provided by tungstate treatment of the cells (Lara et al. 1987 a), which under controlled conditions produces an inactive nitrate reductase (Herrero and Guerrero 1986), or by keeping cells in the dark (Sivak et al. 1989) where the lack of photogenerated reductant leads to negligible nitrate reduction *in vivo*. In either system, intracellular nitrate accumulates at high levels (Fig. 1). In tungstate-treated *A. nidulans* cells assayed under illumination (Fig. 1A) intracellular nitrate increases with time to reach a plateau in about 2 min. Untreated cells assayed in the dark transport nitrate to a similar level although 4 - 5 min are required to reach maximal accumulation (Fig. 1B).

Regarding the plateau as an equilibrium situation, the change in free energy associated with the build up and maintenance of these intracellular nitrate concentrations can be calculated according to the Nernst equation. For a membrane potential of -100 mV (Paschinger 1977; Peschek et al. 1985), an initial external concentration of 50 μM nitrate and an internal concentration of 370 μM (Fig. 1), the corresponding free energy change amounts to about +15 kJ per mol nitrate. *A. nidulans* cells are, therefore, able to accumulate intracellular nitrate against the electrochemical gradient of the ion. The results strongly indicate the operation of an endergonic nitrate transport system in *A. nidulans* (Lara et al. 1987 a). In this regard, preliminary data show that nitrate transport in *A. nidulans* is sensitive to ionophores and dicyclohexylcarbodiimide.

The expression of the nitrate transport system is strongly affected by the nature of the available nitrogen source. High nitrate transport activity is exhibited by nitrate-grown *A. nidulans* cells, whereas only negligible activity is detected in ammonium-

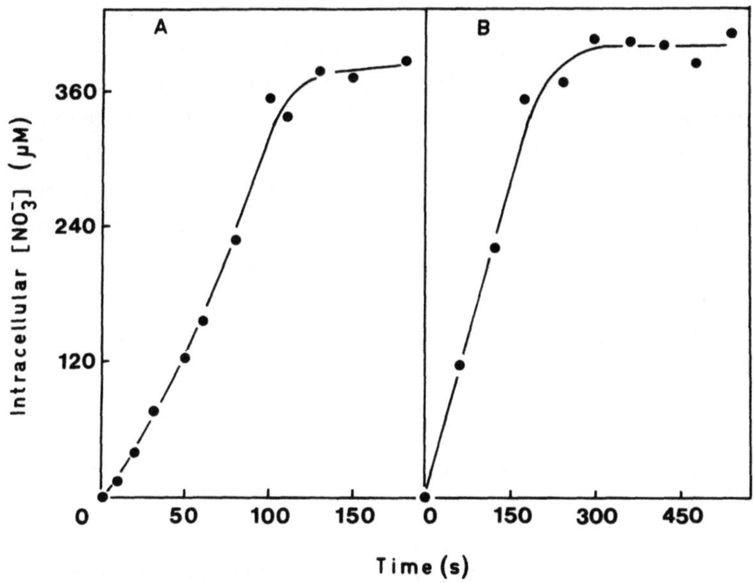

Fig. 1. Time-course of intracellular nitrate accumulation in *Anacystis nidulans* cells. (**A**), tungstate treated cells assayed in the light. (**B**), untreated cells assayed in the dark. In both cases 50 μM KNO_3 was added at zero time. Please note the different time scale on the X-axis. Panel A taken from Lara et al. 1987, with permission

grown cells (Lara et al. 1987 a; Sivak et al. 1989). Nitrate transport activity develops after transfer of ammonium-grown cells to media containing nitrate or no available nitrogen source, but remains absent when the cells are transferred to media containing ammonium or ammonium nitrate. The results indicate that ammonium (or a product of ammonium assimilation) represses the development of nitrate transport activity, nitrate not being required as an inducer (Sivak et al. 1988, 1989). Such a pattern of regulation is analogous to that of the nitrate reducing enzymes of *A. nidulans*, namely nitrate reductase (Herrero et al. 1985) and nitrite reductase (Herrero and Guerrero 1986), suggesting that synthesis of both the nitrate transport and the nitrate reduction systems is subject to the same type of nitrogen control.

Analysis of the polypeptide composition of the cytoplasmic membranes of *A. nidulans* has revealed that the level of a 47-kDa polypeptide changes significantly in response to variations in the available nitrogen source (Sivak et al. 1988, 1989). In nitrate-grown cells, the 47-kDa polypeptide is a major membrane component, representing about 25% of total cytoplasmic membrane protein. Synthesis of the 47-kDa polypeptide is also subject to ammonium repression and is not dependent on nitrate as an inducer, indicating that this adaptive protein is involved in the nitrate assimilation system. The 47-kDa polypeptide might correspond to nitrate or nitrite reductase. Neither enzyme activity was detected, however, in the cytoplasmic membrane fraction. Nitrate reductase activity appeared in the thylakoid fraction, nitrite reductase activity being present in both thylakoids and soluble fraction. Interestingly, a close parallelism exists between the level of the 47-kDa polypeptide in the cytoplasmic membrane and the magnitude of the nitrate transport activity of *A. nidulans* cells in all experimental situations assayed (Sivak et al. 1989). Table 1 illustrates the point, showing the time-course of the development of both the 47-kDa polypeptide and nitrate transport activity of cells, following transfer from ammonium to nitrate medium. Taken together, the data indicate the involvement of the 47-kDa polypeptide in nitrate transport in *A. nidulans*, most probably as a component of the nitrate transporter (Sivak et al. 1988, 1989).

Table 1. Increase in the level of nitrate transport and the 47-kDa cytoplasmic membrane polypeptide in *Anacystis nidulans* following the transfer of ammonium-grown cells to a medium containing KNO_3

Time (h)	Intracellular NO_3^- (μM)	47-kDa polypeptide (area units)
0	< 2	170
1.5	< 2	150
3.0	27	490
4.5	38	570
6.0	76	720
24.0	181	3120

Taken from Sivak et al. (1989), with permission

Further support to this view is provided by data obtained in *Anacystis* R2 (*Synechococcus* PCC 7942) by Omata and coworkers (1988, 1989). The cytoplasmic membranes of this cyanobacterium contain an adaptive 45-kDa polypeptide, which is a major plasma membrane component in nitrate-grown cells but is virtually absent in ammonium-grown cells. A mutant named M45, constructed by inactivating the gene encoding the 45-kDa polypeptide, is specifically devoid of this protein irrespective of the growth conditions. Comparative studies performed with the wild-type and M45 strains on the effect of nitrate concentration on both growth and nitrate-dependent O_2 evolution suggest that the defect in M45 is at the level of nitrate transport (Omata et al. 1989).

The identity of the 45-kDa polypeptide of *Anacystis* R2 with the 47-kDa polypeptide of *A. nidulans* seems most probable. The 45 - 47 kDa protein appears to be an essential component of the nitrate transport system of *Anacystis*, although its exact role in nitrate transport activity remains to be investigated. Otherwise the involvement of other poly-peptides in the nitrate transport system cannot be excluded.

Contribution of Nitrate Transport to the Control of Nitrate Assimilation

Nitrate utilization is efficiently regulated in the short-term through changes in the activity of the system. Nitrate transport appears to be the major rate-limiting step and the first operative control point of the process in cyanobacteria.

In *A. nidulans* and other cyanobacteria, nitrate utilization is rapidly and effectively hampered by the presence of ammonium in the outer medium. This inhibition is mediated by ammonium assimilation products (Flores et al. 1980). In addition, nitrate utilization by *Anacystis* exhibits a tight dependence upon the rate of CO_2 fixation, which is hampered after treatment of the cells with D,L-glyceraldehyde, a selective inhibitor of CO_2 fixation. The negative effect of ammonium and the positive effect of CO_2 on nitrate utilization are somehow related, since ammonium assimilation inhibitors relieve nitrate uptake from both the ammonium inhibition and the CO_2 requirement (Flores et al. 1983; Romero et al. 1985; see Lara et al. 1987 b, for a review). Actually, CO_2 acts as an antagonist of ammonium in the control of nitrate uptake. In cells with moderate ammonium assimilation capacity (after treatment with 5-hydroxylysine) acceleration of CO_2 fixation induced by transition from CO_2-limiting to CO_2-saturating conditions results in abolishment of the negative effect of ammonium (Romero et al. 1987). The nitrate entrance step seems to be the target of the effect of ammonium assimilation and CO_2 fixation products. As shown in Table 2, nitrate transport activity of *Anacystis* cells is effectively inhibited by ammonium addition or by impeding CO_2 fixation with D,L-glyceraldehyde. In either case, the inhibition of nitrate transport is prevented by blocking ammonium assimilation with L-methionine-D,L-sulfoximine, indicating the common basis of the mechanisms underlying the effects of ammonium and CO_2 on nitrate transport. The sensitivity of nitrate transport activity to those factors that influence the overall process of nitrate utilization thus indicates that nitrate assimilation in this cyanobacterium is primarily regulated at the level of substrate supply to the cell (Lara et al. 1987 a).

Table 2. Inhibition of nitrate transport in *Anacystis nidulans* by ammonium or D,L-glyceraldehyde (DLG) and its prevention by L-methionine-D,L-sulfoximine (MSX)

Addition	Mean intracellular nitrate concentration (μM)
None	208
MSX	225
NH_4^+	0
MSX and NH_4^+	211
DLG	0
MSX and DLG	196

Taken from Lara et al.(1987) with permission

The available data thus emphasize the outstanding role played by the nitrate transport step in the regulation of nitrate utilization in cyanobacteria. The situation seems to be shared by eukaryotic microalgae (Syrett 1988), and its validity must be assessed when drawing conclusions on rate-limiting steps in nitrate assimilation by other green cells.

Acknowledgments. Research in the authors' laboratory has been supported by grants from Junta de Andalucía and DGICYT (PB88-0019). We thank Dr. T. Omata for making available unpublished data.

References

Flores E, Guerrero MG, Losada M (1980) Short-term ammonium inhibition of nitrate utilization by *Anacystis nidulans* and other cyanobacteria. Arch Microbiol 128: 137-144

Flores E, Romero JM, Guerrero MG, Losada M (1983) Regulatory interaction of photosynthetic nitrate utilization and carbon dioxide fixation in the cyanobacterium *Anacystis nidulans*. Biochim Biophys Acta 725: 529-532

Guerrero MG, Lara C (1987) Assimilation of inorganic nitrogen. In: Fay P, van Baalen C (eds) The Cyanobacteria. Elsevier, Amsterdam, pp 163-186

Guerrero MG, Vega JM, Losada M (1981) The assimilatory nitrate-reducing system and its regulation. Annu Rev Plant Physiol 32: 169-204

Herrero A, Flores E, Guerrero MG (1981) Regulation of nitrate reductase levels in the cyanobacteria *Anacystis nidulans*, *Anabaena* sp. strain 7119 and *Nostoc* sp. strain 6719. J Bacteriol 145: 175-180

Herrero A, Guerrero MG (1986) Regulation of nitrite reductase in the cyanobacterium *Anacystis nidulans*. J Gen Microbiol 132: 2463-2468

Lara C, Romero JM, Guerrero MG (1987 a) Regulated nitrate transport in the cyanobacterium *Anacystis nidulans*. J Bacteriol 169: 4376-4378

Lara C, Romero JM, Coronil T, Guerrero MG (1987 b) Interactions between photosynthetic nitrate assimilation and CO_2 fixation in cyanobacteria. In: Ullrich WR, Aparicio PJ, Syrett PJ, Castillo F (eds) Inorganic Nitrogen Metabolism. Springer, Berlin Heidelberg New York, pp 45-52

Omata T (1988) Responses to CO_2 and NO_3^- stresses in mutants of *Anacystis nidulans* R2 lacking the major proteins in the cytoplasmic membrane. US-Japan Seminar Environmental Plant Stresses, Alaska, pp 1-3

Omata T, Ohmori M, Arai N, Ogawa T (1989) Genetically engineered mutant of the cyanobacterium *Synechococcus* PCC 7942 defective in nitrate transport. Proc Natl Acad Sci 86: 6612-6616

Paschinger H (1977) DCCD induced sodium uptake by *Anacystis nidulans*. Arch Microbiol 113: 285-291

Pescheck GA, Czerny T, Schmetterer G, Nitschmann WH (1985) Transmembrane proton electrochemical gradients in dark aerobic and anaerobic cells of the cyanobacterium (blue-green alga) *Anacystis nidulans*. Plant Physiol 79: 278-284

Romero JM, Coronil T, Lara C, Guerrero MG (1987) Modulation of nitrate uptake in *Anacystis nidu-lans* by the balance between ammonium assimilation and CO_2 fixation. Arch Biochem Biophys 256: 578-584

Romero JM, Lara C, Guerrero MG (1985) Dependence of nitrate utilization upon active CO_2 fixation in *Anacystis nidulans*: A regulatory interaction between photosynthetic carbon and nitrogen metabolism. Arch Biochem Biophys 237: 396-401

Romero JM, Lara C, Guerrero MG (1989) Determination of intracellular nitrate. Biochem J 259: 545-548

Sivak MN, Lara C, Romero JM, Rodríguez R, Guerrero MG (1988) Relationship between a 47 kDa cytoplasmic membrane polypeptide and nitrate transport in *Anacystis nidulans*. 2nd Int Congr Plant Molec Biol, Jerusalem, Abstract 456

Sivak MN, Lara C, Romero JM, Rodríguez R, Guerrero MG (1989) Relationship between a 47 kDa cytoplasmic membrane polypeptide and nitrate transport in *Anacystis nidulans*. Biochem Biophys Res Commun 158: 257-262

Syrett PJ (1988) Uptake and utilization of nitrogen compounds. In: Rogers LJ, Gallon JR (eds) Biochemistry of the Algae and Cyanobacteria. Clarendon Press, Oxford, pp 23-39

Ullrich WR (1987) Nitrate and ammonium uptake in green algae and higher plants: Mechanism and relationship with nitrate metabolism. In: Ullrich WR, Aparicio PJ, Syrett PJ, Castillo F (eds) Inorganic Nitrogen Metabolism, Springer, Berlin Heidelberg New York, pp 32-38

TARGETED MUTAGENESIS OF THE 45-KDa PROTEIN INACTIVATES NITRATE TRANSPORT IN *SYNECHOCOCCUS* PCC 7942

T. Omata [1], M. Ohmori [2], N. Arai [1] and T. Ogawa [1]

Introduction

Nitrate is taken up into cyanobacterial cells by an active transport system (Flores et al. 1983; Tischner and Schmidt 1984; Lara et al. 1987). Little is known about the mechanism of this transporting system and a protein(s) involved in the transport is not yet identified. Such protein(s) would be located in the cytoplasmic membrane and its synthesis might be suppressed when cells do not utilize nitrate for their growth. We found that ammonium-grown cells of *Synechococcus* PCC 7942 lack the 45-kDa protein, a major component of the cytoplasmic membrane in nitrate-grown cells (Omata and Ogawa 1987). In order to elucidate the function of the 45-kDa protein, we cloned the gene (*nrt*A) encoding this protein and constructed an insertional mutant (M45) of *Synechococcus* that cannot synthesize the 45-kDa protein. Biochemical and physiological characterizations of the mutant indicated that the 45-kDa protein plays an essential role in the nitrate-transporting system.

Materials and Methods

WT (wild-type) and mutant cells of *Synechococcus* PCC 7942 were grown at 30°C in BG-11 medium (Stanier et al. 1971) supplemented with 50 mM KNO_3 (final concentration of nitrate = 67.5 mM), unless otherwise stated, under aeration with 3% CO_2 in air. Continuous illumination was provided by tungsten lamps at 120 µmol of PAR (photosynthetically active radiation) $m^{-2}\,s^{-1}$. "Nitrate-free medium" was a modified BG-11 medium in which $NaNO_3$ had been replaced by NaCl. Media containing different concentrations of nitrate were prepared by adding various amounts of KNO_3 to the nitrate-free medium. To obtain an ammonium-containing medium, 1 mM $(NH_4)_2SO_4$ was added to the nitrate-free medium. All media were buffered with 20 mM TES-NaOH (pH 8.0).

Manipulations and analyses of DNA and RNA were performed according to standard protocols (Maniatis et al. 1982). A DNA fragment carrying a part of the gene for the 45-kDa protein was obtained from a genomic library of *Synechococcus* PCC 7942 constructed in expression vector λgt11 (Young and Davis 1983). The cloned

[1] Solar Energy Research Group, The Institute of Physical and Chemical Research (RIKEN), Wako, Saitama 351-01, Japan;
[2] Ocean Research Institute, University of Tokyo, Minamidai, Nakano-ku, Tokyo 164, Japan

gene fragment was interrupted by the aminoglycoside-3'-phosphotransferase gene (the "kanamycin-resistance [Km^r] cartridge") that originated from transposon Tn 903 (Oka et al. 1981). The plasmid containing the modified gene was used to transform the WT cells of *Synechococcus* PCC 7942 to Km^r through homologous recombination (Williams and Szalay 1983).

Cytoplasmic membranes were prepared as described previously (Omata and Ogawa 1986). SDS-PAGE was performed in the buffer system of Laemmli (1970). Affinity-purified antibody against the 45-kDa protein was obtained using the λgt11 recombinant isolated from the library (Snyder and Davis 1985) and used for immunostaining of Western blots.

Nitrate reductase activity was determined at 30°C according to Herrero et al. (1981). Nitrite reductase activity was determined in a crude extract of cells (Ohmori and Hattori 1970).

When nitrate-dependent O_2 evolution was measured, cells were treated with MSX (L-methionine-DL-sulfoximine) as described by Ohmori and Ohmori (1985), suspended in nitrate-free medium supplemented with 1 mM MSX, 5 mM $NaHCO_3$ and 30 mM DLG (DL-glyceraldehyde) at a chlorophyll concentration of 5 $\mu g\ ml^{-1}$ and then placed in a reaction vessel kept at 30°C for polarographic O_2 measurement (Rank Brothers, Cambridge). The rate of nitrate-dependent O_2 evolution was estimated from the changes of O_2 concentration in the cell suspension upon addition of nitrate in the light (2.3 mmol PAR $m^{-2}\ s^{-1}$). The rate of CO_2-dependent O_2 evolution was less than 5 $\mu mol\ mg\ Chl^{-1}\ h^{-1}$ for both WT and M45 under these conditions.

Results and Discussion

Cytoplasmic membrane of ammonium-grown cells

Nitrate-grown cells of *Synechococcus* PCC 7942 contain a 45-kDa protein as a major protein in the cytoplasmic membrane (Omata and Ogawa 1987). We found that ammonium-grown cells lacked this protein (Fig. 1). The 45-kDa protein appears to be identical with the ammonium-repressible proteins reported recently by Madueño et al. (1988) and Sivak et al. (1989).

Cloning of the gene encoding the 45-kDa protein (nrtA)

A genomic library of *Synechococcus* PCC 7942 constructed in λgt11 phage was screened using antibody against the 45-kDa protein. A recombinant phage that produced peptides recognized by the antibody was cloned and found to contain a 5.8-kbp insert within the *EcoR* I site of the phage. A lysogen of the clone in *E. coli* Y1089 pro-

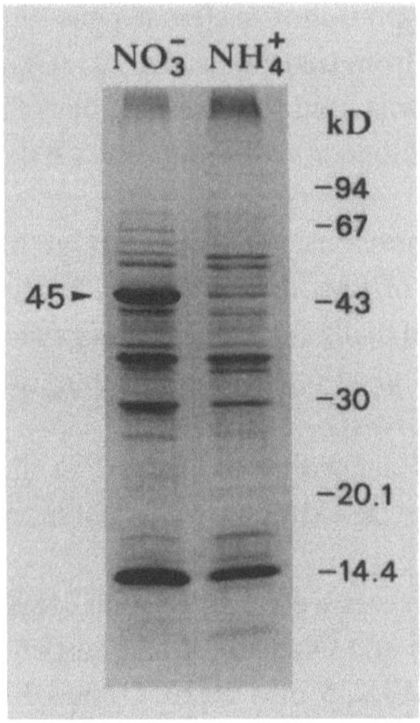

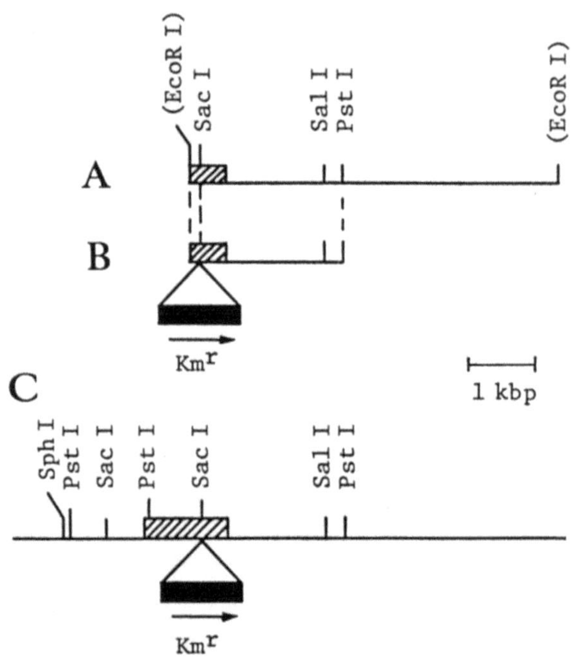

Fig. 1 *(left)*. Electrophoretic profiles showing Coomassie-staining patterns of proteins in the cytoplasmic membrane from WT cells grown in BG-11 medium containing 17.5 mM nitrate (*left*) and in nitrate-deficient medium supplemented with 1 mM $(NH_4)_2SO_4$ (*right*). Samples (25 μg of protein) were solubilized by SDS at 100°C for 5 min and were electrophoresed in an 8-15% gradient SDS-polyacrylamide gel

Fig. 2 *(right)*. (**A**) Restriction map of a 5.8-kbp DNA insert from the recombinant λgt11 which carries a 582-nucleotide open reading frame encoding a portion of the 45-kDa protein (hatched bar). (**B**) Structure of the modified gene fragment constructed by inserting a 1.3-kbp Km^r cartridge (closed bar) at the *Sac* I site. (**C**) Structure of the modified *nrt*A gene obtained by homologous recombination of the construct shown in (**B**) with chromosomal DNA of *Synechococcus* PCC 7942. Transcription of *nrt*A gene (hatched bar) occurs from left to right

duced a β-galactosidase fusion protein, which cross-reacted with the antibody against the 45-kDa protein (not shown). The 5.8-kbp insert was recloned into pUC119 plasmid and analyzed by restriction mapping (Fig. 2A). Analysis of nucleotide sequence revealed an open reading frame of 582 nucleotides (194 amino acids) that had been fused *in frame* to the *lac*Z sequence in the λgt11 recombinant. This indicated that the 5.8-kbp fragment contained about 40% of the structural gene for the 45-kDa protein.

The remainder of the gene was cloned as follows. A 2.4-kbp *Pst* I/*EcoR* I fragment from the 5.8-kbp DNA fragment was subcloned into pUC119 and the 1.3-kbp Km^r cartridge was inserted into the *Sac* I site located in the coding region for the 45-kDa protein (Fig. 2B). The plasmid having the modified gene fragment was used to

```
MSQFSRRKFL LTAGGTAAAA LWLNACGSNN SSTDTTGSTS TPAPSGTSGG  50
DAPEVKGVTL GFIALTDAAP VIIALEKGLF AKYGLPDTKV VKQTSWAVTR 100
DNLELGSDRG GIDGAHILSP MPYLLTAGTI TKSQKPLPMY ILARLNTQGQ 150
GISLSNEFLA EKVQIKDPKL KAIADQKKAS GKLLKAAVTF PGGTHDLWMR 200
YWLAANGIDP NNDADLVVIP PPQMVANMQT GTMDTFCVGE PWNARLVNKK 250
LGYTAAVTGE LWKFHPEKAL TIRADWADKN PKATMALLKA VQEAQIWCED 300
PANLDELCQI TAQDKYFKTS VEDIKPRLQG DIDYGDGRSV KNSDLRMRFW 350
SENASFPYKS HDLWFLTEDI RWGYLPASTD TKALIEKVNR SDLWREAAKA 400
IGREQDIPAS DSRGVETFFD GVTFDPENPQ AYLDGLKFKA IKA        443
```

Fig. 3. Amino acid sequence of the 45-kDa protein as deduced from the nucleotide sequence of *nrt*A

transform WT *Synechococcus* PCC 7942 to Km^r through homologous recombination. This resulted in insertion of the Km^r cartridge into *nrt*A on the chromosome of *Synechococcus* (Fig. 2C). Chromosomal DNA from the transformant (M45) was digested with *Sal* I/*Sph* I and a 5.4-kbp *Sal* I/*Sph* I fragment containing the entire nucleotide sequence of *nrt*A was recovered using the Km^r gene as the marker.

*Nrt*A had an open reading frame of 1329 nucleotides encoding a 443 amino acid polypeptide with a calculated molecular weight of 48,424. The deduced amino acid sequence (Fig. 3) was strongly homologous with that of the 42-kDa cytoplasmic membrane protein from the same cyanobacterium, with 47% of amino acid residues being identical (Omata et al. 1989). No homologous sequence was found in the recent databases of published sequences (the NBRF Protein Sequence Database-Release 20.0).

Characterization of the nrtA insertional mutant

The Km^r transformant (M45) is an insertional mutant carrying the Km^r cartridge in the structural portion of *nrt*A. M45 did not synthesize the 45-kDa protein (Fig. 4). Northern hybridization analysis of total RNA indicated the presence of a 1.4-kb RNA transcript for the 45-kDa protein in WT but not in M45 (not shown). Since the size of the transcript was close to the size of the open reading frame encoding the 45-kDa protein, we assumed that M45 is a simple mutant lacking only the 45-kDa protein.

M45 grew as rapidly as the WT when ammonium (2 mM) was the nitrogen source. M45 did not grow in a medium containing 2 mM KNO_3 as the nitrogen source while the same medium supported a high growth rate in WT. Higher concentrations of nitrate supported the growth of M45, with the optimal concentration being 40 to 70 mM (Fig. 5). When grown in a medium containing nitrate at concentrations lower than 40 mM, the M45 cells became yellowish green due to nitrogen deficiency. The results indicated that the mutant cannot utilize nitrate efficiently.

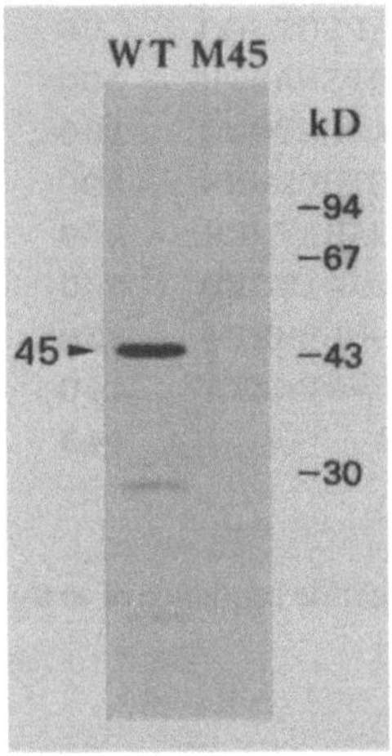

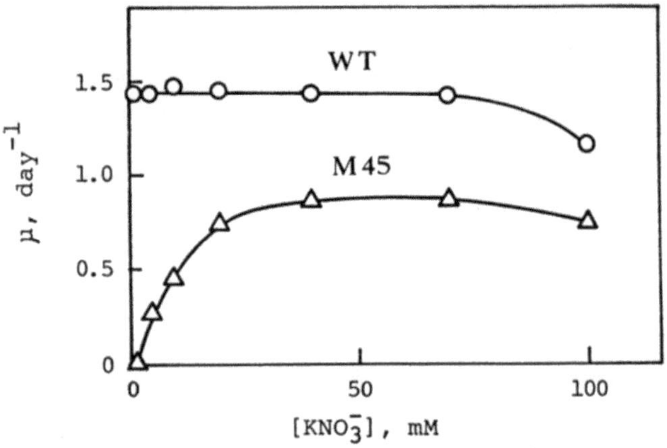

Fig. 4 (left). Immunostaining of the Western blot of the proteins in the cytoplasmic membranes from WT (left) and M45 (right), obtained using antibody against the 45-kDa protein. Cytoplasmic membranes were purified from cells grown at 40 mM nitrate concentration. Samples (5 μg protein) were solubilized as in Fig. 1 and electrophoresed in a 10% SDS-polyacrylamide gel

Fig. 5 *(right)*. Specific growth-rate constant, μ, for WT and M45 as a function of nitrate concentration in the medium. To convert μ to doublings per day one divides by ln 2 (0.693)

Since M45 utilizes ammonium as the nitrogen source, the defect in the mutant must be either at the entrance of nitrate into the cells by an active transport system or at the two-step reduction of nitrate to ammonium catalyzed by nitrate and nitrite reductases. The activity of nitrate reductase in M45 measured *in situ* was 200 μmol mg Chl^{-1} h^{-1}, which was higher than the rate of 70 μmol mg Chl^{-1} h^{-1} obtained for WT. The activity of nitrite reductase in M45 measured in a crude extract was 20 μmol mg $protein^{-1}$ h^{-1},

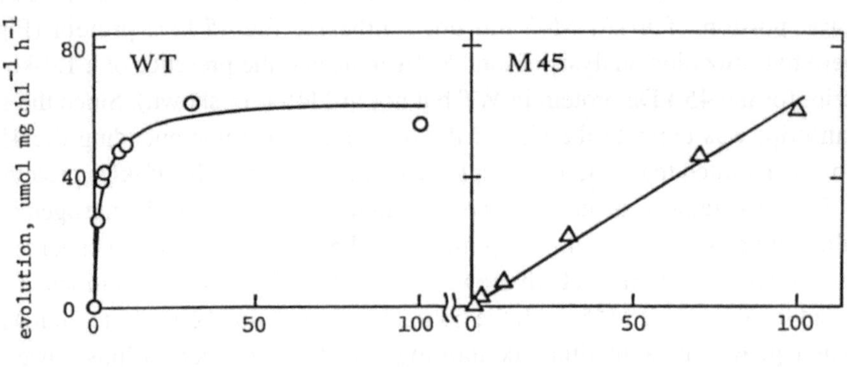

Fig. 6. Rates of nitrate-dependent O_2 evolution by WT and M45 cells as a function of nitrate concentrations in the external medium

which was also higher than the value, 12 μmol mg protein^{-1} h^{-1}, obtained for WT. The high activities of these two enzymes in M45 clearly showed that the defect in the mutant is not on these enzymes.

The rate of nitrate uptake was determined in MSX- and DLG-treated cells by measuring the rate of nitrate-dependent O_2 evolution. The reduction of nitrate to ammonium requires 8 electrons and the amount of O_2 evolved in nitrate-dependent O_2 evolution is twice that of nitrate reduced (Flores et al. 1983). The rate of nitrate-dependent O_2 evolution in WT reached a saturation level at 30 μM nitrate (Fig. 6). The nitrate concentration required for half the saturation was about 1 μM, being much lower than the $K_m(NO_3^-)$ value of 55 μM reported for the nitrate reductase from the cyanobacterium *Plectonema boryanum* (Ida and Mikami 1983). In M45, the nitrate-dependent O_2 evolution could be detected only at nitrate concentrations above 1 mM. The activity increased linearly as the nitrate concentration increased (Fig. 6). These results suggested that active nitrate transport is missing in M45 and that nitrate enters the M45 cells by passive diffusion. We concluded that the 45-kDa protein is an essential component of the nitrate-transporting system in *Synechococcus* PCC 7942.

References

Flores E, Guerrero MG, Losada M (1983) Photosynthetic nature of nitrate uptake and reduction in the cyanobacterium *Anacystis nidulans*. Biochim Biophys Acta 722: 408-416

Herrero A, Flores E, Guerrero MG (1981) Regulation of nitrate reductase levels in the cyanobacteria *Anacystis nidulans*, *Anabaena* sp. strain 7119, and *Nostoc* sp. strain 6719. J Bacteriol 145: 175-180

Ida S, Mikami B (1983) Purification and characterization of assimilatory nitrate reductase from the cyanobacterium *Plectonema boryanum*. Plant Cell Physiol 24: 649-658

Laemmli UK (1970) Cleavage of structural proteins during the assembly of the head of bacteriophage T4. Nature 227: 680-685

Lara C, Romero JM, Guerrero MG (1987) Regulated nitrate transport in the cyanobacterium *Anacystis nidulans*. J Bacteriol 169: 4376-4378

Madueño F, Vega-Palas MA, Flores E, Herrero A (1988) A cytoplasmic-membrane protein repressible by ammonium in *Synechococcus* R2: Altered expression in nitrate-assimilation mutants. FEBS Lett 239: 289-291

Maniatis T, Fritsch EF, Sambrook J (1982) Molecular Cloning: A Laboratory Manual. Cold Spring Harbor Laboratory, New York

Ohmori K, Hattori A (1970) Induction of nitrate and nitrite reductases in *Anabaena cylindrica*. Plant Cell Physiol 11: 873-878

Ohmori M, Ohmori K (1985) Differential effects of methionine sulfoximine on light and dark uptake of nitrate by *Anabaena cylindrica*. Plant Cell Physiol 26: 957-960

Oka A, Sugisaki H, Takanami M (1981) Nucleotide sequence of the kanamycin resistance transposon Tn 903. J Mol Biol 147: 217-226

Omata T, Ogawa T (1986) Biosynthesis of a 42-kDa polypeptide in the cytoplasmic membrane of the cyanobacterium *Anacystis nidulans* strain R2 during adaptation to low CO_2 concentration. Plant Physiol 80: 525-530

Omata T, Ogawa T (1987) Immunochemical studies on the major proteins in cytoplasmic membrane of cyanobacteria. In: Biggins J (ed) Progress in Photosynthesis Research, vol IV. Nijhoff, Dordrecht, p 309

Omata T, Ogawa T, Carlson TJ, Pierce J (1989) Nucleotide sequence of the genes encoding the two major proteins in the cytoplasmic membrane of *Synechococcus* PCC 7942. In: Proceedings of VIIIth International Congress on Photosynthesis. Kluwer, Dordrecht, (in press)
Sivak MN, Lara C, Romero JM, Rodríguez R, Guerrero MG (1989) Relationship between a 47-kDa cytoplasmic membrane polypeptide and nitrate transport in *Anacystis nidulans*. Biochem Biophys Res Comm 158: 257-262
Snyder M, Davis RW (1985) Screening λgt11 expression libraries with antibody probes. In: Springer T (ed) Hybridoma Technology into Bioscience and Medicine. Plenum, New York, p 397
Stanier RY, Kunisawa R, Mandel M, Cohen-Bazire G (1971) Purification and properties of unicellular blue-green algae (order *Chroococcales*). Bacteriol Rev 35: 171-205
Tischner R, Schmidt A (1984) Light mediated regulation of nitrate assimilation in *Synechococcus leopoliensis*. Arch Microbiol 137: 151-154
Williams JGK, Szalay AA (1983) Stable integration of foreign DNA into the chromosome of the cyanobacterium *Synechococcus* R2. Gene 24: 37-51
Young RA, Davis RW (1983) Yeast RNA polymerase II genes: Isolation with antibody probes. Science 222: 778-782

NITROGEN METABOLISM IN HETEROTROPHIC BACTERIA. SIMULTANEOUS AMMONIA INHIBITION OF NITROGEN FIXATION AND NITRATE UPTAKE AND DIVALENT-CATION REGULATION OF NITRATE UPTAKE IN *AZOTOBACTER CHROOCOCCUM*

A. Paneque, F.J. Cejudo, M.T. Ruiz and M.G. de la Vega [1]

Introduction

Nitrogen fixation and nitrate assimilation usually exclude each other because nitrate, through some product(s) generated in the assimilation of the ammonia resulting from its intracellular reduction, represses nitrogenase synthesis (Tubb and Postgate 1973; Postgate 1982). Besides this effect, nitrate exerts a rapid and reversible inhibition of nitrogenase activity that involves also nitrogen assimilation beyond ammonia (Cejudo and Paneque 1986, 1987; Paneque et al. 1987). Moreover, addition of ammonia to *Azotobacter chroococcum* cells either actively taking up nitrate (Revilla et al. 1986) or fixing nitrogen (Cejudo et al. 1984) causes an immediate inhibition of these activities. Whereas nitrogenase activity is the target for the effect of ammonia on nitrogen fixation, the inhibition of nitrate uptake by ammonia is attributed to a direct effect on the nitrate transport system. This transport system acts before the nitrate reduction step and is closely linked to the cell energy metabolism. Thus ammonia, through some compound(s) derived from its metabolism, blocks the first energy-dependent step in the sequence of events involved in either nitrogen fixation or nitrate uptake.

Taking advantage of the fact that *A. chroococcum* cells can transitorily develop the ability to take up nitrate and to fix nitrogen (Cejudo and Paneque 1986) we have studied the effect of exogenously added ammonia on such cells that simultaneously fix nitrogen (measured as acetylene reduction) and utilize nitrate. Ammonia promoted a rapid and total inhibition of both activities but once the added ammonia was consumed nitrate uptake resumed quickly. The subsequent depletion of nitrate in the medium or the interruption of the assimilation beyond the reduction step then restored nitrogenase activity. It is concluded that nitrogen fixation is more sensitive to the true inhibitor than nitrate assimilation

Besides the control of nitrate uptake by ammonia at the energy-dependent nitrate entrance into the cell, we have found that the other step in the assimilatory nitrate reduction that requires also metabolic energy, i. e. the incorporation of the reduced nitrogen into carbon skeletons catalysed by glutamine synthetase, is subject to a diva-

[1] Instituto de Bioquímica Vegetal y Fotosíntesis, Universidad de Sevilla y CSIC, Apartado 1113, 41080 Sevilla (Spain)

lent-cation dependent regulation. We observed that addition of $MnCl_2$ at low concentrations, but not $CaCl_2$, $MgCl_2$ or $ZnCl_2$, to nitrate-grown *A. chroococcum* cells enhanced the rate of nitrate uptake. In the presence of Mn(II), however, the newly reduced anion was released into the medium as ammonia. Mg(II) at a 100-fold concentration with respect to Mn(II) couteracted the effects of Mn(II). The results suggest that Mn(II) exerts an external control over nitrate uptake in *Azotobacter*, glutamine synthetase being the target for the effect of Mn(II) on this process.

Simultaneous Ammonia Inhibition of Nitrogen Fixation and Nitrate Uptake in *Azotobacter*

Figure 1 shows that addition of 0.3 mM NH_4Cl, final concentration, to *A. chroococcum* cells simultaneously reducing acetylene (Fig. 1A) and taking up nitrate (Fig. 1B) caused a rapid and total inhibition of these activities. The NH_4^+-promoted inhibitory effect was reversible since nitrogenase activity and nitrate uptake were restored once ammonia in the medium was consumed, as it occurred when the short-term ammonia inhibition of nitrogen fixation and of nitrate uptake by *A. chroococcum* were studied separately (Paneque et al. 1987). It must be emphasized, however, that while nitrate uptake was completely restored (Fig. 1B) nitrogenase activity showed a rather poor recovery (Fig. 1A). To demonstrate that the scanty recovery of nitrogenase activity was due to nitrate consumption after the added ammonia had been consumed in the medium the following experimental approaches were used. Firstly, the effect of ammonia was investigated in the presence of 0.13 mM KNO_3 and 0.1 mM NH_4Cl, smaller concentrations than those of Figure 1. It was expected that when the medium was depleted of the added ammonia, recovery of nitrogenase activity would occur soon after the low nitrate concentration was also consumed by the cells. It could be confirmed that nitrogenase activity and nitrate uptake were inhibited by supplying the cell suspensions with 0.13 mM NH_4Cl. After the added ammonia and the initially present nitrate were sequentially exhausted in the medium nitrogenase resumed its activity at a relatively high rate. Secondly, the inhibitory effect of ammonia on the activities of nitrogenase and nitrate uptake was studied in cells that, at a given moment, were treated with L-methionine-D,L-sulfoximine (MSX), an inhibitor of glutamine synthetase (Brenchley 1973). As in the previous cases, addition of NH_4Cl inhibited both nitrogenase activity and nitrate uptake. Once the added ammonia was consumed by the cells, 5 mM MSX was applied to the bacterial suspensions. In the presence of the inhibitor nitrate was taken up at a rate even higher than before addition of NH_4Cl. Due to MSX, however, the newly reduced nitrate could not be incorporated into carbon skeletons and was released to the medium as ammonia. Nitrogenase activity was fully restored under these conditions.

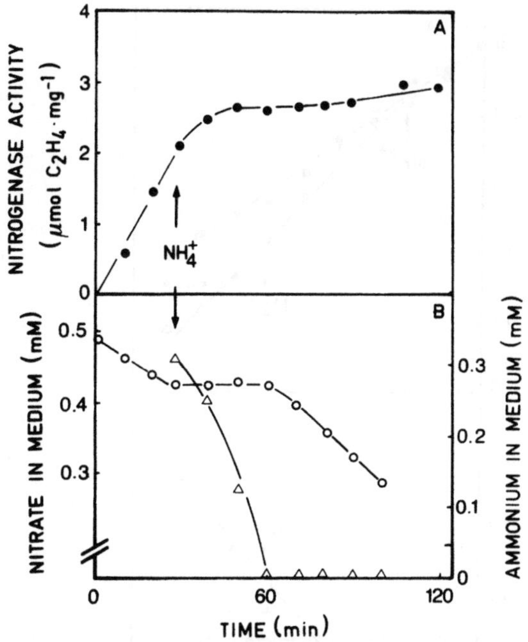

Fig. 1. Simultaneous ammonia inhibition of nitrogenase activity and of nitrate uptake. Two *A. chroococcum* cell suspension samples (80 µg protein ml^{-1}) capable of simultaneously fixing nitrogen and assimilating nitrate were supplied at zero time with both 10 % (v/v) acetylene and 0.5 mM KNO_3 and (**A**) nitrogenase activity (closed circles) and (**B**) nitrate uptake activity (open circles) were determined. When indicated 0.3 mM NH_4Cl was added to each sample and its consumption was followed (triangles)

All these results indicate that the simultaneous inhibition by ammonia of nitrogenase activity and nitrate uptake implies the incorporation of the cation, which leads to an accumulation of an intermediate, the true inhibitor. As soon as ammonia is consumed in the medium, the level of the inhibitory metabolite will decrease and nitrate uptake but not nitrogenase activity will recover. The latter could not be restored because nitrate assimilation did not allow the dilution of the organic nitrogen inhibitor accumulated in the cells during assimilation of the added ammonia.

Divalent-Cation Regulation of Nitrate Uptake Activity in *Azotobacter*

Though it has not been possible so far to decide whether glutamine synthetase can be considered as an enzyme really regulated by divalent cations *in vivo*, Weddler's group has hypothesized that the intracellular concentrations of Mn(II) and Mg(II) controlled

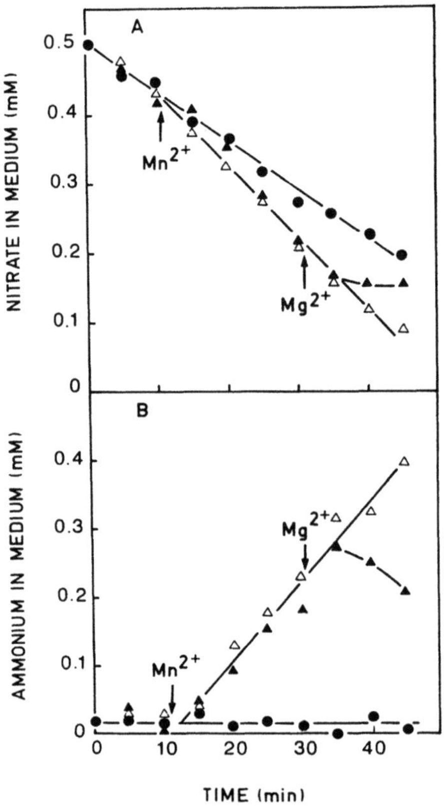

Fig. 2. Effects of Mn(II) and Mg(II) on nitrate uptake activity in *A. chroococcum.* Cell suspension samples (0.45 mg protein ml^{-1}) were assayed for **(A)** nitrate uptake activity in the absence (closed circles) and in the presence of 30 μM $MnCl_2$ (open triangles) and 30 μM $MnCl_2$ plus 3 mM $MgCl_2$ (closed triangles), which were added as indicated by the arrows, and **(B)** for ammonia content in the medium (Ruiz et al., in press)

by hormonal signals and physiological states of glial cells, may be regulators of glutamine synthetase (Weddler and Thomas 1986). We now present data showing that addition of Mn(II) to *A. chroococcum* cells actively taking up nitrate resulted in an immediate increase in nitrate utilization activity, the newly reduced nitrate being released into the external medium as ammonia. The results suggest that glutamine synthetase may be the target for the Mn(II) effect on nitrate uptake.

As shown in Figure 2 A, addition of 30 μM $MnCl_2$ (final concentration) to *A. chroococcum* cells utilizing nitrate produced a rapid increase in the nitrate uptake rate. The addition of Mn(II) also gave rise to an accumulation in the medium of ammonia in amounts accounting for most of the nitrate consumed by the cells. Actually, as illustrated in Figure 2 B, ammonia was formed in stoichiometric amounts to the nitrate taken up in the presence of Mn(II). Figure 2 reveals also that when 3 mM $MgCl_2$ was added to the Mn(II)-treated cells nitrate uptake ceased completely soon thereafter (Fig. 2A), most likely due to the consumption of the ammonia in the medium (Fig. 2B). As a control Mg(II) was added to cell suspensions that had not been pretreated with Mn(II). In this case the rate of nitrate uptake was not affected by the added Mg(II). It could be assessed that the nitrate reducing enzyme system, i. e. nitrate reductase and nitrite reductase, was not affected during the length of the experiment.

These results indicate that Mn(II) did not affect either nitrate transport into the cells or its subsequent reduction to nitrite and ammonia, but did affect the incorporation of the reduced nitrogen into carbon skeletons which became fully inhibited. The further addition of Mg(II) restored ammonia assimilation and thus equally its removal from the medium. When ammonia assimilation was relieved from Mn(II) inhibition, nitrate could not be taken up any more by the cells, in agreement with previous reports that ammonia inhibited nitrate uptake by *A. chroococcum* (Revilla et al. 1985, 1986). It is worth mentioning that the increase in nitrate uptake activity caused by Mn(II), and the concomitant ammonia release into the external medium support the hypothesis that in untreated cells ammonia or a product of its assimilation formed by assimilatory nitrate reduction, exerted a self-control on the process of nitrate assimilation, possibly at the level of nitrate uptake into the bacterium (Revilla et al. 1986; Romero et al. 1987).

In summarizing, the present results confirm the hypothesis that Mn(II) can act as a regulatory signal in metabolism (Williams 1982; Schramm 1982) and they support the assumption that the ratio of Mg(II):Mn(II) may be yet another level in the control of cellular metabolism (Milne and Cook 1979).

Acknowledgments. Research in the authors' laboratory has been supported by grants from Comisión Asesora de Investigación, Comisión Interministerial de Ciencia y Tecnología, and Plan Andaluz de Investigación (Spain).

References

Brenchley JE (1973) Effect of methionine sulfoximine and methionine sulfone on glutamate synthesis in *Klebsiella pneumoniae*. J Bacteriol 144: 667-673

Cejudo FJ, de la Torre A, Paneque A (1984) Short-term ammonium inhibition of nitrogen fixation in *Azotobacter*. Biochem Biophys Res Commun 123: 431-437

Cejudo FJ, Paneque A (1986) Short-term nitrate (nitrite) inhibition of nitrogen fixation in *Azotobacter chroococcum*. J Bacteriol 165: 240-243

Cejudo FJ, Paneque A (1987) Correlation between nitrate uptake rate and nitrate inhibition of nitrogenase activity in *Azotobacter chroococcum*. FEMS Microbiol Lett 43: 5-7

Milne JA, Cook RA (1979) Role of metal cofactor in enzyme regulation. Differences in the regulatory properties of the *Escherichia coli* nicotinamide dinucleotide specific malic enzyme depending whether Mg^{2+} or Mn^{2+} serves as divalent cation. Biochemistry 18: 3604-3618

Paneque A, Cejudo FJ, Revilla E (1987) Nitrogen metabolism in heterotrophic bacteria. A comparative study of the short-term ammonium inhibition of dinitrogen fixation and of nitrate assimilation in *Azotobacter chroococcum*. In: Ullrich WR, Aparicio PJ, Syrett PJ, Castillo F (eds) Inorganic Nitrogen Metabolism, Springer, Berlin Heidelberg New York, pp 53-60

Postgate JR (1982) The Fundamentals of Nitrogen Fixation. Cambridge University Press, Cambridge, UK
Revilla E, Llobell A, Paneque A (1985) The assimilatory nitrate uptake in *Azotobacter chroococcum*. Induction by nitrate and by cyanate. J Plant Physiol 118: 165-176
Revilla E, Cejudo FJ, Llobell A, Paneque A (1986) Short-term ammonium inhibition of nitrate uptake by *Azotobacter chroococcum*. Arch Microbiol 144: 187-190
Romero JM, Coronil T, Lara C, Guerrero MG (1987) Modulation of nitrate uptake in *Anacystis nidulans* by the balance between ammonium assimilation and CO_2 fixation. Arch Biochem Biophys 256: 578-584
Ruiz MT, Cejudo FJ, Paneque A (1989) Role of Mn(II) as regulator of nitrate assimilation in *Azotobacter chroococcum*. Biochim Biophys Acta 993: 36-41
Schramm VL (1982) Metabolic regulation: Could Mn^{2+} be involved? Trends Biochem Sci 7: 369-371
Tubb R, Postgate JR (1973) Control of nitrogenase synthesis in *Klebsiella pneumoniae*. J Gen Microbiol 79: 137-144
Weddler FC, Thomas R (1986) Interactions of Mn(II) with mammalian glutamine synthetase. In: Schramm VL, Weddler FC (eds) Manganese in metabolism and enzyme function. Acad Press, New York London, p 221
Williams RJP (1982) Free manganese (II) and iron (II) cations can act as intracellular cell controls. FEBS Lett 140: 3-10

AMMONIA TRANSPORT SYSTEMS IN CYANOBACTERIA

S. Boussiba [1]

Introduction

Ammonia is a key biochemical compound, being the only molecule which is incorporated into organic linkage, and therefore an obligate intermediate in the utilization of other nitrogen sources. Moreover, ammonia is the inorganic nitrogen source most widely used by microorganisms. The ability to use externally available or internally generated ammonia efficiently is therefore of considerable advantage to free-living cells. This molecule affects also the regulation and/or the activation of some key enzymes of nitrogen metabolism such as nitrogenase, glutamine synthetase and nitrate reductase (Kerby et al. 1987).

Ammonia transport systems were investigated in various microorganisms (see Kleiner 1981, 1985, for reviews), and found to be energy dependent. The organisms studied were proved capable of retaining a large amount of NH_4^+ as well as methylammonium (the commonly used analog of ammonium) gradients across the cell membranes. Since ammonium generally inhibits methylammonium entry completely, it has been assumed that NH_4^+ is in fact the natural substrate of the system.

Two aspects of ammonia translocation in cyanobacteria will be described in this overview: (1) Uptake of NH_4^+/NH_3;(2) Transport of the radioactive ammonium analog $^{14}CH_3NH_3^+$. The term ammonia is used here to cover both the protonated and unprotonated forms of the compound, while NH_3 or NH_4^+ is used where one or the other form is meant specifically.

Uptake of NH_4^+/NH_3

Cyanobacteria appear to use ammonia as a nitrogen source, usually preferring it over nitrate or atmospheric nitrogen. It has been shown repeatedly that ammonia is used first when available at the same time as nitrate (Ohmori et al. 1977). Following its entry ammonia is assimilated via the GS/GOGAT pathway to yield glutamine which is then converted together with α-ketoglutarate to give two molecules of glutamate.

NH_3 is a small uncharged molecule which can be expected to move relatively freely through lipid bilayers. A possible mechanism for an ammonia uptake might therefore involve simple diffusion followed by trapping through protonation, given a membrane permeable to NH_3 but impermeable to NH_4^+. It is therefore expected that the mecha-

[1] Micro-Algal Biotechnology Laboratory, The Jacob Blaustein Institute for Desert Research, Sede Boker, 84993 Israel.

nism of the entry of a molecule such as ammonia whose distribution is pH dependent (pK_a 9.25) could be different in cyanobacteria which can grow in natural habitats over a wide range of pH's. Indeed, during our studies we have provided evidence which may support this assumption (Boussiba et al. 1984; Zimmerman and Boussiba 1987; Boussiba 1989).

One of the first characteristic features of "active" transport of a given solute is its accumulation against its concentration gradient. We have shown (Boussiba et al. 1984) that *Anacystis nidulans* can accumulate ammonium against a concentration gradient of 10^3, but no constant energy was required to maintain it. Ammonia uptake occurred at the same rate over the range of external pH from 6.6 to 8.7 (Table 1), and no obvious changes in apparent K_m (less than 1 μM) were noted.

Table 1. Lack of pH effect on ammonia uptake rate in *Anacystis nidulans* R-2

pH	Uptake rate (nmol min^{-1} mg^{-1} protein)
6.2	26.7
7.0	34.1
8.1	29.7
8.7	31.5

It is reasonable to conclude that the ammonia species involved must be NH_4^+, since its relative concentration is changed by less than 50% due to proton dissociation in the pH range studied, while $[NH_3]$ would represent less than 10^{-3} of the total ammonia at pH 6.2.

The unusual pH insensivity of ammonia uptake found in our studies (Boussiba et al. 1984; Zimmerman and Boussiba 1987) contrasts sharply with the presumed properties of what are thought to be ammonia uptake systems studied using the analog methylamine in other microorganisms (Kleiner 1985). A further point of difference between methylamine uptake in other bacteria and ammonia uptake by cyanobacteria lies in the effect of the nitrogen source during growth. Repression of methylammonium uptake rates by growth in high ammonia concentration was observed in several species (Kleiner 1985), but no significant rate differences associated with the nitrogen source have been detected in the cyanobacteria used in the above studies.

The effect of inhibitors on ammonia uptake (Table 2) suggests that amination may indeed be important in regulating net uptake. Compounds such as ionophores or DCCD, affect primarily photosynthetic energy conversion and ATP formation or utilization, but must affect CO_2 fixation and acceptor regeneration indirectly, as they affect directly reducing energy supply for active membrane transport. Methionine sulfoximine, however, specifically inhibits glutamine synthetase which is the key enzyme in the major ammonia incorporation pathway in cyanobacteria. The sharp inhibition of ammonia uptake observed with this reagent strongly suggests that transport across the cytoplasmic membrane is in fact regulated by the internal pool, and uptake can proceed only through continuous amidation.

Table 2. Effect of inhibitors on ammonia uptake rates in *Anacystis nidulans* R-2

Addition	Final concentration (μM)	Uptake rate (% of control) [a]
DCCD	5	0
CCCP	5	10
SF 6847	5	4
MSX	50	0 [b]

[a] average uptake rate 28.5 nmol min^{-1} mg^{-1} protein
[b] Suspensions (50 μg protein ml^{-1}) were incubated for 10 min with MSX before uptake measurements

We can conclude that in neutrophilic cyanobacteria such as *A. nidulans* or *A. azollae*, NH_4^+ is probably the ammonia form crossing the cytopolasmic membrane when it is used as a nitrogen source. The findings that the entry of this cation is against its concentration gradient and that the process is pH independent, suggests that this cation is penetrating the cell by an energy-dependent transport mechanism.

In alkalophilic strains such as *Spirulina platensis* and *Cyanospira rippkae* the situation is quite different. These microalgae can grow optimally at pH's above 10 (Boussiba 1989) using ammonia as their sole nitrogen source. At those pH values in a growth medium one can expect that NH_3 is the dominant form of ammonia. A simple mechanism of ammonia entry in those cyanobacteria may thus be suggested: diffusion followed by trapping through protonation as a consequence of a pH gradient. The data obtained recently with *S. platensis* (Boussiba 1989) support this suggestion. At pH 10 the uptake of ammonia consisted of two components: a rapid initial disappearance of ammonia from the medium immediately upon its application, which lasted less than 5 seconds (the shortest period measured), and a subsequent slow phase (Fig. 1).

The rapid phase was totally insensitive to MSX and took place even when GS was completely inhibited, while the slow component was totally inhibited by MSX through its effect on GS activity. The rapid component is driven by ΔpH, acidic inside. Preliminary results (S. Boussiba and S. Belkin, unpublished data) indicate that at pH 10.0 the cytoplasmic pH is approximately 8.5. The ΔpH is therefore large enough to support the inward movement of the uncharged ammonia, and leads to its intracellular accumulation (Gaensslen and McCarty 1971; Schuldiner et al. 1972). Indeed, when the cells were toluenized to quench ΔpH the rapid influx of ammonia was totally abolished. Following this initial rapid entry, the continuous activity of glutamine syn-

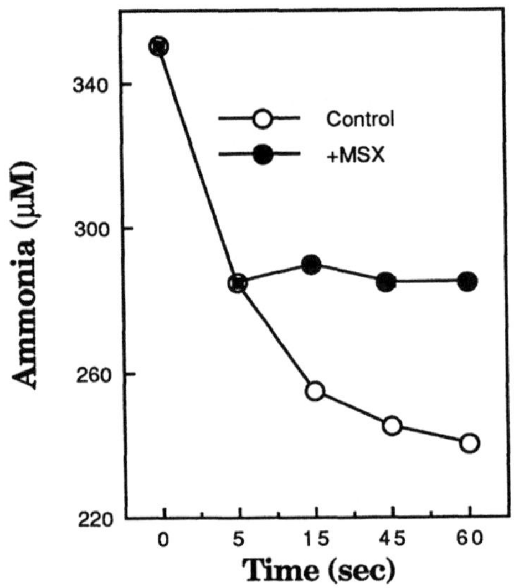

Fig. 1. Effect of MSX on short-term ammonia uptake at pH 10. MSX concentration was 100 µM. Open symbols: without MSX, closed symbols: with MSX

thetase supports further entry by reducing the internal ammonia concentration, thus supplying the driving force for the second, slower phase.

We suggest that under optimal (alkaline) pH the entry of ammonia into alkalophilic cyanobacteria such as *Spirulina* is primarily a diffusion process driven by the pH gradient on the one hand, and by the GS dependent intracellular assimilation on the other hand.

$^{14}CH_3NH_3^+$ Transport

In an effort to distinguish between transport across the cytoplasmic membrane and the subsequent metabolic conversion of the entering molecules, the transport of radioactively labeled methylammonium (MA), an assumed analog of NH_4^+, was studied (Boussiba et al. 1984 b; Rai et al. 1984 ; Boussiba and Gibson 1985, 1987).

The process of methylammonium transport in *A. nidulans* was clearly separated from metabolic conversion of the entered molecules by limiting the time period over which uptake measurements were made or by using selected inhibitors. Unaltered $^{14}CH_3NH_3^+$ could be accumulated to give a concentration gradient of at least 150:1 between illuminated cells and their environment. As might be expected from the very low respiration rate of *A. nidulans*, transport in darkness was, at most, 10% of that seen in light. The driving force for concentration gradient formation appeared to be the membrane potential component of protonmotive force, since additions of high concentrations of K^+ caused a reduction in the methylammonium transport rate and also of light-induced quenching of the fluorescence of a carbocyanine dye, and the penetrant cation triphenylmethyl phosphonium eliminated both. The ΔpH as such did not appear to be involved, since both the rate and extent of transport were unaffected by variations in the external pH over a range that would change not only the magnitude but also the sign (+/–) of ΔpH across the cell boundary (Gibson 1981; Padan and Schuldiner 1978). The ATPase inhibitor dicyclohexylcarbodiimide drastically reduced transport and fluorescence changes showing that ATP is an essential intermediate in energizing transport. This emphasizes the difference in membrane functional organization between *Anacystis* and *Plectonema* sp., in which dicyclohexylcarbodiimide does not affect glucose along transport (Raboy and Padan 1978).

A careful comparison of the uptake of $CH_3NH_3^+$ and NH_4^+ in *A. nidulans* indicates that $CH_3NH_3^+$ is concentrated within the cyanobacterial cell by a transport system which also handles NH_4^+ and has a higher affinity for the latter, and presumably natural, substrate. It does not follow, however, that this system is the only mechanism involved in establishing and maintaining NH_4^+ gradients in these cells, and some apparently major differences were found between the transport of $^{14}CH_3NH_3^+$ and that of NH_4^+ (Table 3).

Table 3. Comparison between NH_4^+ uptake and $^{14}CH_3NH_3^+$ transport

	NH_4^+	$^{14}CH_3NH_3^+$
1.	no Na^+ requirement	requires 5 mM Na^+ optimum
2.	not inhibited by 200 mM KCl	inhibition of the first phase by 200 mM KCl
3.	unaffected when culture grown on either NO_3^- or NH_4^+	severely inhibited when culture grown on 5 mM NH_4^+
4.	high affinity, $K_m < 1$ μM	low affinity, $K_m = 7.2$ μM

The Na^+ requirement, repressibility by growth in NH_4^+, and inhibition by high KCl concentrations could be demonstrated only in measurements of $^{14}CH_3NH_3^+$ accumulation, whereas NH_4^+ uptake was unaffected by these factors.

At least two explanations for these differences can be suggested. *A. nidulans* may possess two distinct systems responsible for NH_4^+ transport, of which one is highly specific for NH_4^+ and constitutive and the other recognizes both NH_4^+ and $CH_3NH_3^+$ and is repressible. It is also possible that there is in fact only one system, and that the differences observed experimentally can be explained with kinetic considerations. $^{14}CH_3NH_3^+$ transport can be separated from conversion to a metabolic product both in time and by the use of inhibitors, so that the entry rate can be clearly visualized and measured. The net rate of NH_4^+ entry, on the other hand, may be governed by the amidation rate, as suggested by the need for bicarbonate during the incubation period preceding uptake measurements. The potential transport rate may thus be very much higher than the net rate actually observed. If this is so, even extensive repression of the system may go unnoticed in the type of measurements made here. The obvious difference in Na^+ requirement for transport of the two compounds could have a comparable explanation, and even 90% reduction in potential activity may still fail to become apparent in the overall rate limiting of the uptake process. This suggestion implies that the inward transport capability of *A. nidulans* is much greater than that needed to maintain the growth rate and may account for the finding that NH_4^+ is taken up in the presence of much higher concentrations of NO_3^- as rapidly as when it constitutes the only available nitrogen source. It could also support the suggestion that an important normal function of an NH_4^+ transport system is to recapture NH_4^+ generated internally from other inorganic nitrogen sources, such as NO_3^- or N_2.

The $NH_4^+/CH_3NH_3^+$ transport system is regulated in different ways in prokaryotes of different habitats. At one extreme, in *Azotobacter vinelandii*, the system appears to be expressed under all growth conditions, and its activity is modulated by the internal pool of glutamine (Jayakumar and Barnes 1984). In *E. coli*, modulation of both gene expression and of existing activity has been reported: expression is under the control of the general regulatory system (Jayakumar et al. 1986), but expressed activity also affected by the glutamine pool (Jayakumar et al. 1987). In *Synechococcus* changes in $CH_3NH_3^+$ transport depend on new protein synthesis or turnover of membrane components (Boussiba and Gibson 1987). Since prolonged exposure to $CH_3NH_3^+$, metabolized only as far as γ-glutamic methylamide in this organism (Boussiba et al. 1984), as well as to NH_4^+ brings about repression, it is probable that the internal pool of glutamine or its analog act as signals for the change in gene expression.

Conclusions

The need of simplified systems such as plasma membrane vesicles which are devoid of metabolic activities, and genetic analysis will be needed to establish unequivocally whether one or several transport systems are involved in NH_4^+ transport in neutrophilic cyanobacteria. In alkalophilic ones, a simple mechanism for NH_3 uptake is suggested: a diffusion process driven by the pH gradient on the one hand, and by the GS dependent intracellular assimilation on the other hand.

References

Boussiba S (1989) Ammonia uptake in the alkalophilic cyanobacterium *Spirulina platensis*. Plant Cell Physiol 30: 303-308

Boussiba S, Dilling W, Gibson J (1984) Methylammonium transport in *Anacystis nidulans* R-2. J Bacteriol 160: 204-210

Boussiba S, Gibson J (1985) The role of glutamine synthetase activity in ammonium and methylammonium transport in *Anacystis nidulans* R-2. FEBS Lett 180: 13-16

Boussiba S, Gibson J (1987) Regulation of methylammonium/ammonium transport in the unicellular cyanobacterium *Synechococcus* R-2 (PCC 7942). FEMS Microbiol Lett 43: 289-293

Boussiba S, Resch CM, Gibson J (1984) Ammonium uptake and retention in some cyanobacteria. Arch Microbiol 138: 287-292

Gaensslen RE, McCarty RE (1971) Amine uptake in chloroplasts. Arch Biochem Biophys 147: 55-65

Gibson J (1981) Movement of acetate across the cytoplasmic membrane of the unicellular cyanobacteria *Synechococcus* and *Aphanocapsa*. Arch Microbiol 130: 175-179

Jayakumar A, Barnes EM (1984) The role of glutamine in regulation of ammonium transport in *Azotobacter vinelandii*. Arch Biochem Biophys 231: 95-101

Jayakumar A, Hong JS, Barnes EM (1987) Feedback inhibition of ammonium (methylammonium) ion transport in *E.coli* by glutamine and glutamine analogs. J Bacteriol 169: 553-557

Jayakumar A, Schulman I, MacNeil D, Barnes EM (1986) Role of the *E. coli glnALG* operon in regulation of ammonium transport. J Bacteriol 166: 281-286

Kleiner D (1981) The transport of NH_3 and NH_4^+ across biological membranes. Biochem Biophys Acta 639: 41-52

Kleiner D (1985) Bacterial ammonium transport. FEMS Microbiol Rev 32: 87-100

Padan E, Schuldiner S (1978) Energy transduction in the photosynthetic membranes of the cyanobacterium (blue-green alga) *Plectonema boryanum*. J Biol Chem 253: 3281-3286

Raboy B, Padan E (1978) Active transport of glucose and α-methylglucoside in the cyanobacterium *Plectonema boryanum*. J Biol Chem 253: 3287-3291

Rai AN, Rowell P, Stewart WDP (1984). Evidence for an ammonium transport system in free-living and symbiotic cyanobacteria. Arch Microbiol 137: 241-246

Schuldiner S, Rotenberg H, Avron M (1972) Determination of ΔpH in chloroplasts. Eur J Biochem 25: 64-70

AMMONIA UPTAKE AND ASSIMILATION BY MUTANT STRAINS OF THE CYANOBACTERIUM *ANABAENA VARIABILIS* RESISTANT TO AMMONIA-ANALOGUES

N.W. Kerby [1], P. Rowell [1] and A. Reglinski [2]

Introduction

The utilization of nitrogenous compounds by microalgae and cyanobacteria is a series of metabolic processes which include membrane transport, reduction of dinitrogen, nitrate or nitrite, and ammonia assimilation. Most studies on the uptake of nitrogenous compounds have considered the summation of these processes and the existence and mechanism of transport into cells has received relatively little attention.

Ammonia is taken up in preference to other nitrogen sources by cyanobacteria (Ohmori et al. 1977; Guerrero et al. 1981). Ammonia transport is of particular interest since ammonia is a powerful repressor of nitrogenase activity and synthesis in N_2 fixing cyanobacteria and also of nitrate uptake and assimilation in cyanobacteria (see Kerby et al. 1989). The terms ammonia and methylamine are used unless the charged (NH_4^+ or $CH_3NH_3^+$) or uncharged (NH_3 or CH_3NH_2) species are meant specifically. The uptake of ammonia can be measured by the depletion of exogenous ammonia, uptake of ^{15}N-ammonia and uptake of ammonia analogues (e.g. methylamine). The first two methods tend to measure transport together with assimilation, which appear to be tightly coupled. Analogues can be used to measure transport as distinct from uptake when they are not assimilated. Methylammonium ($CH_3NH_3^+$) has been used as a substrate for NH_4^+ transport (see Kleiner 1981) and for the measurement of intracellular pH (Kashket and Wilson 1973) in microorganisms. $CH_3NH_3^+$ has been used to study NH_4^+-transport in *Anacystis nidulans* (Boussiba et al. 1984 a,b) and various N_2-fixing cyanobacteria (Kashyap and Johar 1984; Rai et al. 1984; Singh et al. 1985; Kerby et al. 1986 a; Reglinski et al. 1989).

NH_3 is a small uncharged molecule which can pass freely through lipid bilayers by passive diffusion. A mechanism for the accumulation of ammonia at external pH values higher than that of the cytoplasm (pH 6.5 to 7.5), such as those found during bloom conditions, could involve simple diffusion and ion trapping given a membrane permeable to NH_3 but not NH_4^+. At pH values near neutral, as found in many naturally occurring water bodies, a specific energy-dependent uptake process is essential. An NH_4^+ transport system may also be required for the retention of ammonia generated in-

[1] A.F.R.C. Research Group on Cyanobacteria, Department of Biological Sciences, University of Dundee, Dundee DD1 4HN, UK

[2] Present Address: Scottish Crop Research Institute, Invergowrie, Dundee DD2 5DA, UK

ternally by N_2-fixation, nitrate reduction, photorespiration and protein and nucleotide catabolism.

We have isolated mutant strains of the N_2-fixing cyanobacterium *Anabaena variabilis* which are resistant to the ammonia analogues ethylenediamine (1,2-diaminoethane) and methylamine and in this paper we consider some of the properties of these strains.

Methylamine-Resistant Mutant Strains

The uptake of methylamine by cyanobacteria is typically biphasic (see Kerby et al. 1987) at external pH values of 7.0 and 9.0. The initial rapid phase (1 min) is attributed at pH 7.0 to uptake via a $CH_3NH_3^+$ (NH_4^+) transport system and at pH 9.0 to passive diffusion and trapping by protonation. The initial accumulation is greater at pH 9.0 than at pH 7.0 and addition of ammonia displaces the internal free pool of methylamine, inhibiting further uptake. The initial phase is followed by a second slower phase which is sensitive to inhibition by L-methionine-D,L-sulphoximine (MSX), an inhibitor of the primary ammonia assimilating enzyme glutamine synthetase (GS), and is considered to represent assimilation of methylamine to γ-methylglutamine by GS (Rai et al. 1984; Kerby et al. 1986 a). This biphasic pattern has been observed in various cyanobacteria (see Kerby et al. 1987). Studies using uncouplers and ATPase inhibitors have shown that transport of NH_4^+ and $CH_3NH_3^+$ is an energy-dependent, electrogenic process (Boussiba et al. 1984 a,b; Rai et al. 1984).

Methylamine-resistant mutant strains of *Anabaena variabilis* were obtained following chemical mutagenesis and selection for resistance to 1 mM methylamine, a concentration which totally inhibits growth of the parent strain (for details see Reglinski et al. 1989). Figure 1 shows the biphasic pattern of ^{14}C-methylamine uptake by the parent strain at pH 7.0 and pH 9.0 together with the kinetics of uptake by a mutant strain (4m3) resistant to methylamine. At pH 7.0 the mutant strain 4m3, grown in N-free medium, showed a much reduced first phase of methylamine uptake and a reduced second slower phase as compared to the parent strain, whereas at pH 9.0 the kinetics of uptake were similar to the parent strain. The reduced first phase of uptake at pH 7.0 indicates a deficiency in $CH_3NH_3^+$ transport which is not apparent at pH 9.0 where CH_3NH_2 can enter by passive diffusion and accumulate by ion trapping in response to a pH gradient. Growth of strain 4m3 in the presence of 3mM methylamine restored the rate of the first phase of uptake but not the rate of the second slower phase indicating the presence of a methylamine-induced $CH_3NH_3^+$ transport system (Fig. 1). The uncoupler carbonylcyanide-m-chlorophenylhydrazone (CCCP) reduced the first phase of $CH_3NH_3^+$ at pH 7.0 by 75% in the parent strain but only reduced the first phase by 30% in strain 4m3 inferring that the first phase of $CH_3NH_3^+$ transport by strain 4m3 largely represents non-specific binding and carry-over of $^{14}CH_3NH_3^+$.

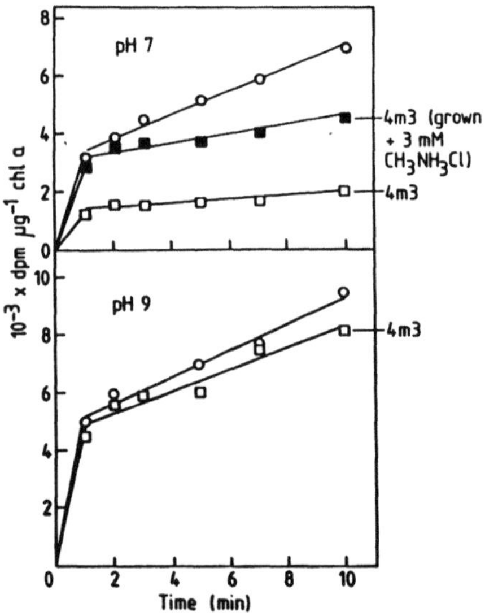

Fig. 1

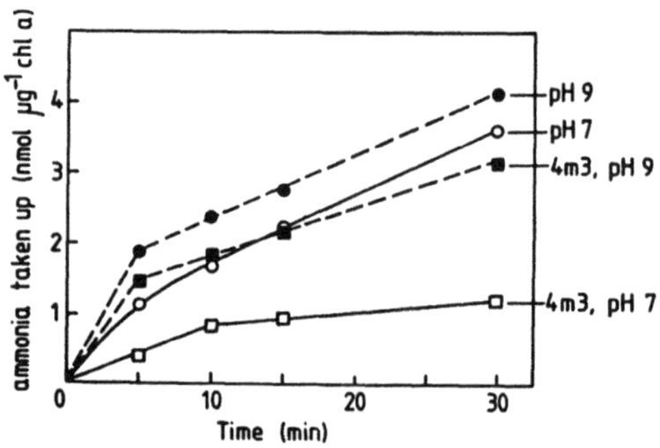

Fig. 2

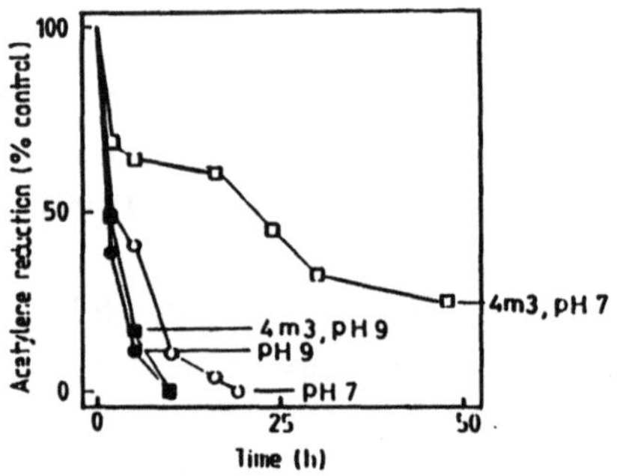

Fig. 3

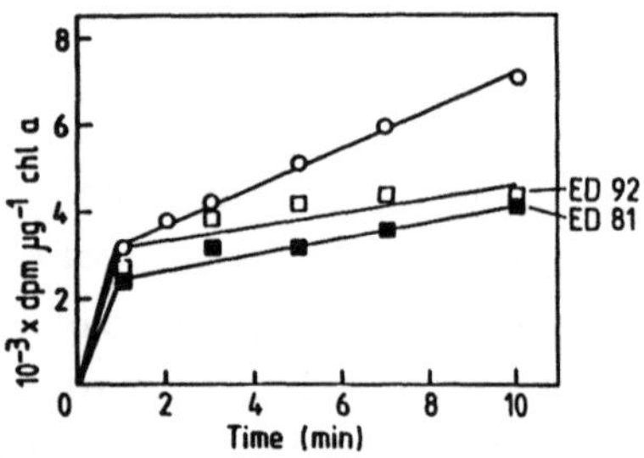

Fig. 4

Fig. 1. ^{14}C-methylamine uptake by parent strain *Anabaena variabilis* (o) and mutant strain 4m3, grown in the presence and absence of 3 mM methylamine at pH 7.0 and pH 9.0

Fig. 2. Ammonia uptake by parent strain and mutant strain 4m3 at pH 7.0 and pH 9.0 measured as depletion from the media using an ion-specific electrode

Fig. 3. Effect of 3 mM NH_4Cl on nitrogenase activity in parent strain and mutant strain 4m3 at pH 7.0 and pH 9.0

Fig. 4. Uptake of $^{14}CH_3NH_3^+$ by parent strain (o) and mutant strains resistant to EDA at pH 7.0. For experimental details see Reglinski et al. (1989)

Figure 2 shows the rate of ammonia uptake at pH 7.0 and pH 9.0 by the parent strain and mutant strain 4m3 measured using an ion-specific electrode. Uptake in this case represents both transport and subsequent assimilation. At pH 9.0, where uptake of NH_3 can occur by diffusion, both strains showed similar kinetics. However, at pH

7.0 the rate of ammonia uptake was much reduced in strain 4m3 confirming the data on $CH_3NH_3^+$ transport which indicated the lack of a $CH_3NH_3^+$ (NH_4^+) transport system. Growth of strain 4m3 in the presence of methylamine did not enhance the rate of ammonia uptake at pH 7.0.

The rate of γ-methylglutamine formation by both strains, as measured by amino acid analysis, following addition of methylamine at pH 7.0 corresponded to the rate of the second phase of uptake of methylamine and was reduced in the mutant strain 4m3. This confirms that the second slower phase of uptake depends on metabolism of methylamine to γ-methylglutamine.

Since strain 4m3 appears to lack an NH_4^+ transport system the effect of exogenous NH_4^+ on nitrogenase activity was determined. Figure 3 shows that addition of 3mM ammonia to N_2-fixing cultures at pH 7.0 caused a more rapid inhibition of nitrogenase activity in the parent strain than in strain 4m3, whereas at pH 9.0 a similar pattern of inhibition was observed in both strains. It should be noted that, in contrast to the effects of ammonia, addition of nitrate to N_2-fixing cultures at pH 7.0 resulted in a slightly greater inhibition of nitrogenase activity in strain 4m3 than in the parent strain. Ammonia was not liberated into the external medium by strain 4m3 unless treated with MSX to inhibit assimilation via GS. This is in contrast to the findings for an NH_4^+-transport deficient mutant of *Klebsiella pneumoniae* (Castorph and Kleiner 1984) but in agreement with the findings for an NH_4^+-transport deficient mutant of *Chlamydomonas reinhardtii* (Franco et al. 1987).

Ethylenediamine-Resistant Mutant Strains

Mutant strains of *Anabaena variabilis* (ED81 and ED92) resistant to lethal concentrations of ethylenediamine (EDA) have been selected (Polukhina et al. 1982) and characterized (Kerby et al. 1986 b). EDA is not taken up by the $CH_3NH_3^+$ (NH_4^+) transport system but enters the cells by passive diffusion at high pH values and is metabolized by GS to a glutamine analogue (aminoethylglutamine) (Kerby et al. 1985). Mutant strains ED81 and ED92 liberate ammonia into their growth medium due to their low levels of GS biosynthetic activity (Kerby et al.1986 b) and can fix nitrogen at high rates in the presence of exogenous ammonia. The mutations responsible for the low levels of GS biosynthetic activity in strains ED81 and ED92 appear to be different since strain ED92 has low levels of GS protein, transferase and biosynthetic activities (Kerby et al. 1986 b) which result from low levels of GS mRNA (Hien et al. 1988), whereas strain ED81 has normal GS mRNA and protein levels, normal GS transferase activity but re-duced GS biosynthetic activity.

Further evidence confirming that the second slower phase of $CH_3NH_3^+$ uptake is attributable to metabolism of methylamine was obtained using strains ED81 and ED92 (Fig. 4). The first rapid phase was unaltered in these strains indicating no deficiency in the $CH_3NH_3^+(NH_4^+)$ transport system but the second phase of uptake was reduced by about 60%, corresponding to their reduced GS activities.

Characterization of Glutamine Synthetase from Methylamine-Resistant and Ethylenediamine-Resistant Mutant Strains

Strain ED81 has reduced GS biosynthetic activity but normal transferase activity and strain 4m3 has a reduced second phase of $CH_3NH_3^+$ and γ-methylglutamine accumulation. Therefore, GS was purified from these strains and compared to purified GS from the parent strain. GS was purified by $(NH_4)_2SO_4$ fractionation, FPLC ion-exchange chromatography (Mono Q) and FPLC gel-filtration (Superose 6 Prep grade) to homogeneity as estimated by polyacrylamide gel electrophoresis (PAGE) using silver staining. This method resulted in a yield of 11 to 23% and a purification of 55 to 68 fold. The specific GS biosynthetic activity of strain ED81 was reduced at all stages of purification as compared to the parent strain and strain 4m3, which had similar activities.

Properties of GS purified from the three strains are given in Table 1. No significant differences in the native relative molecular mass (M_r) (estimated using 10 - 15% Phast gradient gels), subunit M_r (estimated by SDS-PAGE) or isoelectric point of GS from the three sources were detected. The results are similar to those obtained for GS purified from other species of *Anabaena* (Sampaio et al. 1979; Orr et al. 1981). Kinetic properties were also similar for GS purified from these strains. It should be noted that the apparent K_m for methylamine is substantially higher than that obtained for ammonia. An increase in the apparent K_m for glutamate was found in strains ED81 and 4m3 as compared to the parent. However, since the internal free pool of glutamate in cyano-

Table 1. Properties of purified glutamine synthetase from parent strain *Anabaena variabilis* and mutant strains 4m3 and ED81

	Strain		
	Parent	4m3	ED81
Native M_r	596	582	591
Subunit M_r	58.5	57.0	57.0
pI	4.55	4.55	4.55
K_m Ammonia	11	7	2
K_m Methylamine	36,700	32,600	11,200
K_m Glutamate	730	1,850	1,710
K_m ATP	91	96	127
V_{max}	7.2	5.6	0.9

M_r = relative molecular mass x 10^{-3}, apparent K_m = μM, V_{max} = mol min^{-1} (mg protein $)^{-1}$

bacteria is about 20 mM the observed increase in K_m may not be of significance *in vivo*. No other significant differences in kinetic constants were observed between the three sources of enzyme.

Concluding remarks

Strain 4m3 has a reduced uptake of both ammonia and methylamine at pH 7.0 due to a defective NH_4^+ ($CH_3NH_3^+$) transport system but uptake by passive diffusion at pH 9.0 was unaffected. Since ammonia was not detected in the growth media of strain 4m3 it appears that an active NH_4^+ transport system may not be required for the retention of internally generated ammonia in cyanobacteria. The transport of ammonia and its assimilation appear to be tightly coupled in cyanobacteria. No significant differences in the properties of GS that may have accounted for the reduction in GS biosynthetic activity of strain ED81 and for the reduction in the second phase of $CH_3NH_3^+$ uptake by strain 4m3, were observed. However, GS purified from ED81 has a much reduced V_{max}.

Acknowledgments . We thank the Agricultural and Food Research Council, The Science and Engineering Research Council and the Royal Society for financial support.

References

Boussiba S, Resch CM, Gibson J (1984 a) Ammonia uptake and retention in some cyanobacteria. Arch Microbiol 138: 28-292

Boussiba S, Dilling W, Gibson J (1984 b) Methylammonium transport in *Anacystis nidulans* R2. J Bact 160: 204-210

Castorph H, Kleiner D (1984) Some properties of a *Klebsiella pneumoniae* ammonium transport negative mutant (Amt). Arch Microbiol 139: 245-247

Franco AR, Cárdenas J, Fernández E (1987) A mutant of *Chlamydomonas reinhardtii* altered in the transport of ammonium and methylammonium. Mol Gen Genet 206: 414-418

Guerrero MG, Vega JM, Losada M (1981) The assimilatory nitrate reducing system and its regulation. Annu Rev Plant Physiol 32: 169-204

Hien NT, Kerby NW, Machray GC, Rowell P, Stewart WDP (1988) Expression of glutamine synthetase in mutant strains of the cyanobacterium *Anabaena variabilis* which liberate ammonia. FEMS Microbiol Lett 56: 337-342

Kashket ER, Wilson TH (1973) Proton-coupled accumulation of galactosidase in *Streptococcus lactis* 7962. Proc Natl Acad Sci USA 70: 2866-2869

Kashyap AK, Johar G (1984) Genetic control of ammonium transport in the nitrogen-fixing cyanobacterium *Nostoc muscorum*. Mol Gen Genet 197: 509-512

Kerby NW, Rowell P, Stewart WDP (1985) Ethylenediamine uptake and metabolism in the cyanobacterium *Anabaena variabilis*. Arch Microbiol 141: 244-248

Kerby NW, Rowell P, Stewart WDP (1986 a) The uptake and metabolism of methylamine by N_2-fixing cyanobacteria. Arch Microbiol 143: 353-358

Kerby NW, Musgrave SC, Rowell P, Shestakov SV, Stewart WDP (1986 b) Photoproduction of ammonium by immobilised mutant strains of *Anabaena variabilis*. Appl Microbiol Biotechnol 24: 42-46

Kerby NW, Rowell P, Stewart WDP (1987) Cyanobacterial ammonium transport, ammonium assimilation and nitrogenase regulation. NZ J Mar Freshwater Res 21: 447-455

Kerby NW, Rowell P, Stewart WDP (1989) The transport, assimilation and production of nitrogenous compounds by cyanobacteria and microalgae. In: Cresswell RC, Rees TAV, Shah N (eds) Algal and Cyanobacterial Biotechnology. Longman, London, pp 50-90

Kleiner D (1981) The transport of NH_3 and NH_4^+ across biological membranes. Biochim Biophys Acta 639: 41-52

Ohmori M, Ohmori K, Strotmann H (1977) Inhibition of nitrate uptake by ammonia in a blue-green alga, *Anabaena cylindrica*. Arch Microbiol 114: 225-229

Orr J, Keefer LM, Keim P, Nguyen TD, Wellems T, Heinrikson L, Haselkorn R (1981) Purification, physical characterization and NH_2-terminal sequence of glutamine synthetase from the cyanobacterium *Anabaena* 7120. J Biol Chem 256: 13091-13098

Polukhina LE, Sakhurieva GN, Shestakov SV (1982) Ethylenediamine resistant *Anabaena variabilis* mutants with derepressed nitrogen fixing system. Microbiol 51: 90-95

Rai AN, Rowell P, Stewart WDP (1984) Evidence for an ammonium transport system in free-living and symbiotic cyanobacteria. Arch Microbiol 137: 241-246

Reglinski A, Rowell P, Kerby NW, Stewart WDP (1989) Characterization of methylammonium/ammonium transport in mutant strains of *Anabaena variabilis* resistant to ammonium analogues. J Gen Microbiol 135: 1441-1451

Sampaio MJAM, Rowell P, Stewart WDP (1979) Purification and some properties of glutamine synthetase from the nitrogen-fixing cyanobacteria *Anabaena cylindrica* and a *Nostoc* sp. J Gen Microbiol 111: 181-191

Singh DT, Rai AN, Singh HN (1985) Methylammonium (ammonium) uptake in a glutamine auxotroph of the cyanobacterium *Anabaena cycadae*. FEBS Lett 186: 51-53

Part 3 Carbon-Nitrogen Relationship

NITROGEN METABOLISM OF MUTANTS OF BARLEY LACKING KEY ENZYMES OF AMMONIA ASSIMILATION

P.J. Lea [1], K.W. Joy [2] and R.D. Blackwell [1]

The photorespiratory cycle as outlined in Figure 1 was originally proposed by Keys et al. (1978) (see Givan et al. 1988, for a review). Ammonia is released in the mitochondria following the conversion of glycine to serine at a rate ten times that generated by the reduction of nitrate. It is now considered probable that the majority of the ammonia passes directly into the chloroplast where it is converted to glutamine by the chloroplastic isoenzyme of glutamine synthetase (GS) (Lea et al. 1990 a,b). In most C_3 plants the cytosolic form of GS is not present in sufficient amounts to carry out the high rates of assimilation required (McNally et al. 1983). ^{15}N-labelling studies have clearly shown that the glutamine is converted to glutamate (Woo et al. 1982, 1987 a; Rhodes et al. 1989) by ferredoxin-dependent glutamate synthase which is localized solely in the chloroplast (Wallsgrove et al. 1983). Although the mitochondria of plant leaves contain high levels of glutamate dehydrogenase, there is little to suggest that the enzyme may play an important role in ammonia reassimilation (Singh et al. 1985; Yamaya and Oaks 1987; Rhodes et al. 1989; Lea et al. 1990 a). An essential prerequisite of the pathway shown in Figure 1 is that ammonia and 2-oxoglutarate must enter the chloroplast at the same rate and that glutamate must be exported in order to balance the cycle. Woo and his colleagues have studied the system in detail and have proposed both two (Woo et al. 1987 b) and three (Yu and Woo 1988) translocator models for the transport of metabolites across the chloroplast envelope.

It is, however, the isolation and characterization of mutants lacking enzymes of the photorespiratory nitrogen cycle that have gone furthest in establishing the key role played by nitrogen in photosynthesis. Somerville and Ogren devised a method to isolate mutant lines of *Arabidopsis thaliana* that were able to grow only at a sufficiently high level of CO_2 that would competitively inhibit RuBP oxygenase and thus prevent the synthesis of phosphoglycollate (Ogren 1984). The mutant lines of *A. thaliana* have been fully discussed by Somerville (1984, 1986). The isolation and characterization of a similar extensive range of mutant lines of barley (*Hordeum vulgare*) has also been the subject of a number of comprehensive review articles (Blackwell et al. 1988; Lea et al. 1989; Lea and Blackwell 1990).

Barley mutants lacking the chloroplast form of GS have been isolated by research groups at Rothamsted (Wallsgrove et al. 1987) and the University of Lancaster (Blackwell et al. 1987 a). The leaves of the mutant lines accumulated high levels of

[1] Division of Biological Sciences, University of Lancaster, LA1 4YQ, UK
[2] Department of Biology, Carleton University, Ottawa, Canada, K1S 5B6

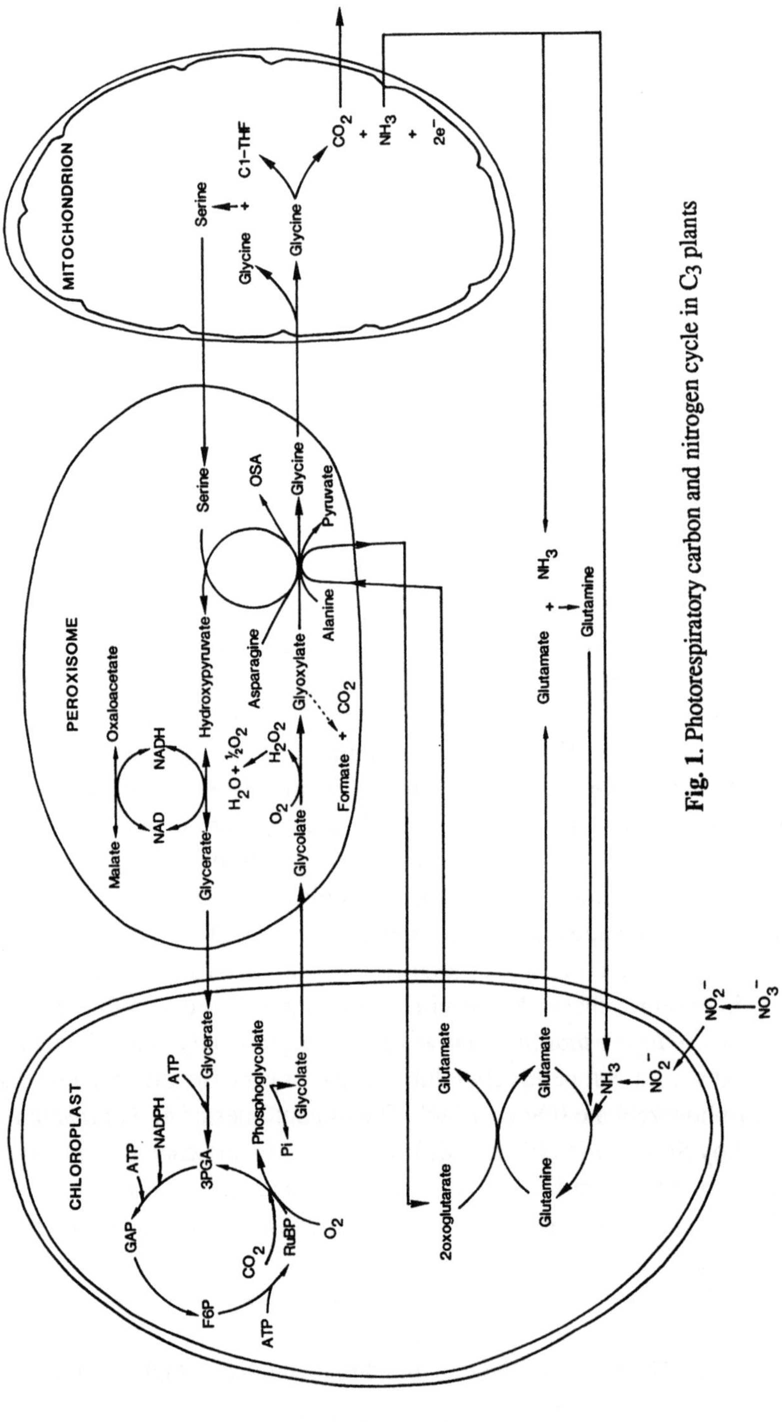

Fig. 1. Photorespiratory carbon and nitrogen cycle in C_3 plants

ammonia when placed in air in the light and there was a steady decline in the soluble amino acid content (Fig. 2 c). The rate of photosynthetic CO_2 fixation also fell following exposure to air; this process could be reversed by the feeding of either alanine, asparagine or glutamine (Lea et al. 1989; Lea and Blackwell 1990).

The GS deficient mutants contained wild-type levels of the leaf cytoplasm enzymes and normal root activity. This information backs up the evidence that the chloroplast form of GS is coded for by a separate single gene (see Cullimore et al., this volume). At the present time fourteen GS deficient lines of barley have been isolated to contain varying concentrations of enzyme protein by Western blot analysis (Wallsgrove et al. 1987; Blackwell 1988). No such mutants have been isolated in *Arabidopsis thaliana*, which contains only the chloroplastic isoenzyme.

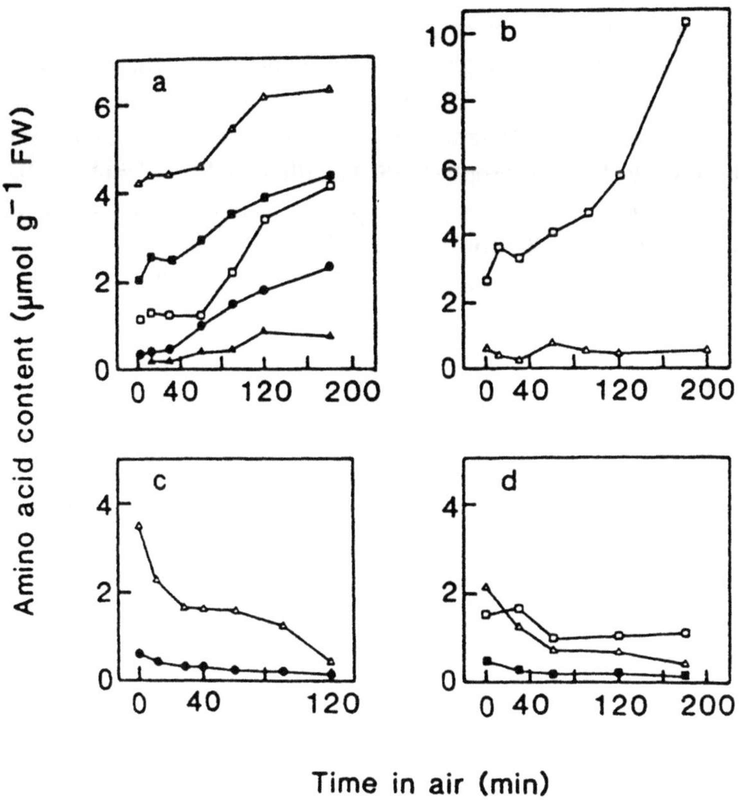

Fig. 2. Change of leaf amino acid content with time following transfer of wild type (**a**), GS deficient (**b**), glutamate synthase deficient (**c**) and double mutant (**d**) from 0.7% CO_2 to air. Levels of glutamic acid (△), glutamine (□), alanine (■), glycine (▲) and serine (o). Amino acids not shown in the mutants were below the levels of detection of the sensitive HPLC technique used

Mutant lines deficient in glutamate synthase were initially isolated from *A. thaliana* (Somerville and Ogren 1980; Grumbles 1987; Morris et al. 1989). Similar lines have also been isolated from barley (Kendall et al. 1986; Blackwell et al. 1987 b, 1988 b) and pea (Blackwell et al. 1987 b). The chloroplastic ferredoxin dependent enzyme is virtually absent in the leaves and roots of the barley mutants. However, an NADH dependent enzyme (5% of the leaf ferredoxin dependent activity) is present in all mutant lines so far tested at normal wild type levels. The NADH and ferredoxin dependent enzymes are under separate gene control, the latter being coded for by a single nuclear gene (Kendall et al. 1986; Blackwell et al. 1988 b).

The rate of photosynthesis in the glutamate synthase deficient mutants of barley falls rapidly following transfer from 1% O_2 to air. This inhibition is far greater than seen with the GS deficient mutants under the same conditions (Kendall et al. 1986; Blackwell et al. 1988 b). The level of glutamine increases rapidly in the leaves following exposure to air, with a corresponding decrease in the other amino acids (Fig. 2 b). Alanine and, to a lesser extent, glutamate were able to prevent the sharp decline in photosynthetic rate (Blackwell et al. 1988 b). Ammonia levels are elevated in the leaves of the glutamate synthase deficient mutants, but to a much lesser extent than determined in those lacking chloroplast GS. Detailed studies of mutants lacking the enzymes of ammonia assimilation and the use of GS inhibitors have shown that the high levels of ammonia are not initially responsible for the drop in the photosynthetic rate (Blackwell et al. 1988 a; Lea et al. 1989; Lea and Ridley 1989; Morris et al. 1989).

Table 1. ^{15}N-labelling of amino acids following addition of [^{15}N-amide]-glutamine to leaves of barley

Amino Acid	Wild Type				Glutamate Synthase Deficient Mutant			
	2% O_2		Air		2% O_2		Air	
	2 h	4 h	2 h	4 h	2 h	4 h	2 h	4 h
Asp	7.9	22.5	130.5	98.2	0.10	0.15	0.64	0.68
Glu	48.7	112.8	275.5	280.9	0.88	0.94	1.8	3.1
Asn (Amino)	11.1	18.4	7.7	6.2	0.06	0.12	0.06	0.72
Asn (Amide)	2.8	4.6	*NA*	1.2	*NA*	0.03	*NA*	0.18
Ser	46.0	146.9	68.3	168.8	0.10	0.10	0.20	0.36
Gln (Amino)	10.8	51.6	74.3	121.2	65.6	78.2	199.8	396.7
Gln (Amide)	51.7	130.7	*NA*	272.9	*NA*	195.9	*NA*	785.1
Gly	4.4	16.1	7.0	18.0	0.19	1.2	2.4	2.9
Ala	5.2	19.9	73.3	110.2	0.06	0.56	0.01	0.09

Leaves from plants grown in elevated CO_2 were detached and placed in air or 2% O_2 and supplied with 2 mM [^{15}N-amide]-glutamine. Leaves were illuminated for the time shown at 190 μEinstein m^{-2} s^{-1}. Results are expressed as nmol ^{15}N g^{-1} fresh weight; - *NA* - not assayed

By normal genetic crossing techniques it has been possible to construct mutant lines of barley lacking both chloroplast GS and ferredoxin dependent glutamate synthase

(Kendall et al. 1987; Blackwell et al. 1988 b). The double mutant plants are just capable of growth in elevated levels of CO_2, but show severe stress symptoms immediately following transfer to air. The plants have the properties of both of the parent lines, exhibiting very low rates of CO_2 fixation combined with high levels of ammonia when exposed to air. The amino acid levels of the double mutant are low and slowly decrease in photorespiratory conditions (Fig. 2 d).

The feeding of [^{15}N]-amide labelled glutamine to wild type barley leaves (Table 1) clearly shows the rapid metabolism to a range of amino acids (Bauer et al. 1977). In the glutamate synthase deficient mutant, label was concentrated solely in glutamine as would be predicted from Figure 2 b. However, the distribution of ^{15}N in the amino group of glutamine clearly shows that the NADH dependent glutamate synthase has been able to convert some glutamine to glutamate, but in the presence of excess GS and ammonia this is rapidly converted back to glutamine.

In addition to the mutants described in this article lacking the enzymes of ammonia assimilation, barley lines lacking six other key enzymes of photorespiratory metabolism have also been isolated at the University of Lancaster. These have been discussed in detail in recent review articles (Blackwell et al. 1988 a; Lea et al. 1989; Lea and Blackwell 1990).

References

Bauer A, Joy KW, Urquhart AA (1977) Amino acid metabolism of pea leaves - labelling studies on utilization of amides. Plant Physiol 59: 920-924

Blackwell RD (1988) Isolation and characterisation of mutants of higher plants unable to carry out photorespiration. PhD Thesis, University of Lancaster

Blackwell RD, Murray AJS, Joy KW, Lea PJ (1987 a) Inhibition of photosynthesis in barley with decreased levels of chloroplastic glutamine synthetase. J Exp Bot 38: 1799-1809

Blackwell RD, Murray AJS, Lea PJ (1987 b) The isolation and characterisation of photorespiratory mutants of barley and pea. Progr Photosynth Res III: 625-628

Blackwell RD, Murray AJS, Lea PJ, Kendall AC, Hall NP, Turner JC, Wallsgrove RM (1988 a) The value of mutants unable to carry out photorespiration. Photosynth Res 16: 155-176

Blackwell RD, Murray AJS, Lea PJ, Joy KW (1988 b) Photorespiratory amino donors, sucrose synthesis and the induction of CO_2 fixation in barley deficient in glutamine synthetase and/or glutamate synthase. J Exp Bot 39: 845-858

Givan CV, Joy KW, Kleczkowski LA (1988) A decade of photorespiratory nitrogen cycling. Trends in Biochem Sci 13: 433-437

Grumbles RM (1987) The effect of glutamate synthase deficiency and ammonia on *Arabidopsis* metabolism. J Plant Physiol 130: 363-371

Kendall AC, Wallsgrove RM, Hall NP, Turner JC, Lea PJ (1986) Carbon and nitrogen metabolism in barley (*Hordeum vulgare* L.) mutants lacking ferredoxin dependent glutamate synthase. Planta 168: 316-323

Kendall AC, Bright SWJ, Hall NP, Keys AJ, Lea PJ, Turner JC, Wallsgrove RM (1987) Barley photorespiration mutants. Progr Photosynth Res III: 629-632

Keys AJ, Bird IF, Cornelius MJ, Lea PJ, Wallsgrove RM, Miflin BJ (1978) Photorespiratory nitrogen cycle. Nature 275: 741-743

Lea PJ, Blackwell RD (1990) Photorespiratory mutants. In: Zelitch I (ed) Perspectives in Biochemical and Genetic Regulation of Photosynthesis. Alan R Liss, New York, in press

Lea PJ, Ridley SM (1989) Glutamine synthetase and its inhibition. In: Dodge A (ed) Herbicides and Plant Metabolism. Cambridge Univ Press, Cambridge, pp 137-167

Lea PJ, Blackwell RD, Murray AJS, Joy KW (1989) The use of mutants lacking glutamine synthetase and glutamate synthase to study their role in plant nitrogen metabolism. In: Conn EE (ed) Plant Nitrogen Metabolism. Plenum Press, New York, pp 157-190
Lea PJ, Blackwell RD, Chen F-L, Hecht U (1990 a) The enzymes of ammonia assimilation. In: Methods in Plant Biochemistry, vol 3. Acad Press, London New York, pp 129-144
Lea PJ, Robinson SA, Stewart GR (1990 b) The enzymology and metabolism of glutamine and glutamate. In: Miflin BJ, Lea PJ (eds) The Biochemistry of Plants, Acad Press, New York London, in press
McNally SF, Hirel B, Gadal P, Mann AF, Stewart GR (1983) Glutamine synthetase of higher plants. Evidence for a specific isoform content related to their possible physiological role and their compartmentation within the leaf. Plant Physiol 72: 22-25
Morris PF, Layzell DB, Canvin DT (1989) Photorespiratoy ammonia does not inhibit photosynthesis in glutamate synthase mutants of *Arabidopsis*. Plant Physiol 89: 498-500
Ogren WL (1984) Photorespiration: Pathways, regulation and modification. Annu Rev Plant Physiol 35: 415-442
Rhodes D, Brunk DG, Magalhães JR (1989) Assimilation of ammonia by glutamate dehydrogenase. In: Conn EE (ed) Plant Nitrogen Metabolism. Plenum Press, New York, pp 191-226
Singh P, Kumar PA, Abrol YP, Naik MS (1985) Photorespiratory nitrogen cycle - A critical evaluation. Physiol Plant 66: 169-176
Somerville CR (1984) The analysis of photosynthetic carbon dioxide fixation and photosynthesis by mutant selection. In: Miflin BJ (ed) Oxford Surveys of Plant Molecular and Cell Biology, vol 1. Oxford Univ Press, Oxford, pp 103-131
Somerville CR (1986) Analysis of photosynthesis with mutants of higher plants and algae. Annu Rev Plant Physiol 37: 467-507
Somerville CR, Ogren WL (1980) Inhibiton of photosynthesis in *Arabidopsis* mutants lacking leaf glutamate synthase activity. Nature 286: 257-259
Wallsgrove RM, Keys AJ, Lea PJ, Miflin BJ (1983) Photosynthesis, photorespiration and nitrogen metabolism. Plant Cell Environ 6: 301-309
Woo KC, Morot-Gaudry JF, Summons RE, Osmond BC (1982) Evidence for the glutamine synthetase/glutamate synthase pathway during the photorespiratory nitrogen cycle in spinach leaves. Plant Physiol 70: 1514-1517
Woo KC, Boyle FA, Flügge IU, Heldt HW (1987 a) ^{15}N-ammonia assimilation, 2-oxoglutarate transport and glutamate export in spinach chloroplasts. Plant Physiol 85: 621-625
Woo KC, Flügge IU, Heldt HW (1987 b) A two translocator model for the transport of 2-oxoglutarate and glutamate in chloroplasts during ammonia assimilation in the light. Plant Physiol 84: 624-632
Yamaya T, Oaks A (1987) Synthesis of glutamate by mitochondria - an anaplerotic function for glutamate dehydrogenase. Physiol Plant 70: 749-756
Yu J, Woo KC (1988) Glutamine transport and the role of the glutamine translocator in chloroplasts. Plant Physiol 88: 1048-1054

THE GREATER ^{13}C NATURAL ABUNDANCE IN NITRATE-GROWN THAN IN AMMONIUM-GROWN *RICINUS COMMUNIS* IS MAINLY A FUNCTION OF THE LOWER RATIO OF CO_2-TRANSPORT LIMITATION TO BIOCHEMICAL LIMITATION OF PHOTOSYNTHESIS IN NITRATE-GROWN PLANTS, WITH THEIR HIGHER ORGANIC ANION CONTENT AS A LESS SIGNIFICANT FACTOR

J.A. Raven [1] and G.D. Farquhar [2]

Introduction

The $^{13}C/^{12}C$ ratio of terrestrial C_3 plants is determined by the $^{13}C/^{12}C$ ratio of the CO_2 used for growth, the temperature and photon flux density under which the plants were grown, the relative limitation of photosynthesis by CO_2 transport and by biochemical reactions of photosynthesis, the fraction of plant C derived from carboxylations other than that catalysed by RUBISCO (ribulose bisphosphate carboxylase-oxygenase) and by the genotype used (Farquhar et al. 1989).

In an investigation of the growth of *Ricinus communis* L. in water culture with nitrate or ammonium as N source the plant genotype, the carbon isotope ratio of the source CO_2, the temperature and the photon flux density were the same for the plants growing under the two N regimes (Allen and Raven 1987; Raven et al. 1984). The observed higher $^{13}C/^{12}C$ ratio of nitrate-grown than of ammonium-grown plants was interpreted in terms of a lower CO_2 transport conductance relative to biochemical conductance of photosynthesis for the nitrate-grown plants (Raven et al. 1984). The rationale was that the water lost per unit dry matter increase was lower for the nitrate-grown than for the ammonium-grown plants (Allen and Raven 1984, 1987). The most reasonable explanation of this difference in water use efficiency is a lower CO_2 transport conductance relative to biochemical conductance during photosynthesis in nitrate-grown plants rather than a lower respiratory loss of C in nitrate-grown plants (Allen and Raven 1984, 1987; Raven and Farquhar, in preparation). However, Raven et al. (1984) ignored the possibility that the higher organic anion content in nitrate-grown than in ammonium grown plants (Allen and Raven 1987) could be at least partly responsible, via an increased fraction of organic C derived from PEPc (phosphoenol-pyruvate carboxylase) activity in the nitrate-grown plants, for the higher $^{13}C/^{12}C$ ratio (Farquhar et al. 1989). This paper presents some of the results of calculations (Raven and Farquhar, in preparation) of the quantitative significance of PEPc activity related to

[1] Department of Biological Sciences, University of Dundee, Dundee DD1 4HN, UK
[2] Plant Environmental Biology Group, Research School of Biological Sciences, Australian National University, GPO Box 475, Canberra, ACT 2601, Australia

the higher rate of net organic acid synthesis in accounting for the higher $^{13}C/^{12}C$ ratio of nitrate-grown than of ammonium-grown *Ricinus communis*.

Computations

The calculations resulted in predicted $^{13}C/^{12}C$ ratios in plant material expressed in terms of Δ, a measure of the $^{13}C/^{12}C$ ratio of the plant material relative to that of the CO_2 from which it was derived (Farquhar et al. 1989). A larger value of Δ means a lower $^{13}C/^{12}C$ ratio in the plant relative to the source CO_2.

The composition of the plant was used to compute the fraction of organic C in the plant at harvest which was contributed by RUBISCO and the residual portion contributed by other carboxylases (predominantly PEPc). The data of Allen and Raven (1987) show a larger fraction of plant C in organic anions whose synthesis involves PEPc activity, in nitrate-grown than in ammonium-grown *Ricinus*: this is related to regulation of acid-base balance in the plant (Raven and Smith 1976). The lower discrimination against the ^{13}C in the inorganic C dissolved in cell water by PEPc than by RUBISCO (after corrrection for the different immediate inorganic C substrates used by the two enzymes) when the CO_2 supply conditions are constant means that a larger fraction of plant organic C in organic acid anions leads to a higher $^{13}C/^{12}C$ in the plant organic matter and thus to a lower value of Δ (Farquhar et al. 1989). Application of these principles to the data of Allen and Raven (1987) shows that the differences in organic acid anion content between nitrate-grown and ammonium-grown *Ricinus* should result in a value of nitrate-grown plants which was 0.11×10^{-3} lower than that of ammonium-grown plants. The observed difference in Δ is in the same direction as the one computed for the effect of different fractional contributions of PEPc activity to plant C at harvest, but the magnitude is substantially greater: nitrate-grown plants have a Δ value 2.1×10^{-3} lower than that of ammonium-grown plants. Since many plants have a larger difference in organic anion content resulting from growth on nitrate rather than growth on ammonium than does *Ricinus*, it would be expected that the increment in Δ between nitrate- and ammonium-grown plants which is attributable to differences in the fractional contribution of PEPc to whole-plant C would be greater than the 0.11×10^{-3} computed for *Ricinus*.

Conclusions

The lower Δ value for nitrate-grown than that for ammonium-grown *Ricinus communis* can be mainly attributed to differences in CO_2 transport limitations relative to biochemical limitations of photosynthesis as a function of N source; the influence of changes in the fraction of C assimilated via PEPc is in the same direction as the observed effect but is much smaller. The lower water use efficiency of growth on ammonium than on nitrate for *Ricinus communis* (Allen and Raven 1984, 1987) is also found for *Phaseolus vulgaris* (Allen et al. 1988), while a lower water loss per unit net photosynthesis was found for *Ceratonia siliqua* L. (Martins-Loução 1985), suggesting that a lower Δ value resulting from a lower value of CO_2 transport conductance relative to biochemical conductance for nitrate-grown than for ammonium-grown plants may not be restricted to *Ricinus*, although data for Δ are not available for the other species as a function of N source.

The finding of differences in the fractional limitation of photosynthesis by CO_2 transport and by biochemistry as a function of N source is of considerable interest: what is the mechanism and what are the ecological and agronomic consequences? At all events, the changes in Δ as a function of N source can be of the same magnitude as differences in Δ caused by growth at different temperatures or photon flux densities, or in environments yielding different water use efficiencies by changing root water supply relative to shoot evaporation potential (Farquhar et al. 1989). Variations in N source should be added to the factors which might alter the Δ value of plants growing in the field.

References

Allen S, Raven JA (1984) NH_4^+ as N source increases water uptake by *Ricinus*. In: Cram WJ, Janácek K, Rybová R, Sigler K (eds) Membrane Transport in Plants. Academia, Praha, pp 105-106

Allen S, Raven JA (1987) Intracellular pH regulation in *Ricinus communis* grown with ammonium or nitrate as N source: the role of long distance transport. J exp Bot 38: 580-586

Allen S, Raven JA, Sprent JI (1988) The role of long-distance transport in intracellular pH regulation in *Phaseolus vulgaris* grown with ammonium or nitrate as N source, or nodulated. J. exp Bot 39: 513-528

Farquhar GD, Ehleringer JR, Hubick KT (1989) Carbon isotope discrimination and photosynthesis. Annu Rev Plant Physiol Plant Mol Biol 40: 503-537

Martins-Loução MA (1985) Estudos fisiológicos e microbiológicos da associação da alfarrobeira (*Ceratonia siliqua* L.) com bacterias de *Rhizobiaceae*. PhD Thesis, Lisboa

Raven JA, Smith FA (1976) Nitrogen assimilation and transport in vascular land plants in relation to intracellular pH regulation. New Phytol 76: 415-431

Raven JA, Allen S, Griffiths H (1984) N source, transpiration rate and stomatal aperture in *Ricinus*. In: Cram WJ, Janácek K, Rybová R, Sigler K (eds) Membrane Transport in Plants. Academia, Praha, pp 161-162

INTERACTIONS BETWEEN RESPIRATION AND NITROGEN ASSIMILATION DURING PHOTOSYNTHESIS

D.H. Turpin, H.G. Weger, R.G. Smith, W.C. Plaxton and M. Lin [1]

Introduction

The assimilation of inorganic nitrogen into amino acids requires ATP, reducing power and carbon skeletons in the form of ketoacids. The carbon skeletons required in amino acid synthesis are primarily respiratory intermediates. Therefore, net amino acid synthesis requires respiratory carbon flow (Syrett 1953, 1956 a,b) even during photosynthesis (Weger and Turpin 1989). The purpose of this contribution is to examine some of these interactions as illustrated by our work with the green alga *Selenastrum minutum* (Naeg.) Collins (UTEX 2459).

In N sufficient algae the molar rate of N assimilation is approximately 1/6 to 1/10 the rate of carbon assimilation. Consequently the changes in metabolism accompanying the onset of N assimilation in N-sufficient cells are difficult to observe. In natural environments, algae often grow under N-deficiency. Under these conditions the resupply of inorganic nitrogen results in N assimilation rates which are equivalent to, and in some cases greater than, the rates of photosynthetic carbon fixation. Consequently, the interactions between N assimilation and respiration during photosynthesis are much easier to observe with these microalgae. To capitalize on this we have used N-limited chemostat cultures of the green alga *Selenastrum minutum.* The techniques employed have been outlined in detail elsewhere (Elrifi and Turpin 1986, 1987; Elrifi et al. 1988; Guy et al. 1989; Lin et al. 1989; Weger and Turpin 1989).

The Provision of Carbon for Amino Acid Synthesis

Ammonium supplied to the N-limited green alga *Selenastrum minutum* is rapidly assimilated via the GS/GOGAT pathway into glutamate and glutamine. This is reflected by a rapid drop in glutamate and increase in glutamine (Fig. 1). The demand for 2-oxoglutarate (2-oxo) in the GOGAT reaction results in its rapid decrease. Within one minute of NH_4^+ addition the 2-oxo levels stabilized, yet the rate of net gln accumulation continues. This infers an increase in the rate of carbon flow through the TCA cycle to 2-oxo which is then used in ammonium assimilation.

This increase in carbon flow to support N assimilation occurs in both the light and

[1] Department of Biology, Queen's University, Kingston, Ontario, Canada K7L 3N6

the dark. In the dark this is reflected by an increase in respiratory CO_2 release during N assimilation. This phenomenon has been known for several decades (Syrett 1953, 1956 a,b). Unmasking respiratory CO_2 release during photosynthesis is somewhat more difficult and requires the use of two different carbon isotopes. Use of a membrane inlet mass spectrometer and $^{13}CO_2$ enrichment has allowed us to resolve the effects of N assimilation on respiratory CO_2 release in the light (Weger and Turpin 1989). Figure 2 shows that NO_3^-, NO_2^- and NH_4^+ all cause an increase in respiratory CO_2 release during photosynthesis. It is also interesting to note that the rate of CO_2 release is highest when NO_3^- is being assimilated and lowest during NH_4^+ assimilation. The ratios of CO_2 released/N assimilated for NO_3^-, NO_2^- and NH_4^+ are 1.25, 0.80 and 0.36 in the light. Even larger differences were found in the dark due to the lack of photosynthetic fixation of respired CO_2 (Weger and Turpin 1989). These results imply that the assimilation of 1 molecule of NO_3^- requires more respiratory carbon flow than 1 molecule of NH_4^+. If the only role of the TCA cycle during N assimilation was the provision of carbon skeletons, then there should be an equivalent increase in TCA cycle carbon flow during the assimilation of all inorganic N sources. The observation that NO_3^- and NO_2^- evoke higher rates of CO_2 release suggests that the TCA cycle not only provides carbon skeletons for N assimilation but also reducing power for the reduction of NO_3^- and NO_2^- during photosynthesis, as well as in the dark.

Regardless of whether the TCA cycle is functioning to provide carbon skeletons or reducing power, maintenance of carbon flow requires the continued oxidation of NADH and $FADH_2$. A scheme showing how such interactions may occur is given in Figure 3.

During the NH_4^+ assimilation-induced enhancement in TCA cycle activity, there is an accompanying increase in the rate of mitochondrial electron transport reflected by an increase in respiratory O_2 consumption. However, when NO_3^- or NO_2^- are assimilated there is little or no increase in respiratory O_2 consumption (Weger and Turpin 1989). This implies that there is an oxidizing agent other than O_2 supporting TCA cycle carbon flow during NO_3^- and NO_2^- assimilation, possibly NO_3^- and NO_2^- themselves (Fig. 3).

If the mitochondrial electron transport chain is responsible for oxidizing TCA cycle reductant during NH_4^+ assimilation, but NO_3^- and NO_2^- fulfill this role during their assimilation (Fig. 3), then two predictions can be made. The first is that inhibition of the mitochondrial electron transport chain during photosynthesis should result in a greater inhibition of respiratory CO_2 release during NH_4^+ assimilation than during the assimilation of NO_3^- or NO_2^-. Second, inhibition of the mitochondrial electron transport chain should inhibit NH_4^+ assimilation but not NO_3^- and NO_2^- assimilation.

Previous work with *S. minutum* has shown that inhibition of the mitochondrial electron transport chain via anaerobiosis during photosynthesis inhibits both control and NH_4^+-stimulated CO_2 release (Weger and Turpin 1989). Anaerobiosis, however,

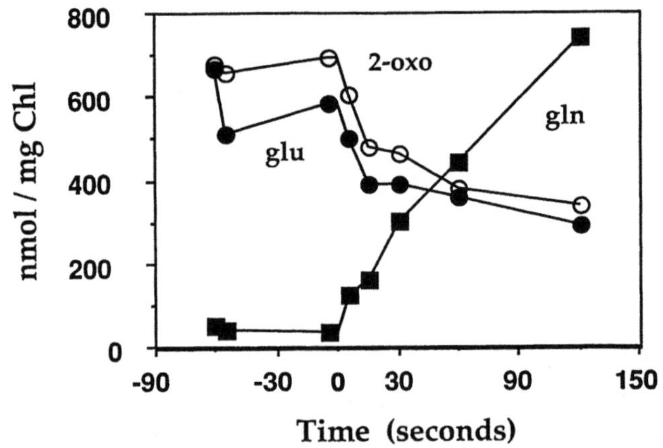

Fig. 1. Changes in cellular pool sizes of glutamine, glutamate and 2-oxoglutarate in response to ammonium addition at time 0. The data points represent the means of three separate experiments (modified from Smith et al. 1989)

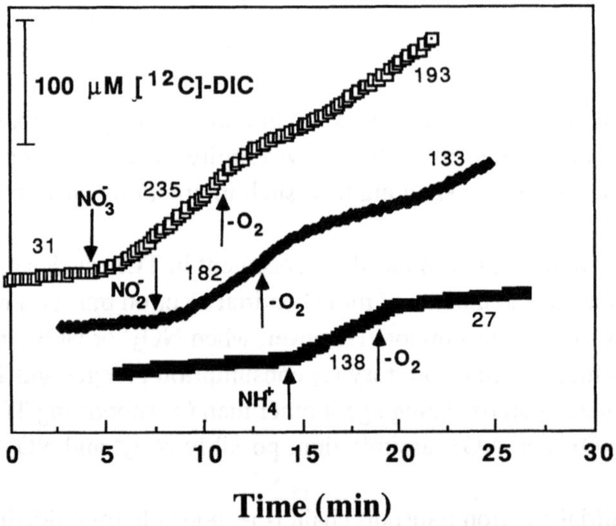

Fig. 2. Effects of different N sources and anaerobiosis on respiratory $^{12}CO_2$ release during photosynthesis into medium containing [^{13}C]-DIC (dissolved inorganic carbon). The first arrow indicates addition of 1 mM N, and the second arrow indicates the onset of anaerobiosis. Similar results were obtained in the presence of DCMU. Rates are expressed as µmol CO_2 mg^{-1} Chl·h^{-1} (modified from Weger and Turpin 1989)

allowed much greater rates of respiratory CO_2 release during NO_3^- and NO_2^- assimilation (Fig. 2). These results are consistent with NO_3^- and NO_2^- acting as an electron

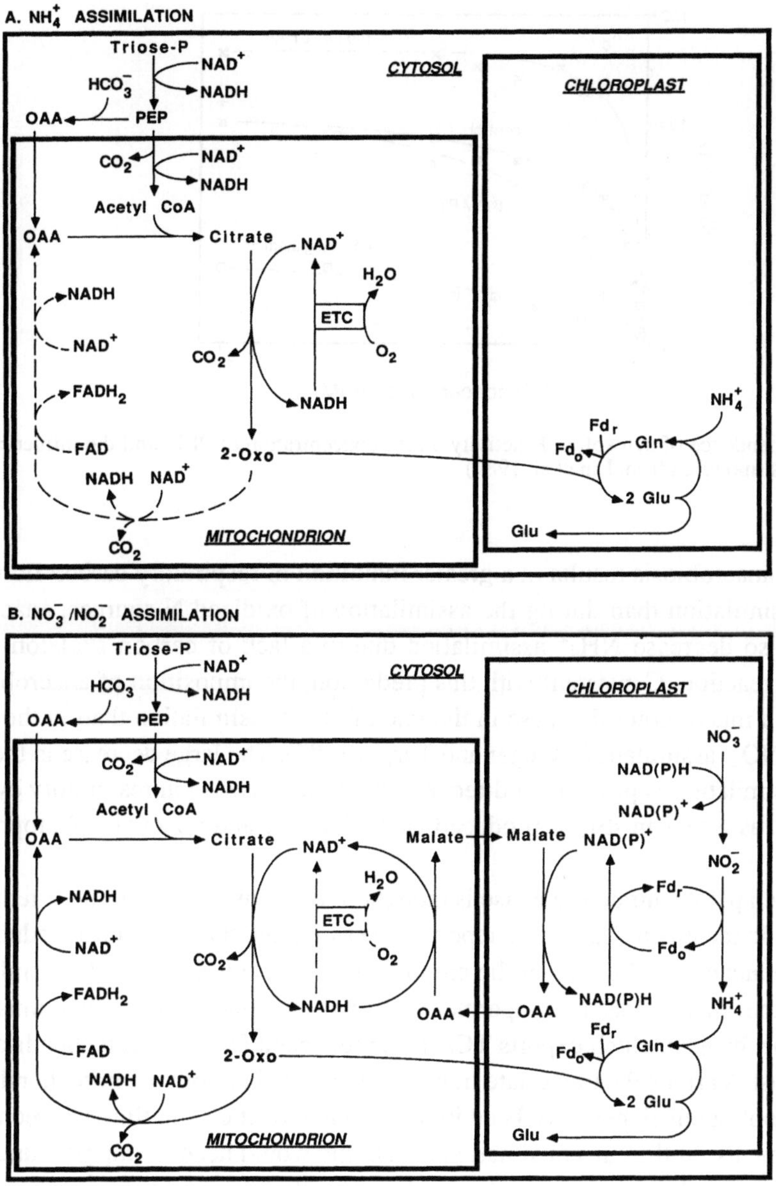

Fig. 3. Proposed pathways of TCA cycle electron and carbon flow during transient N assimilation in the light by N-limited *S. minutum*. Cytosolic triose phosphate would be provided by the chloroplast, from the Calvin cycle and from starch breakdown in the light. The actual mechanism of reductant shuttling during N assimilation is unknown, and has been represented in (B) as malate/OAA exchange (modified from Weger and Turpin 1989)

sink for respiratory reductant, thereby maintaining respiratory carbon flow and associated CO_2 release.

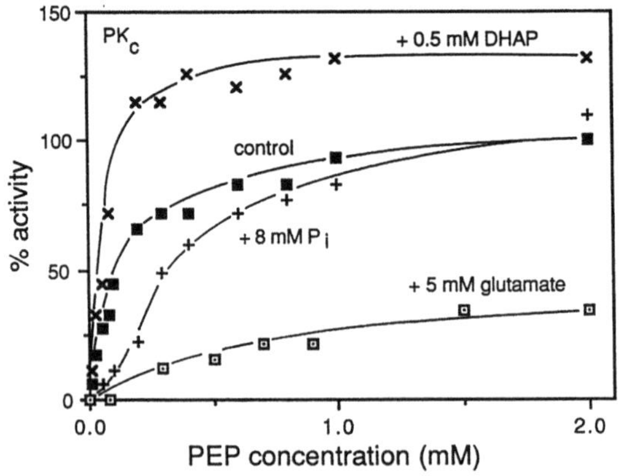

Fig. 4. Dependence of cytosolic PK activity on the concentration of PEP, and the influence of other metabolites (modified from Lin et al. 1989)

Since anaerobiosis results in a greater inhibition of respiratory carbon flow during NH_4^+ assimilation than during the assimilation of oxidized N sources, anaerobiosis should also decrease NH_4^+ assimilation due to a lack of carbon skeletons for the GOGAT reaction. Consistent with this prediction, the imposition of anaerobiosis resulted in a much larger decrease in the rate of NH_4^+ assimilation than in the rates of NO_3^- or NO_2^- assimilation (Weger and Turpin 1989). The large decrease in the rate of NH_4^+ assimilation is probably a direct effect on the decrease in respiratory cycle carbon flow, as ATP can still be produced via both cyclic and noncyclic photophosphorylation.

In green plants nitrate reductase is thought to be cytosolic while nitrite reductase is plastidic. In the green algae some reports have suggested that both nitrite reductase and nitrate reductase are located in the chloroplast. Since the mitochondrial and chloroplastic inner membranes are impermeable to pyridine nucleotides, there must exist a shuttle mechanism which exports TCA cycle reductant from the mitochondrion to the chloroplast. Various dicarboxylate transporters located in both the mitochondrion and the chloroplast envelopes have been implicated in reductant shuttling between cellular compartments. In the current situation the shuttle would need to import reductant into the chloroplast during photosynthesis, the opposite of most of the proposed shuttle mechanisms. Such a system would enable respiration to supply both 2-oxo and reductant to the chloroplast during photosynthesis. Thus the reductant supplied by the photosynthetic electron transport chain could be supplemented by respiratory reductant during NO_3^- and NO_2^- assimilation. Even if nitrate reductase was located in the cytosol, as is thought for higher plants, reductant import by the chloroplast for NO_2^- reduction must still be occurring. The actual shuttle mechanism is unknown, but may be based upon an OAA/malate exchange system.

Replenishment of TCA Cycle Intermediates Used in Biosynthesis

Use of TCA cycle intermediates in amino acid synthesis requires the operation of anaplerotic reactions to maintain the TCA cycle intermediates. In the case of *S. minutum* this function is served by the PEPcarboxylase (PEPc) reaction. The fixation of $H^{14}CO_3^-$ into OAA via PEPc, and the condensation of OAA with acetyl-CoA would eventually result in labeled carbon appearing in the α-carboxyl of 2-oxo and therefore the α-carboxyl of glu and gln. Radio-carbon pulse chase experiments in the light show this to be the case, demonstrating the role of PEP carboxylation in amino acid biosynthesis (Elrifi and Turpin 1987). Using $^{13}C/^{12}C$ isotopic discrimination techniques, Guy et al. (1989) have shown that about 90% of carbon fixed by *S. minutum* during transient NH_4^+ assimilation in the light is actually supplied via PEPc.

This increased flux of carbon into the TCA cycle through PEPc must be met with an increase in carbon flow through pyruvate kinase (PK). Hence both PEPc and PK must be regulated by the demand for TCA cycle intermediates in biosynthesis (Smith et al. 1989). Recently we have purified the plastid and cytosolic isozymes of PK from this organism (Lin et al. 1989) and found that the cytosolic form is allosterically inhibited by glu and P_i and activated by DHAP (Fig. 4) (Lin et al. 1989). Addition of NH_4^+ to photosynthesizing cells results in a rapid decrease in glu (Fig. 1) and an increase in DHAP (Smith et al. 1989). Thus changes in metabolite concentration during NH_4^+ assimilation would serve to activate PK and increase carbon flow to the TCA cycle in support of amino acid biosynthesis.

Summary

Respiration plays an important role in support of photosynthetic nitrogen assimilation. Carbon skeletons required for amino acid synthesis are for the most part respiratory intermediates. Therefore the onset of N-assimilation increases respiratory carbon flow. During the assimilation of NO_3^- and NO_2^- by N-limited *Selenastrum minutum*, respiration serves to provide not only carbon for amino acid biosynthesis, but also reducing power for NO_3^- and NO_2^- reduction.

References

Elrifi IR, Turpin DH (1986) Nitrate and ammonium induced photosynthetic suppression in N-limited *Selenastrum minutum*. Plant Physiol 81: 273-279

Elrifi IR, Turpin DH (1987) The path of carbon flow during NO_3^--induced photosynthetic suppression in N-limited *Selenastrum minutum*. Plant Physiol 83: 97-104

Elrifi IR, Holmes JJ, Weger HG, Mayo WP, Turpin DH (1988) RuBP limitation of photosynthetic carbon fixation during NH_3 assimilation: Interactions between photosynthesis, respiration and ammonium assimilation in N-limited green algae. Plant Physiol 87: 395-401

Guy RD, Vanlerberghe GC, Turpin DH (1989) Significance of phosphoenol pyruvate carboxylase during ammonium assimilation: Carbon isotope discrimination in photosynthesis and respiration by the N-limited green alga *Selenastrum minutum*. Plant Physiol 89: 1150-1157

Lin M, Turpin DH, Plaxton WC (1989) Pyruvate kinase isozymes from the green alga, *Selenastrum minutum* II. Kinetic and regulatory properties. Arch Biochem Biophys 269: 228-238

Smith RG, Vanlerberghe GC, Stitt M, Turpin DH (1989) Short-term metabolite changes during transient ammonium assimilation by the N-limited green alga *Selenastrum minutum*. Plant Physiol 91: 749-755

Syrett PJ (1953) The assimilation of ammonia by nitrogen-starved cells of *Chlorella vulgaris*. Part I. The correlation of assimilation with respiration. Ann Bot 65: 1-19

Syrett PJ (1956 a) The assimilation of ammonia and nitrate by nitrogen-starved cells of *Chlorella vulgaris*. II. The assimilation of large quantities of nitrogen. Physiol Plant 9: 19-27

Syrett PJ (1956 b) The assimilation of ammonia and nitrate by nitrogen starved cells of *Chlorella vulgaris*. IV. The dark fixation of carbon dioxide. Physiol Plant 9: 165-171

Weger HG, Turpin DH (1989) Mitochondrial respiration can support NO_3^- and NO_2^- reduction during photosynthesis: Interactions between photsynthesis, respiration and N-assimilation in the N-limited green alga *Selenastrum minutum*. Plant Physiol 89: 409-415

CARBON METABOLISM AND AMMONIUM ASSIMILATION UNDER LIGHT OR DARK CONDITIONS IN N-SUFFICIENT AND N-LIMITED UNICELLULAR ALGAE

C. Rigano, V. Di Martino Rigano, V. Vona, S. Esposito and C. Di Martino [1]

Introduction

The assimilation of ammonium by N-sufficient cells of unicellular algae is strictly dependent on light and CO_2. However, cells previously kept under conditions of nitrogen starvation (Thacker and Syrett 1972; Syrett et al. 1986), or cells grown in chemostat culture under conditions of nitrogen limitation (Di Martino Rigano et al. 1986), can assimilate ammonium in darkness. Moreover, nitrogen deprivation or nitrogen limitation leads to an increase in the rate at which ammonium is absorbed when supplied to the cells. Besides, as demonstrated in several algae the supply of NH_4^+ to N-limited or N-starved cells stimulates dark CO_2 fixation (Syrett 1956; Ohmori et al. 1984; Elrifi and Turpin 1986) and starch breakdown (Miyachi and Miyachi 1985), and affects photosynthetic and respiratory metabolism to a much greater extent than in N-sufficient cells. This effect on carbon metabolism is stricly correlated to the increased demand for carbon skeletons suitable for ammonium incorporation and amino acid synthesis. Thus, results obtained with cells grown under different culture conditions may provide clues about the regulatory mechanisms operating in the pathways of nitrogen assimilation. In this article observations concerning regulatory aspects of carbon and nitrogen metabolism interaction in the green alga *Chlorella sorokiniana*, a facultative autotroph, and in the red alga *Cyanidium caldarium*, an obligate autotroph, are discussed.

Pattern of Ammonium Assimilation by N-Sufficient Cells of *Cyanidium caldarium* and *Chlorella sorokiniana*

Assimilation of ammonium by cells of *C. caldarium* (Fig. 1) and *C. sorokiniana* (not shown) grown in batch under conditions of excess ammonium (N-sufficient cells) was almost immediately inhibited by removal of light and CO_2, and was immediately restored upon light and CO_2 resupply, which supports the view that ammonium assimilation requires an active photosynthesis. This is in agreement with findings (Kanazawa et al. 1970; Elrifi and Turpin 1986) that the Calvin cycle, through an involvement of dark CO_2 fixation, can provide a source of triose phosphate for replen-

[1] Dipartimento di Biologia Vegetale, Università di Napoli, via Foria 223, 80139 Napoli, Italy

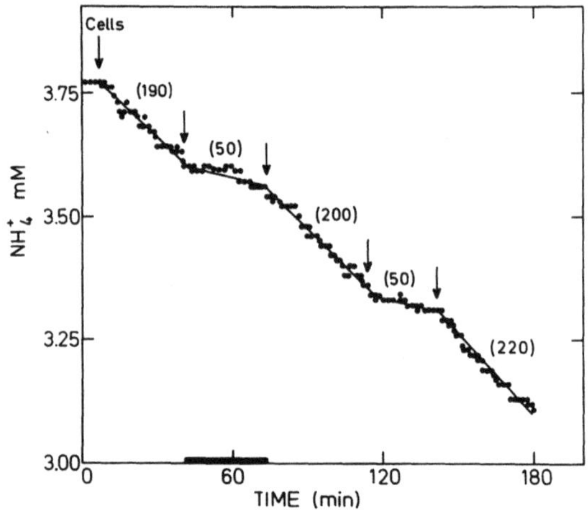

Fig. 1. Ammonium uptake by N-sufficient cells of *Cyanidium caldarium* (grown in batch culture with 20 mM ammonium). The cell suspension was kept in the light at 42°C, and was gassed with air containing 5% CO_2. Experiments started with the addition of the cells (first arrow). At the time indicated by the second arrow, the cell suspension was transferred to dark, and thereafter again to light (third arrow). At the time indicated by the fourth arrow, CO_2 was removed and was again supplied successively (fifth arrow). The numbers in parentheses indicate the uptake rates expressed as µmol NH_4^+ h^{-1} ml packed cell volume $(pcv)^{-1}$. Ammonium disappearance from the external medium was monitored continuously with an ammonia electrode as described previously (Di Martino Rigano et al. 1986)

ishment of C_4 and C_5 carbon compounds suitable for ammonium assimilation. The strict dependence on light and CO_2 indicates that, by this way, in N-sufficient cells the Calvin cycle constitutes the sole source of carbon for ammonium assimilation.

Patterns of Ammonium Assimilation by N-Limited Cells of *Chlorella sorokiniana* and *Cyanidium caldarium*

Cells of *C. sorokiniana* grown in chemostat under conditions of nitrate limitation (N-limited cells) were able to assimilate NH_4^+ both in the light and dark at similar rates. Under either conditions the N-limited cells assimilated ammonium at about a 3-fold higher rate than did N-sufficient cells in the light (not shown).

These results are consistent with the contention that in N-limited cells the major source of carbon for ammonium assimilation must be starch breakdown. On the other hand, like in other algae, respiration of N-limited cells of both *C. sorokiniana* and *C. caldarium*, not of N-sufficient cells, is stimulated 3.5-fold by the addition of ammonium (not shown).

Also N-limited chemostat cells of *C. caldarium* assimilated ammonium either in the light or dark at higher rates than did N-sufficient cells in the light. However, different

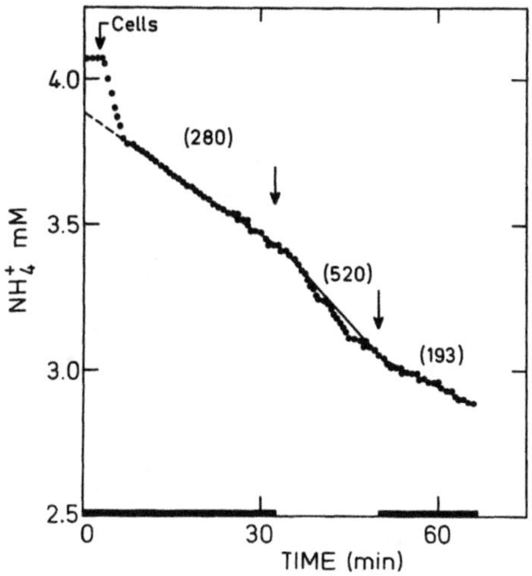

Fig. 2. Ammonium uptake by N-limited cells of *Cyanidium caldarium* (grown in chemostat under conditions of N limitation), kept at 42°C and gassed with air containing 5% CO_2. The cells were initially kept in the light, and were then subjected to alternating dark and light periods as indicated by the arrows and the black bars (dark) on the abscissa. Activity (numbers in parentheses) and NH_4^+ assay as indicated in Fig. 1

from *C. sorokiniana*, the rate of assimilation in the light was higher than that in the dark (Fig 2). It is noteworthy that (i) the difference between the rate of ammonium assimilation measured in the light and that measured in the dark, was similar to the rate at which ammmonium was assimilated in the light by N-sufficient cells, and (ii) this light-dependent part of ammonium assimilation in N-limited cells, as the sole light-dependent activity of N-sufficient cells, was suddenly inhibited upon removal of light and was immediately restored upon light resupply. This suggests (i) that in N-limited cells carbon skeletons for ammonium assimilation must be either from the Calvin cycle (the light-dependent assimilation) or from starch breakdown (the dark assimilation), (ii) that in *C. caldarium* both ways can operate simultaneously, and (iii) that each of these metabolic pathways alone is inadequate to supply all the carbon needed for full ammonium assimilation.

Effect of Removal of CO_2 on NH_4^+ Assimilation in N-Limited Cells

As shown previously (Di Martino et al. 1986) in cell suspensions of N-limited chemostat cells of *C. caldarium*, in contrast to N-sufficient cells, ammonium assimilation was not immediately inhibited by CO_2 deprivation, but inhibition behaved like a time-dependent phenomenon both in light and darkness. During an initial period of about

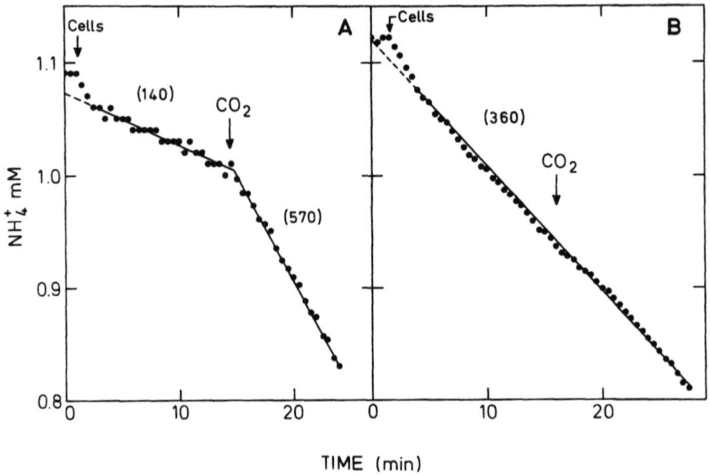

Fig. 3. Effect of CO_2 deprivation on initial ammonium uptake by N-limited cells of *Cyanidium caldarium* in light and dark. The cell suspensions were kept at 42°C in the light (**A**) or in darkness (**B**), and were gassed with CO_2-free air. To remove CO_2 the air was passed through 10 mM NaOH before it reached the cell suspensions. At the time indicated by the arrows CO_2 (5%) was added again. Activity (number in parentheses) and NH_4^+ assay as indicated in Fig. 1

30 min, however, CO_2 deprivation was without effect on ammonium assimilation in darkness, but it produced a 75 % inhibition in light which suggests a light-dependent inhibition promoted by the absence of CO_2 (Fig. 3). Thus, under CO_2-free air, not only light-dependent ammonium assimilation (which, as referred above, is strictly dependent also on CO_2) was prevented, but there occurred also a light-dependent decrease of dark ammonium assimilation. This light-dependent inhibition of dark ammonium assimilation occurs also in *C. sorokiniana* (not shown). Resupply of CO_2 stimulated ammonium assimilation in the light but was without effect on ammonium assimilation in the dark. Thus, as a working hypothesis, we are investigating whether the light dependent decline of ammonium assimilation in the light in the absence of CO_2 could be promoted by photogenerated ATP and reductants which, since they are not being used in the Calvin cycle, attain intracellular concentrations suitable to trigger control mechanisms operating at the level of starch breakdown. Another level of control would be at the glycolytic pathway which ultimately constitutes the primary source of substrates for dark respiration and, consequently, for the generation of carbon substrates for ammonium assimilation. This would be consistent with the finding that dark respiration of leaves and algae is partially inhibited in the light as a consequence of a light-dependent increase in the cytoplasmic ratio of ATP/ADP (Heber 1974). If so, it could provide a control mechanism by which the cell would be able to economize carbohydrate reserves not replenished in the absence of CO_2. However, the consistent stimulation of respiration after the supply of ammonium suggests that some

endogenous CO_2 is available. Thus, alternatively, it might be suggested that under conditions of limited CO_2 supply, in the light the photosynthetic carboxylation reaction will compete with the dark carboxylation reaction for the available CO_2. Hence it is likely that the production of C_4 and C_5 precursors for amino acid biosynthesis will be slower in the light than in the dark and, as a consequence, ammonium assimilation will be slower, too.

Glutamine Synthetase Activity

In *C. caldarium* N-limitation brings about a 10-fold increase of glutamine synthetase activity (Rigano et al. 1981), and in *C. sorokiniana* about an 8 fold increase (Tischner and Hüttermann 1980), which accounts for the increased capacity of N-limited cells to assimilate ammonium. However, in *C. caldarium*, the rate of ammonium assimilation which strictly depends on light/CO_2, as shown by the difference between ammonium assimilation by N-limited cells in light and darkness, did not increase upon N-limitation. Thus the increase in capacity to assimilate ammonium upon N-limitation, besides derepression of glutamine synthetase and of other enzymes of nitrogen metabolism (Menacho and Vega 1989) is, apparently, based also on derepression of regulatory enzyme(s) of dark respiratory metabolism. This enables the N-limited cells to obtain from mobilization of polysaccharide reserves the major amount of carbon skeletons for NH_4^+ assimilation. However, it is noteworthy that with the fungus *Neurospora* (Hernández and Mora 1986) it was proposed that a coordinated regulation of carbon and nitrogen utilization is achieved by sensing the carbon and nitrogen flows through glutamine synthesis, mediated by changes in the intracellular ATP content. Thus, ultimately the variation in the nitrogen status and, in turn, the nitrogen demand of the cell, through a modulation of glutamine synthetase activity and, perhaps, of other enzymes of nitrogen metabolism, could provide a control signal for coordinated regulation of carbon and nitrogen metabolism.

References

Di Martino Rigano V, Vona V, Di Martino C, Rigano C (1986) Effect of darkness and CO_2 starvation on NH_4^+ and NO_3^- assimilation in the unicellular alga *Cyanidium caldarium*. Physiol Plant 68: 34-38

Elrifi IR, Turpin DH (1986) Nitrate and ammonium induced photosynthetic suppression in N-limited *Selenastrum minutum*. Plant Physiol 81: 273-279

Heber U (1974). Metabolite exchange between chloroplasts and cytoplasm. Annu Rev Plant Physiol 25: 393-421

Hernández G, Mora J (1986) Glutamine synthesis regulates sucrose catabolism in *Neurospora crassa*. J Gen Microbiol 132: 3315-3323

Kanazawa T, Distefano M, Bassham JA (1983) Ammonia regulation of intermediary metabolism in photosynthesizing and respiring *Chlorella pyrenoidosa*: Comparative effects of methylamine. Plant Cell Physiol 24: 979-986

Menacho A, Vega JM (1989). Effect of nitrogen starvation on ammonium assimilation by *Chlamydomonas reinhardtii*. Physiol Plant 75: 285-289
Miyachi S, Miyachi S (1985). Ammonia induces starch degradation in *Chlorella* cells. Plant Cell Physiol 26: 245-252
Ohmori M, Miyachi S, Okabe K, Miyachi S (1984). Effects of ammonia on respiration, adenylate levels, amino acid synthesis and CO_2 fixation in cells of *Chlorella vulgaris* 11h in darkness. Plant Cell Physiol 25: 749-756
Rigano C, Di Martino Rigano V, Vona V, Fuggi A (1981). Nitrate reductase and glutamine synthetase activities, nitrate and ammonia assimilation, in the unicellular alga *Cyanidium caldarium*. Arch Microbiol 129: 110-114
Syrett PJ (1956) The assimilation of ammonia and nitrate by nitrogen-starved cells of *Chlorella vulgaris*. IV. The dark fixation of carbon dioxide. Physiol Plant 9: 165-171
Syrett PJ, Flynn KJ, Molloy CJ, Dixon GK, Peplinska AM, Cresswell RC (1986) Effects of nitrogen deprivation on rates of uptake of nitrogenous compounds by the diatom *Phaeodactylum tricornutum* Bohlin. New Phytol 102: 39-44
Thacker A, Syrett PJ (1972). The assimilation of nitrate and ammonium by *Chlamydomonas reinhardii*. New Phytol 71: 423-433
Tischner R, Hüttermann A (1980). Regulation of glutamine synthetase by light and during nitrogen deficiency in synchronous *Chlorella sorokiniana*. Plant Physiol 66: 805-808

THE ROLE OF AMINO ACIDS IN THE REGULATION OF CYANOBACTERIAL GLUCOSE-6-PHOSPHATE DEHYDROGENASE

P. Rowell and G.G. Simpson [1]

Introduction

At present we have considerable knowledge of the pathways of carbon and nitrogen metabolism in photosynthetic microorganisms, although relatively little is known about the precise biochemical mechanisms involved in their regulation or the interrelationships between them (see Smith 1982; Lara et al. 1987; Turpin et al. 1988). In this paper we consider the mechanisms by which the enzyme glucose-6-phosphate dehydrogenase (G6PDH) of cyanobacteria is regulated. The oxidative pentose phosphate (OPP) pathway, is the main route of fixed carbon dissimilation in cyanobacteria and it is considered that metabolic control is exerted at the first step of the pathway, catalysed by G6PDH (Pelroy and Bassham 1972; Pelroy et al. 1972). This pathway is important, not only in respiratory metabolism, but has also been implicated in the provision of reductant for N_2-fixation in the light (see Stewart and Rowell 1986). Notably, the activities of G6PDH and 6-phosphogluconate dehydrogenase are several fold higher in heterocysts than in vegetative cells (Winkenbach and Wolk 1973). Previous studies (Grossman and McGowan 1975; Schaeffer and Stanier 1978; Udvardy et al. 1983, 1984; Cossar et al. 1984) have demonstrated the complex regulatory properties of G6PDH. It is generally considered that effectors, such as $NADP^+$, ribulose-1,5-bisphosphate, NADPH, ATP and reduced thioredoxin, and changes in the cytoplasmic pH may be involved in inhibition/deactivation of the enzyme in the light (Duggan and Anderson 1975). This may prevent the oxidation of glucose in cells which are assimilating CO_2 via the Calvin cycle. However, it is not known precisely how a high activity of the enzyme is maintained in N_2-fixing heterocysts in the light.

Effects of Amino Acids on the Activity of G6PDH from *Anabaena variabilis*

Schaeffer and Stanier (1978) reported that glutamine, like the substrate glucose-6-phosphate (G6P), is a positive effector of G6PDH from *Anabaena* 7120 and suggested that it might be involved in maintaining a high activity of the enzyme in heterocysts. We have further investigated the effects of glutamine and other amino acids,

[1] A.F.R.C. Research Group on Cyanobacteria and Department of Biological Sciences, University of Dundee, Dundee DD1 4HN, UK

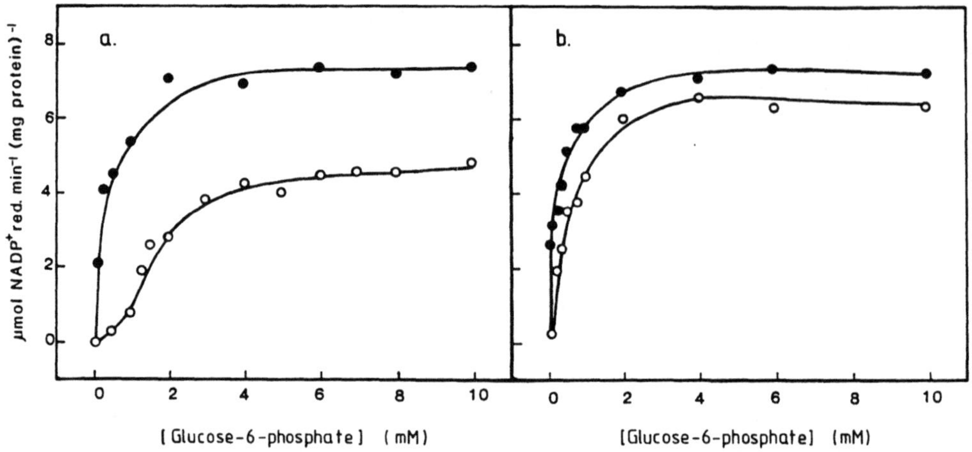

Fig. 1. Effects of glucose-6-phosphate (G6P) concentration on the activity of G6PDH from *A. variabilis* in the presence (●) and absence (○) of glutamine at (**a**) pH 6.9 and (**b**) pH 7.4.
Enzyme used for this Figure and Figs. 2 and 3 was prepared immediately prior to assay by the following partial purification procedure. *A. variabilis* ATCC 29413 cells were suspended in 20 mM bis-tris propane, 2.5 mM G6P, 10 mM $MgCl_2$, 1 mM EDTA, 10 mM 2-mercaptoethanol (pH 6.5) and disrupted by sonication. Following centrifugation at 150,000 g for 1 h streptomycin sulphate was added to 1.6% (w/v). After incubating at 4°C for 30 min, the precipitate was removed by centrifugation at 36,000 g and the supernatant was subjected to FPLC anion exchange chromatography (mono Q, HR 5/5, Pharmacia) eluting with a gradient of 0.125 M - 0.3 M KCl in 20 mM bis-tris propane, 1 mM G6P, 10 mM $MgCl_2$, 1 mM EDTA, 1 mM 2-mercaptoethanol (pH 6.5). The resulting preparations had a specific activity of 5 to 7 µmol $NADP^+$ reduced min^{-1} (mg protein)$^{-1}$, were free of 6-phosphogluconate dehydrogenase activity and did not reduce $NADP^+$ in the absence of G6P. The enzyme had a native relative molecular mass (M_r) of approximately 250,000 as determined by FPLC gel filtration chromatography (Superose 6 Prep grade). Activity was determined at 28°C in a volume of 1 ml, containing 10 µmol $MgCl_2$, 0.5 µmol $NADP^+$, G6P (as indicated) and bis-tris propane buffer (I = 0.05M) at the appropriate pH, and absorbance was monitored at 340 nm

using partially purified G6PDH preparations from *A. variabilis*.

Figure 1 shows the different kinetic properties of G6PDH when assayed at pH 7.4 and 6.9, values which approximate to the cytoplasmic pH values in the light and dark respectively (see Hawkesford et al. 1983). When the enzyme was assayed at pH 7.4, there was an apparently sigmoidal relationship between the initial activity and the G6P concentration, although in the presence of 1 mM glutamine activity was higher and a hyperbolic curve was obtained. At pH 6.9, in the presence or absence of glutamine, the relationship between G6PDH activity and G6P concentration was similar to that at pH 7.4 in the presence of glutamine. The sigmoidal curve obtained at pH 7.4 in the absence of glutamine may be explained, partly, by the rapid loss of activity which occurs at this pH, particularly at low concentrations of G6P. This rapid loss of activity at pH 7.4 in the absence of glutamine and at a low concentration of G6P is illustrated in Figure 2. In contrast to the effect of the low concentration of G6P used here, 1 mM glutamine promoted a marked reactivation of the enzyme. Therefore, like higher con-

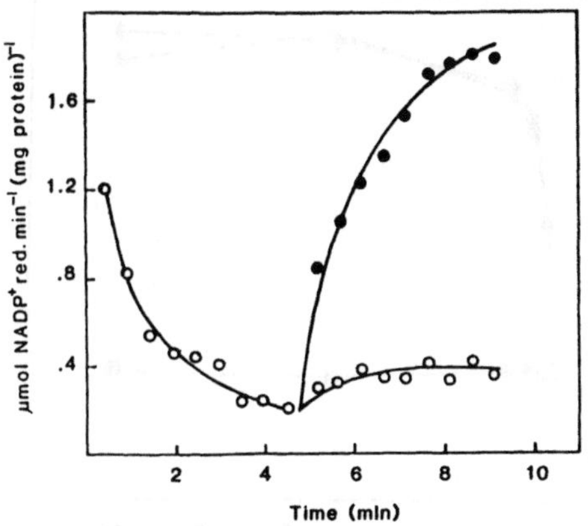

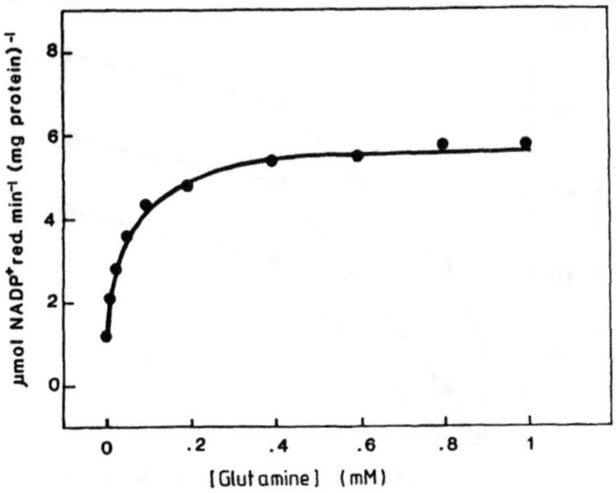

Fig. 2 *(left)*. Rate of $NADP^+$ reduction by G6PDH from *A. variabilis* at intervals after addition of enzyme to reaction mixtures (1 ml final volume) containing 10 µmol $MgCl_2$, 0.5 µmol $NADP^+$ and 1 µmol G6P at pH 7.4. After 5 min, either 1 µmol glutamine (●) or a further 1 µmol G6P (○) was added

Fig. 3 *(right)*. Effect of glutamine concentration on the activity of G6PDH from *A. variabilis*. Enzyme was added to reaction mixtures (1 ml final volume) containing 10 µmol $MgCl_2$, 0.5 µmol $NADP^+$, 1 µmol G6P and glutamine (as indicated) at pH 7.4

centrations of G6P (Schaeffer and Stanier 1978), glutamine not only prevented deactivation of the enzyme, but also promoted reactivation. Figure 3 shows that concentrations of glutamine, as low as 0.2 mM, were effective in preventing deactivation of the enzyme. Such concentrations fall within the range of values for the intracellular free glutamine pool of *A. variabilis* (our unpublished data; see also Reglinski et al. 1989), indicating that the effects may be physiologically significant.

This effect was not, however, restricted to glutamine. When tested at a concentration of 1 mM, many other protein amino acids were found to have an effect similar to that of glutamine. Alanine, glycine, serine, methionine and cysteine were the most effective, whereas some, such as glutamate, lysine and arginine were ineffective. Of those amino acids which were identified as positive effectors of G6PDH, glutamine, alanine, serine and glycine are all major free pool amino acids in *A. variabilis* and are effective at concentrations within the cytoplasmic concentration range. Notably, these amino acids are important intermediates in nitrogen metabolism, being either products of primary nitrogen assimilation or intermediates of the photorespiratory nitrogen cycle.

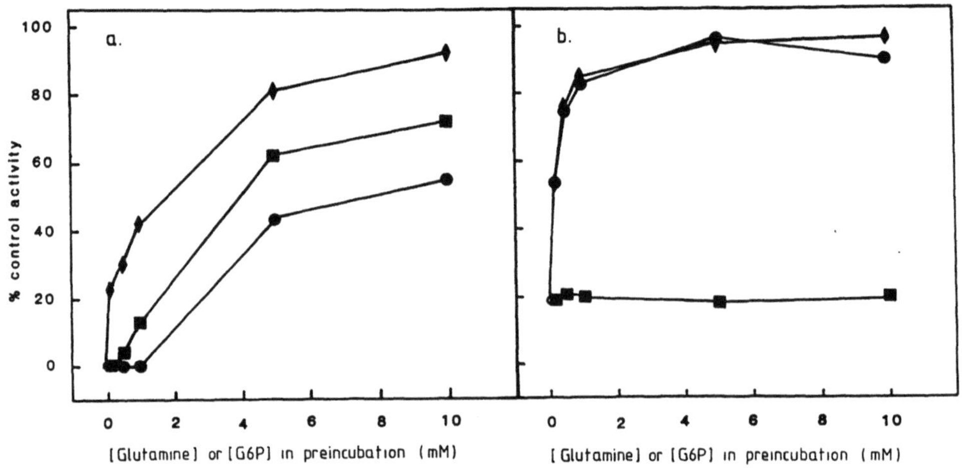

Fig. 4. G6PDH activity in cell-free extracts of (**a**) *A. variabilis* and (**b**) *Synechococcus* PCC 7943 following preincubation for 10 min at pH 6.5 with 20 mM DTT plus glutamine (■) G6P (●) or G6P plus 1 mM glutamine (◆) (at the concentrations indicated). Initial activities were then determined in a reaction mixture (1 ml final volume) containing 10 μmol $MgCl_2$, 0.5 μmol $NADP^+$ and 10 μmol G6P at pH 6.5, and are expressed as % of controls treated in the same way, but with DTT omitted (activity = 29.9 and 26.5 nmol $NADP^+$ reduced min^{-1} for enzyme from *A. variabilis* and *Synechococcus* 7943, respectively)

Effects of Thioredoxin on G6PDH Activity

Those cyanobacteria so far examined possess a thioredoxin (M_r approximately 12,000) which is structurally similar to *Escherichia coli* thioredoxin and chloroplast thioredoxin *m* (see Crawford et al. 1988; Gleason and Holmgren 1988; Rowell et al. 1988). A single gene encoding this thioredoxin has been isolated and cloned from *Anabaena* PCC 7119 (Lim et al. 1986) and *Anacystis nidulans* R2 (Muller and Buchanan 1989). In contrast to *E. coli*, where thioredoxin is not essential for growth, this thioredoxin is apparently essential for the photosynthetic growth of *Anacystis nidulans* R2 (Muller and Buchanan 1989) although the physiological function of the protein has not been firmly established. The rapid reduction of thioredoxin *m*, on transfer of *Anabaena variabilis* and *Anabaena cylindrica* from dark to light (Darling et al. 1986; Rowell et al. 1988), is consistent with reduction via ferredoxin-thioredoxin reductase (Droux et al. 1987) and photosynthetically reduced ferredoxin. Although a protein with properties similar to those of chloroplast thioredoxin *f* has been identified in *Anabaena* 7119 (Crawford et al. 1984; Whittaker and Gleason 1984) its significance remains to be determined. Using G6PDH as a probe when isolating thioredoxin from *Anabaena* species only a single protein with the properties of thioredoxin *m* has been isolated (Ip et al. 1984; Cossar et al. 1984; Udvardy et al.1984).

G6PDH of *A. variabilis* (Cossar et al. 1984) and *Anabaena* PCC 7120 (Udvardy et al. 1984) is rapidly deactivated by low concentrations of reduced thioredoxin *m*. Cossar et al. (1984) demonstrated that deactivation of *A. variabilis* G6PDH by dithiothreitol (DTT) was dependent on thioredoxin and could be prevented by the amino acid glutamine. Figure 4a shows that, when an extract of *A. variabilis* was preincubated with DTT, deactivation could be prevented by G6P or glutamine. When added together at low concentrations (e.g. 1 mM) the degree of protection afforded was greater than the sum of their effects when added separately. These experiments were carried out at pH 6.5, so there was no loss of activity in the absence of DTT. Much higher concentrations of glutamine were required to prevent deactivation by DTT than were required to prevent deactivation at pH 7.4 in the absence of DTT (Fig. 3). When the effect of DTT on G6PDH from the unicellular cyanobacterium *Synechococcus* 7943 was examined in the same way (Fig. 4 b), glutamine was found to have no protective effect and, in contrast to the *A. variabilis* enzyme, G6P afforded protection at concentrations in the region of 1 mM. The lack of an effect of glutamine, in this case, indicates that under these conditions, at least, the effect of glutamine may not represent a widespread regulatory property of cyanobacterial G6PDH. Whether it is a specific property of N_2-fixing or heterocystous cyanobacteria remains to be determined.

Effects of 6-Phosphogluconate on Phosphoribulokinase Activity

Phosphoribulokinase (PRuK) has been suggested to be a potential regulatory enzyme of the Calvin cycle (Pelroy and Bassham 1973; Pelroy et al. 1976). Evidence has been obtained for a light-dependent activation of the enzyme in cyanobacteria (Duggan and Anderson 1975). Although reduced thioredoxin has been shown to activate the enzyme and has been implicated in the light-dark regulation of its activity, the structure and regulatory properties of the enzyme have not been extensively investigated (Marsden and Codd 1984; Rowell et al. 1988; Serra et al. 1989). The enzyme purified from the heterocystous cyanobacterium *Anabaena* 7119 (Rowell et al. 1988) and *Anabaena cylindrica* (Serra et al. 1989) was apparently composed of two identical subunits of M_r approximately 42,000 and could be activated by reduced thioredoxin.

Of the other potential regulators tested (including amino acids) the enzyme purified to homogeneity from *Anabaena* 7119 or partially purified from *A. variabilis* was inhibited by 6-phosphogluconate (6PG) (Fig. 5). It should be noted that inhibition of cyanobacterial ribulose-1,5-bisphosphate carboxylase/oxygenase (RuBisCO) by 6PG has also been reported (see Smith 1982; Tabita 1987) and it is therefore possible that 6PG, accumulated as a result of high G6PDH activity (Pelroy and Bassham 1972), inhibits these enzymes *in vivo*.

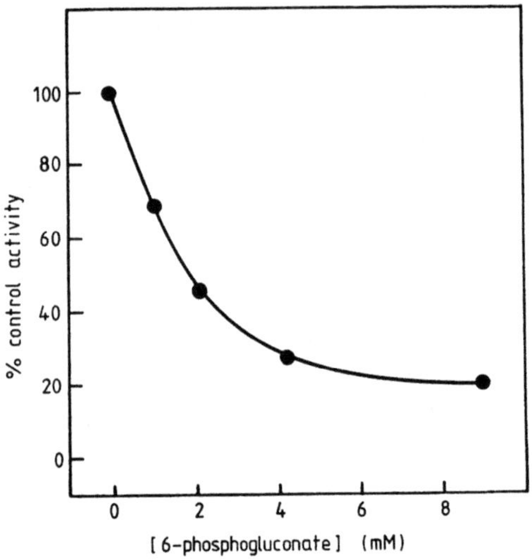

Fig. 5. Effect of 6-phosphogluconate concentration on the activity of phosphoribulokinase from *Anabaena* 7119. PRuK was purified to electrophoretic homogeneity (silver staining) by ammonium sulphate fractionation, chromatography on blue sepharose CL-6B and FPLC anion exchange chromatography (mono Q, HR 5/5, Pharmacia). All buffers were supplemented with 5 mM DTT. Activity was assayed in the presence of the substrates Ru5P (0.46 mM), ATP (1.6 mM) and $MgCl_2$ (37 mM) by coupling the formation of ADP to NADH oxidation. Specific activity was 61 µmol NADH oxidized min^{-1} (mg protein)$^{-1}$

Concluding Remarks

It is clear from the data presented above that, for at least some cyanobacteria, the effects of amino acids must be taken into account when considering the mechanisms by which the key regulatory enzyme G6PDH is controlled. Certain major pool amino acids are powerful positive effectors which override the regulatory effects of pH and $NADP^+$, being more effective in this respect than equivalent concentrations of G6P. In the case of *A. variabilis* G6PDH these amino acids, although capable of preventing deactivation by reduced thioredoxin under certain conditions (e.g. pH 6.5; see Fig. 4), are effective only at relatively high concentrations and are even less effective at higher pH values (e.g. pH 7.4) (García-González, Simpson and Rowell, unpublished data). Without a detailed knowledge of the levels of and changes in the many effectors involved, caution must be counselled in predicting their relative contributions to modulation of enzyme activity *in vivo*. However, the interaction of intermediates of pathways of nitrogen metabolism with a regulatory enzyme of carbon dissimilation may represent an important point of interaction between these pathways. The increase in 6PG levels that may result from the activation of G6PDH and thus the OPP pathway (Pelroy and Bassham 1972) may result in inhibition of PRuK and RuBisCO. Although we have found no direct effect of the amino acids tested on PRuK, for example, their effect on G6PDH may indirectly affect CO_2 fixation via the Calvin cycle.

Acknowledgments. We thank the Natural Environment Research Council, the Agricultural and Food Research Council, the Royal Society and the Society for General Microbiology for financial support.

References

Cossar JD, Rowell P, Stewart WDP (1984) Thioredoxin as a modulator of glucose-6-phosphate dehydrogenase in a N_2-fixing cyanobacterium. J Gen Microbiol 130: 991-998

Crawford NA, Sutton CW, Yee BC, Johnson TC, Carlson C, Buchanan BB (1984) Contrasting modes of photosynthetic enzyme regulation in oxygenic and anoxygenic prokaryotes. Arch Microbiol 139: 124-129

Crawford NA, Yee BC, Droux M, Carlson DE, Buchanan BB (1988) Ferredoxin/thioredoxin system. Methods in Enzymology 167: 415-427

Darling AJ, Rowell P, Stewart WDP (1986) Light induced transitions in the redox state of thioredoxin in the N_2-fixing cyanobacterium *Anabaena cylindrica.* Biochim Biophys Acta 850: 116-120

Droux M, Jacquot J-P, Miginac-Maslow M, Gadal P, Huet JC, Crawford NA, Yee BC, Buchanan BB (1987) Ferredoxin-thioredoxin reductase, an iron-sulfur enzyme linking light to enzyme regulation in oxygenic photosynthesis: Purification and properties of the enzyme from C_3, C_4 and cyanobacterial species. Arch Biochem Biophys 252: 426-439

Duggan JX, Anderson LE (1975) Light regulation of enzyme activity in *Anacystis nidulans* (Richt.). Planta 122: 293-297

Gleason FK, Holmgren A (1988) Thioredoxin and related proteins in procaryotes. FEMS Microbiol Revs 54: 271-298

Grossman A, McGowan RE (1975) Regulation of glucose-6-phosphate dehydrogenase in blue-green algae. Plant Physiol 55: 658-662

Hawkesford MJ, Rowell P, Stewart WDP (1983) Energy transduction in cyanobacteria. In: Papageorgiou GC, Packer L (eds) Photosynthetic Prokaryotes: Cell Differentiation and Function. Elsevier, Amsterdam New York, pp 199-218

Ip SM, Rowell P, Aitken A, Stewart WDP (1984) Purification and characterization of thioredoxin from the N_2-fixing cyanobacterium *Anabaena cylindrica.* Eur J Biochem 141: 497-504

Lara C, Romero JM, Coronil T, Guerrero MG (1987) Interactions between photosynthetic nitrate assimilation and CO_2 fixation in cyanobacteria. In: Ullrich WR, Aparicio PJ, Syrett PJ, Castillo F (eds) Inorganic Nitrogen Metabolism. Springer, Berlin Heidelberg New York, pp 45-52

Lim C-J, Gleason FK, Fuchs JA (1986) Cloning, expression and characterization of the *Anabaena* thioredoxin gene in *Escherichia coli.* J Bacteriol 168: 1258-1264

Marsden WJN, Codd GA (1984) Purification and molecular properties of phosphoribulokinase from the cyanobacterium *Chlorogloeopsis fritschii.* J Gen Microbiol 130: 999-1006

Muller EGD, Buchanan BB (1989) Thioredoxin is essential for photosynthetic growth. The thioredoxin *m* gene of *Anacystis nidulans.* J Biol Chem 264: 4008-4014

Pelroy RA, Bassham JA (1972) Photosynthetic and dark carbon metabolism in unicellular blue-green algae. Arch Mikrobiol 86: 25-38

Pelroy RA, Bassham JA (1973) Kinetics of glucose incorporation by *Aphanocapsa* 6714. J Bacteriol 115: 943-948

Pelroy RA, Levine GA, Bassham JA (1976) Kinetics of light-dark CO_2 fixation and glucose assimilation by *Aphanocapsa* 6714. J Bacteriol 127: 663-643

Pelroy RA, Rippka R, Stanier RY (1972) Metabolism of glucose by unicellular blue-green algae. Arch Mikrobiol 87: 303-322

Reglinski A, Rowell P, Kerby NW, Stewart WDP (1989) Characterization of methylammonium /ammonium transport in mutant strains of *Anabaena variabilis* resistant to ammonium analogues. J Gen Microbiol 135: 1441-1451

Rowell P, Darling AJ, Amla DV, Stewart WDP (1988) Thioredoxin and enzyme regulation. In: Rogers LJ, Gallon JR (eds) Biochemistry of the Algae and Cyanobacteria. Clarendon Press, Oxford, pp 201-216

Schaeffer F, Stanier RY (1978) Glucose-6-phosphate dehydrogenase of *Anabaena* sp. Kinetic and molecular properties. Arch Microbiol 116: 9-19

Serra JL, Llama MJ, Rowell P, Stewart WDP (1989) Purification and characterization of phosphoribulokinase from the N_2-fixing cyanobacterium *Anabaena cylindrica*. Plant Science 59: 1-9

Smith AJ (1982) Modes of cyanobacterial carbon metabolism. In: Carr NG, Whitton BA (eds) The Cyanobacteria. Blackwell, London, pp 47-85

Stewart WDP, Rowell P (1986) Biochemistry and physiology of nitrogen fixation with particular emphasis on nitrogen-fixing phototrophs. Plant and Soil 90: 167-191

Tabita FR (1987) Carbon dioxide fixation and its regulation in cyanobacteria. In: Fay P, Van Baalen C (eds) The Cyanobacteria. Elsevier, Amsterdam New York, pp 95-117

Turpin DH, Elrifi IR, Birch DG, Weger HG, Holmes JJ (1988) Interactions between photosynthesis, respiration and nitrogen assimilation in microalgae. Can J Bot 66: 2083-2097

Udvardy J, Juhasz A, Farkas GL (1983) Interaction between hysteric regulation and redox modulation of glucose-6-phosphate dehydrogenase from *Anacystis nidulans*. FEBS Lett 152: 97-100

Udvardy J, Borbely G, Juhasz A, Farkas GL (1984) Thioredoxins and the redox modulation of glucose-6-phosphate dehydrogenase in *Anabaena* sp. strain PCC 7120 vegetative cells and heterocysts. J Bacteriol 157: 681-683

Whittaker MM, Gleason FK (1984) Isolation and characterization of thioredoxin *f* from the filamentous cyanobacterium *Anabaena* sp. 7119. J Biol Chem 259: 14088-14093

Winkenbach F, Wolk CP (1973) Activities of enzymes of the oxidative and the reductive pentose phosphate pathways in heterocysts of a blue-green alga. Plant Physiol 52: 480-483

REGULATION OF INORGANIC NITROGEN METABOLISM IN THE PHOTOTROPHIC BACTERIUM *RHODOBACTER CAPSULATUS* E1F1

C. Moreno-Vivián, F.J. Caballero, M. Martínez-Luque, J. Cárdenas and F. Castillo [1]

Introduction

The phototrophic bacterium *Rhodobacter capsulatus* E1F1 utilizes oxidized, inorganic nitrogen compounds such as dinitrogen, nitrate or nitrite, as well as reduced nitrogen sources such as amino acids, urea or ammonium for phototrophic growth. Although nitrogen fixation is wide-spread among *Rhodospirillaceae*, only few strains are able to reduce and/or assimilate nitrate (Ferguson et al. 1987). *R. capsulatus* E1F1 reduces nitrate (Moreno-Vivián et al. 1982) and nitrite (Kerber et al. 1981) to ammonium only under anaerobic conditions in the light and exhibits an ammonium-sensitive nitrate reductase (Moreno-Vivián et al. 1982). Ammonium is also the end product of nitrogen fixation, a process regulated by light intensity, nitrogen compounds and oxygen. In *R. capsulatus* E1F1, assimilation of ammonium derived from nitrate or dinitrogen reduction takes place mainly by the GS/GOGAT cycle (Moreno-Vivián et al. 1983), and the nitrogen flux through this pathway is regulated at the level of both glutamine synthetase activity (Caballero et al. 1985) and synthesis (Caballero et al. 1989 b).

Alternative routes for ammonium assimilation have been described in *R. capsulatus* (Johansson and Gest 1976; Moreno-Vivián et al. 1983). Excess of extracellular ammonium is assimilated in the presence of pyruvate by ADH coupled to a specific aminotransferase. These enzymes are induced by pyruvate or alanine with simultaneous inactivation of the GS/GOGAT pathway (Caballero et al. 1989 b). ADH from *R. capsulatus* E1F1 has been purified to electrophoretic homogeneity. It has an M_r of 240,000 and consists of six identical subunits of 42,000 each, requires SH groups for maximal activity and exhibits a high apparent K_m for ammonium (16 mM). Its physiological role is either the reductive amination of pyruvate in the presence of high ammonium concentrations, or the oxidative deamination of alanine at high concentrations of this amino acid, thus providing carbon skeletons for phototrophic growth (Caballero et al. 1989 a). The enzyme can play an important role in bacterial growth in some natural environments such as contaminated fresh waters rich in ammonium. Also,

[1] Departamento de Bioquímica, Biología Molecular y Fisiología, Facultad de Ciencias, Universidad de Córdoba, 14071 Córdoba, Spain

Abbreviations: ADH: alanine dehydrogenase; AOAT: alanine:2-oxoglutarate aminotransferase; Asnase: asparaginase; AT: aminotransferase; GOGAT: glutamate synthase; GOAT: glycine:oxaloacetate aminotransferase; GOT: aspartate aminotransferase; GS: glutamine synthetase; MSX: L-methionine-DL-sulfoximine; N_2ase: nitrogenase; NR: nitrate reductase

R. capsulatus E1F1 is able to grow on other amino acids, by using them either as nitrogen sources alone or as both nitrogen and carbon sources. Several amino acids induce specific aminotransferases and other auxiliary enzymes, which drive nitrogen flux to glutamate biosynthesis and provide carbon skeletons for phototrophic growth. Under these conditions the GS/GOGAT pathway is turned off (Caballero et al. 1989 b).

In the following paragraphs, we propose a unified mechanism controlling inorganic nitrogen assimilation through the regulation of the key steps of nitrogen photometabolism, i.e. that in photosynthetic bacteria, nitrate transport and reduction, nitrogen fixation and glutamine biosynthesis are all regulated by the actual intracellular C/N ratio.

Regulation of Nitrate Assimilation

The proteins responsible for nitrate transport and reduction were induced by nitrate or nitrite (Moreno-Vivián et al. 1982; Caballero et al. 1986) and repressed by ammonium. Induction depended on carbon availability in the extracellular medium. In media with nitrate and low concentrations of malate (0.5 mM), nitrate uptake was almost undetectable and NR levels dropped abruptly after reaching a maximum of activity (Fig. 1A). However, in media with 5 mM malate, a significant nitrate uptake was measured and NR activity remained high for a longer time (Fig. 1B). In the presence of ammonium, neither nitrate uptake nor NR was induced in C-starved cells (0.5 mM malate, Fig. 2A), while under high malate (5 mM) conditions, both nitrate transport and NR were induced after malate consumption (Fig. 2B), when a high concentration of ammonium still remained in the extracellular medium. This indicates that nitrate assimilation is repressed not by ammonium *per se*, but by a product of ammonium assimilation whose concentration would drop sharply after the carbon supply is exhausted (Fig. 2B). This product is formed through the GS/GOGAT pathway since, in the presence of MSX, a strong inhibitor of glutamine synthetase, ammonium formed from nitrate reduction is stoichiometrically released to the medium. After a while with MSX, the cells developed an alternative route for ammonium assimilation (the ADH/ AOAT pathway), and simultaneous uptake of ammonium and nitrate, and growth resumption took place (Moreno-Vivián et al. 1983).

The effect of the intracellular C/N status on both nitrate transport and NR induction was corroborated by culturing *R. capsulatus* E1F1 with nitrate, L-amino acids, or urea as nitrogen sources in the presence or absence of malate. When either glutamine or glutamate was used as carbon source, excess of nitrogen was released as ammonium, and both nitrate uptake and NR remained repressed. When these amino acids were used as nitrogen source in the presence of malate, ammonium was not released and nitrate assimilation was derepressed. As expected, urea inhibited nitrate uptake and

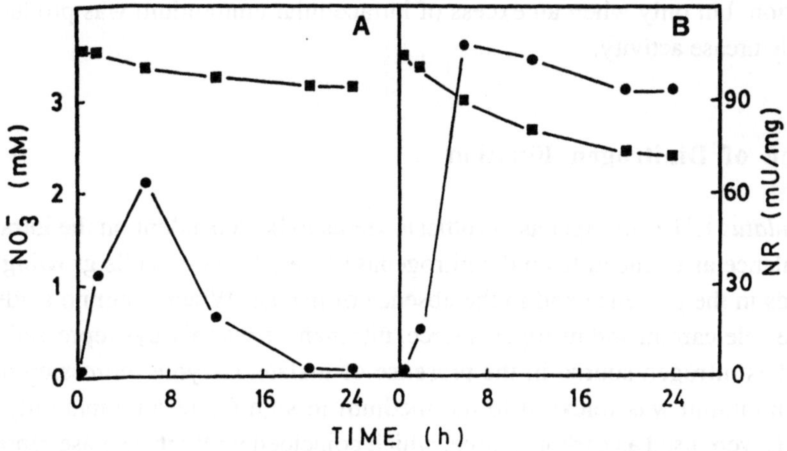

Fig. 1. Effect of carbon source availability on NR derepression in *R. capsulatus* E1F1 cells cultured with KNO_3. **A**: 0.5 mM malate; **B**: 5 mM malate. Cells grown on glutamate were inoculated in media containing 3.5 mM KNO_3. Culture conditions, NR assay (●) and nitrate determination in the medium (■) as described by Moreno-Vivián et al. (1982)

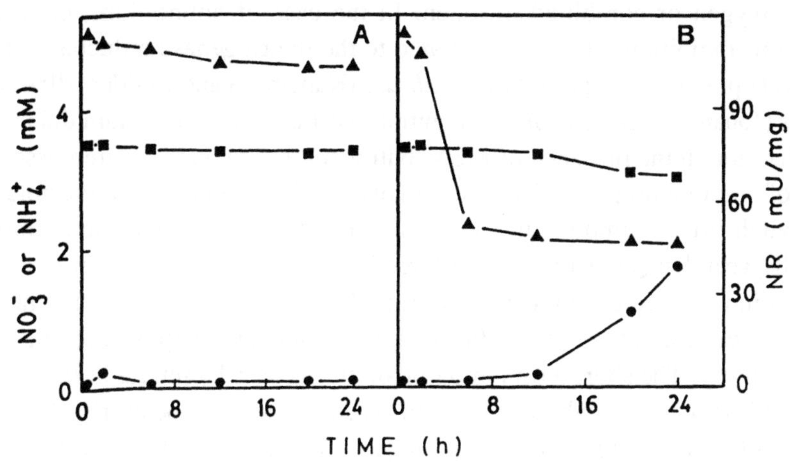

Fig. 2. Effect of carbon source availability on NR derepression in *R. capsulatus* E1F1 cells cultured with KNO_3 and NH_4Cl. **A**: 0.5 mM malate; **B**: 5 mM malate.Cells grown on glutamate were inoculated in media containing 3.5 mM KNO_3 and 5 mM NH_4Cl. Culture conditions, NR assay (●) nitrate (■) and ammonium (▲) determination in the medium as described previously (Moreno-Vivián et al. 1982)

NR induction, but only when an excess of intracellular ammonium was produced by an inducible urease activity.

Regulation of Dinitrogen Fixation

In *R. capsulatus* E1F1, nitrogenase synthesis seems to be dependent on the intracellular C/N balance as deduced from the nitrogenase levels found in cells growing on L-amino acids in the presence and in the absence of malate. When L-amino acids were used as the sole carbon and nitrogen source, nitrogenase was always repressed, while when used as nitrogen source in the presence of malate, enzyme repression did not occur. Ammonium was released to the medium in significant amounts only when amino acids were used as carbon source, which coincided with nitrogenase repression. These results, together with those discussed above, indicate that amino acids did not affect inorganic nitrogen assimilation *per se*, since when they were used in the presence of a suitable carbon source that brings about a well-balanced C/N status, neither nitrate transport, nor nitrate reduction nor dinitrogen fixation was repressed.

In *Rhodospirillaceae*, nitrogenase is regulated by a short-term reversible inhibition process which depends on environmental factors such as dark conditions or the presence of dioxygen or combined nitrogen. In the case of ammonium, dioxygen or darkness, this inhibition seems to be related to the dinitrogenase reductase interconversion cycle previously reported in *R. rubrum* (Kanemoto and Ludden 1984) and *R. capsulatus* (Jouanneau et al. 1988). Like nitrate assimilation, ammonium did not affect nitrogen fixation in the presence of MSX. Nitrogenase synthesis is derepressed in the presence of ammonium plus MSX, and *in vivo* MSX prevented the reversible nitrogenase switch-off by ammonium. Besides, when MSX was added to ammonium-treated cells, acetylene reduction was restored immediately even at higher rates than in non-ammonium treated cells (Arp and Zumft 1983). This indicates that ammonium affected nitrogenase activity only during ammonium assimilation through the GS/GOGAT pathway. The short-term inhibition of nitrogenase by nitrate in *R. capsulatus* cells or by nitrite either in whole cells or in crude extracts can be ascribed to a direct inactivation of the catalytic centers of the enzyme (Moreno-Vivián et al. 1986).

Regulation of Ammonium Assimilation

Ammonium formed from inorganic nitrogen reduction was used for the biosynthesis of glutamate through the GS/GOGAT pathway, with GS playing an important regulatory role. GS is fully active in cells growing on dinitrogen, nitrate or nitrite and malate as carbon source, while the enzyme is partially repressed in cells growing on ammonium/malate, ammonium/succinate, ammonium/pyruvate, glutamine/malate, and strongly re-

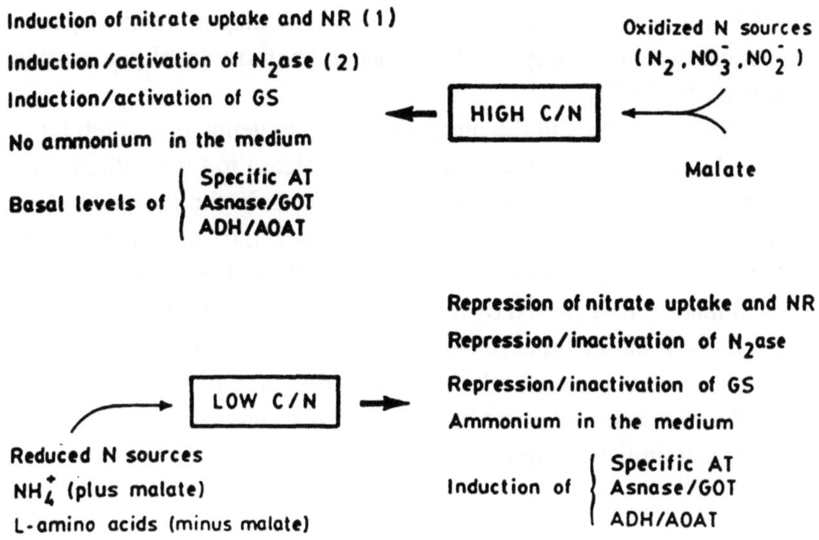

Fig. 3. Effect of C/N status on regulation of inorganic nitrogen assimilation in *R. capsulatus* E1F1. (1) Only in the presence of nitrate or nitrite (2) Only in the absence of nitrate or nitrite

pressed in alanine- or asparagine-grown cells. These nutritional conditions also bring forth a high adenylylation of GS. By contrast, glutamine seems to induce GOGAT activity either under phototrophic or dark-heterotrophic growth conditions (Caballero et al. 1989 b).

In the presence of amino acids, induction of alternative routes of ammonium assimilation took place, with the exception of glutamate dehydrogenase activity which was never detected in *R. capsulatus* E1F1. ADH and AOAT were coinduced in cells growing on pyruvate/ammonium or L-alanine either as carbon and nitrogen source or as nitrogen source in the presence of malate (Caballero et a l. 1989 a, b). A lesser induction of ADH was observed in cells grown on ammonium/malate (Caballero et al. 1989 b) or in nitrate/malate-grown cells after MSX addition (Moreno-Vivián et al. 1983). Induction of ADH activity was independent of illumination and took place to the same extent either in phototrophically or dark-heterotrophically grown cells. Other amino acids, such as glycine or L-aspartate, induced specific aminotransferases which always exhibited basal levels, as in the case of GOAT or GOT (Caballero et al. 1989 b).

A high growth yield was obtained when *R. capsulatus* E1F1 was grown under light-anaerobic conditions on L-asparagine, either as unique nitrogen source in the presence of malate, or as the sole carbon and nitrogen source. Bacterial growth on asparagine was due to the light-dependent induction of asparaginase and aspartate aminotransferase activities (Caballero et al. 1989 b). These growth conditions caused an excess of intracellular nitrogen which was released to the medium as ammonium that, in turn, inactivated nitrate transport, nitrogen fixation and glutamine biosynthesis, and repressed NR, nitrogenase and GS. Figure 3 shows diagrammatically the steps of nitrogen metabolism regulated by a C/N balance in phototrophically anaerobic-grown cells of *R. capsulatus* E1F1.

Oxidized nitrogen sources in the presence of malate generated a high intracellular C/N ratio as deduced from the lack of ammonium release to the medium. This metabolic state activates nitrate transport and NR induction in the presence of nitrate or nitrite and is also responsible for the activation of nitrogen fixation and glutamine biosynthesis. Nitrogenase is fully active only if both nitrate and nitrite are absent, because of the chemical inactivation caused by nitrite. On the other hand, when nitrogen supply produces an excess of intracellular nitrogen (as ammonium or L-amino acids when the latter are used as carbon sources), all the above processes, nitrate uptake and reduction, nitrogen fixation and glutamine biosynthesis become inactivated and alternative routes of nitrogen assimilation are induced.

Acknowledgments. The authors thank Comisión Asesora de Investigación (CAICYT, Grant no. PB86-0167-CO3-02, Spain) and Alexander von Humboldt Foundation (West Germany) for financial support to F.C. The secretarial assistance of C. Santos and I. Molina is also gratefully acknowledged.

References

Arp DJ, Zumft WG (1983) L-methionine-SR-sulfoximine as a probe for the role of glutamine synthetase in nitrogenase switch-off by ammonia and glutamine in *Rhodopseudomonas palustris*. Arch Microbiol 134: 17-22

Caballero FJ, Cejudo FJ, Florencio FJ, Cárdenas J, Castillo F (1985) Molecular and regulatory properties of glutamine synthetase from the phototrophic bacterium *Rhodopomonas capsulata* E1F1. J Bacteriol 162: 804-809

Caballero FJ, Moreno-Vivián C, Castillo F, Cárdenas J (1986) Nitrite uptake system in photosynthetic bacterium *Rhodopseudomonas capsulata* E1F1. Biochim Biophys Acta 848: 16-23

Caballero FJ, Cárdenas J, Castillo F (1989 a) Purification and properties of L-alanine dehydrogenase of the phototrophic bacterium *Rhodobacter capsulatus* E1F1. J Bacteriol 171: 3205-3210

Caballero FJ, Igeño MI, Cárdenas J, Castillo F (1989 b) Regulation of reduced nitrogen assimilation in *Rhodobacter capsulatus* E1F1. Arch Microbiol (in press)

Ferguson SJ, Jackson JB, McEwan AG (1987) Anaerobic respiration in the *Rhodospirillaceae*: Characterization of pathways and evaluation of roles in redox balancing during photosynthesis. FEMS Microbiol Rev 46: 117-143

Johansson BC, Gest H (1976) Inorganic nitrogen assimilation by the photosynthetic bacterium *Rhodopseudomonas capsulata*. J Bacteriol 128: 683-688

Jouanneau Y, Meyer C, Vignais PM (1988) Regulation of nitrogenase activity in *Rhodobacter capsulatus*: ADP-ribosylation of the Fe protein. In: Bothe H, de Bruijn FJ, Newton WE (eds) Nitrogen Fixation: Hundred Years After. Gustav Fischer, Stuttgart New York, pp 173

Kanemoto RH, Ludden PW (1984) Effect of ammonia, darkness and phenazine methosulfate on whole-cell nitrogenase activity and Fe protein modification in *Rhodospirillum rubrum*. J Bacteriol 158: 713-720

Kerber NL, Caballero FJ, Cárdenas J (1981) Assimilatory nitrite reductase from *Rhodopseudomonas capsulata* E1F1. FEMS Microbiol Lett 11: 249-252

Moreno-Vivián C, Castillo F, Cárdenas J (1982) Effect of light and darkness on nitrate assimilation by *Rhodopseudomonas capsulata* E1F1. Photosynth Res 3: 313-319

Moreno-Vivián C, Cejudo FJ, Cárdenas J, Castillo F (1983) Ammonia assimilation pathways in *Rhodopseudomonas capsulata* E1F1. Arch Microbiol 136: 147-151

Moreno-Vivián C, Cárdenas J, Castillo F (1986) *In vivo* short-term inhibition of nitrogenase by nitrate in *Rhodopseudomonas capsulata* E1F1. FEMS Microbiol Lett 34: 105-109

THE REGULATION OF THE REDUCTION OF INORGANIC NITROGEN IN CHLOROPHYLLOUS TISSUE: UPTAKE AND REDUCTION OF NITRITE BY INTACT CHLOROPLASTS

C.F. Cresswell, M.P. Watt, A.M. Amory and A. Whittaker [1]

Introduction

It is well established that plants reduce nitrate in the leaves, and that the primary products of the light reaction are directly involved in the reduction of nitrite within the chloroplast. The products of the carbon reduction cycle, reduced triose phosphate (3-phosphoglyceraldehyde or dihydroxyacetone phosphate), or the reduction of oxaloacetate within the chloroplast to malate, have been suggested to participate in the reduction of nitrate to nitrite.

As early as 1959, Anisimov observed that both nitrogen and phosphorus stimulated the translocation of assimilates from the leaves of both wheat (*Triticum aestivum*) and sugar beet (*Beta vulgaris*). Heldt et al. (1977) proposed that inorganic phosphorus has an important role in influencing the starch content in chloroplasts, and that a phosphorus translocator regulates the internal levels of triose-phosphate and inorganic phosphate. Work with isolated chloroplasts (Heldt et al. 1977) demonstrated that starch is the major photosynthetic product only when the stromal ratio of phosphoglycerate (PGA) to inorganic phosphate (P_i) is high, a result of low external P_i. These studies indicated that the export of phosphate esters from the chloroplast stroma and the import of P_i into the stroma is regulated. The outer chloroplast membrane appears to be non-specifically permeable, whereas the inner membrane does not permit unidirectional transfer of either P_i or phosphate esters (Heber and Walker 1979). This controlled movement regulates the control of the internal ratio of PGA/P_i and thus starch levels.

Ariovich and Cresswell (1983), using the C_4 photosynthetic grass *Panicum maximum* grown with different forms of nitrogen supply and phosphorus, observed that in the presence of nitrate and phosphorus little or no starch accumulated. In the absence of nitrate (phosphorus present), and in the absence of phosphorus (nitrate present) there was substantial starch accumulation. The presence of ammonium (with phosphorus) had little effect in preventing starch accumulation. These findings suggested that under limiting conditions the products of the Reductive Pentose Phosphate Shunt are not readily transported out of the chloroplast, but instead are converted to, and accumulate as, starch. The discrimination between the forms of inor-

[1] Foundation for Research and Development/University of Natal Photosynthetic Nitrogen Metabolism Research Unit, Biology Department, University of Natal, Durban, 4001, South Africa

ganic nitrogen supplied,shown by the extent of the starch accumulation, may be due to a need for a selective anion-dependent translocator system (NO_2^- and PO_4^{3-}) for the transport of the trioses out of the chloroplasts (see Klepper et al. 1971). Evidence does exist for the presence of a nitrite reductase within the bundle sheath chloroplasts of *P. maximum* (Rathnam and Edwards 1976), the site of starch accumulation.

Under dark anaerobic conditions nitrite accumulates in green tissue. Studies in our laboratory have shown that the capacity of *Zea mays* leaf tissue to reduce dark-anaerobically accumulated nitrite was dependent on the carbohydrate status of the tissue (Watt et al. 1986, 1987). Maize leaves, grown from seedlings with and without nitrate, were supplied with nitrate and/or glucose through the transpiration stream four hours prior to the experiment. The accumulation of nitrite in the tissue under dark-N_2 conditions and its utilization upon transfer to oxygen were investigated. In nitrate-starved leaves, the supply of glucose induced a relatively small (16%) increase in the accumulation of nitrite but had no effect on the subsequent aerobic assimilation of the accumulated nitrite. In contrast, in leaves of nitrate-grown plants the capacity to reduce nitrite on transfer to oxygen was greatly diminished unless exogenous glucose was supplied. There was, however, a marked increase (100%) in the level of nitrite accumulation in response to dark anaerobiosis, with supply of exogenous nitrate. This supports the role of nitrate as a regulator of triose-phosphate translocation out of the chloroplast (Klepper et al. 1971) for the necessary production of reductants for the assimilation of nitrate in the cytoplasm.

The present study was initiated to elucidate the interrelationship between energy supply and nitrite reduction by intact viable chloroplasts in the presence and absence of an exogenous carbon source. Although the control of nitrite reduction has been well studied, the possible controlling influence of nitrite permeation into the chloroplast has received less attention. Brunswick and Cresswell (1988) observed nitrite uptake into intact *Pisum sativum* chloroplasts to be closely dependent upon stromal pH and to nitrite accumulation and its reduction within the organelle. Furthermore, net nitrite uptake into chloroplasts was found to be profoundly influenced by plant nitrate nutrition: urea-fed plants receiving less than 0.01 mM nitrate exhibited only 20% of the net chloroplastic nitrite uptake rate compared to plants receiving 10 mM nitrate. A five minute incubation of the chloroplasts in the dark resulted in the complete cessation of nitrite uptake even at high (0.65 mM) nitrite concentrations, supporting the view that nitrite uptake is not a simple diffusion process.

The aim of the present study was to investigate the effect of an external supply of PGA on the uptake and reduction of nitrite in isolated chloroplasts of *Pisum sativum* under light and dark conditions. An attempt was also made to try to distinguish between the events occurring during nitrite uptake into, and nitrite reduction within, the chloroplast.

Materials and Methods

Pisum sativum var. Dark Skin Perfection, grown in vermiculite, was supplied with Long Ashton nutrient containing 10 mM KNO_3. Chloroplasts, exhibiting 92 to 97% intactness, were isolated from shoots of two to three week old seedlings (Brunswick and Cresswell 1988). Chloroplasts (0.4 to 0.5 mg Chl ml^{-1}) were incubated in 2 mM KNO_2 plus 0 to 20 mM PGA (final concentrations) in an incubation mixture containing 0.66 M sorbitol, 50 mM Hepes-KOH, 4 mM Na-pyrophosphate, 0.8 mM KH_2PO_4, 2 mM $MgCl_2$, 2 mM $MnCl_2$, 4 mM EDTA, and 2 mM ATP, at pH 7.8, for 35 min at 600 μmol m^{-2} sec^{-1} light.

To determine nitrite loss from the system, one ml aliquots were sampled, at different time intervals, and added to the reagents for the colorimetric nitrite assay (Hageman and Reed 1980). After a 10 minute centrifugation at 1000 x g the supernatant was read at 540 nm. Nitrite uptake and reduction was determined by separating the chloroplasts from the incubation medium (centrifugation, 1000 x g for 5 min) and assaying the two sample components for nitrite. Chlorophyll was determined by the method of Arnon (1949) and viability of the intact chloroplasts was observed by $^{14}CO_2$ fixation.

Results and Discussion

Under dark aerobic conditions the total nitrite loss from the system (chloroplasts + medium) (Fig. 1) was found to be directly related to, and enhanced by, the concentration of supplied PGA. Chloroplasts receiving 0 or 5 mM PGA only utilized exogenous nitrite after a 15 min incubation period. This limited nitrite utilization was not recorded in the study of Brunswick and Cresswell (1988) as the incubation time was restricted to this lag period. Addition of higher (10 and 20 mM) PGA concentrations overcame the initial lag phase and increased the final rates of nitrite loss from the assay system. This data is in agreement with previous work (Watt et al. 1987) who showed that in the dark and in the presence of oxygen glucose enhanced the utilization of nitrite.

The opposite pattern emerged when chloroplasts were supplied with 0 to 20 mM PGA in the light. Nitrite loss from the system decreased with increasing levels of exogenous PGA; no initial lag phase was observed (Fig. 2). Under light anaerobic conditions nitrite reduction was decreased by approximately 27%, but the pattern of inhibition by PGA was similar to that observed under air (Fig. 3). Under anaerobic conditions in the dark there was a total cessation of nitrite utilization irrespective of the level of PGA supplied (results not shown).

In an attempt to obtain a picture of the uptake and reduction of nitrite by the chloroplasts, the total loss of nitrite from the medium, and the nitrite content of chloroplasts,

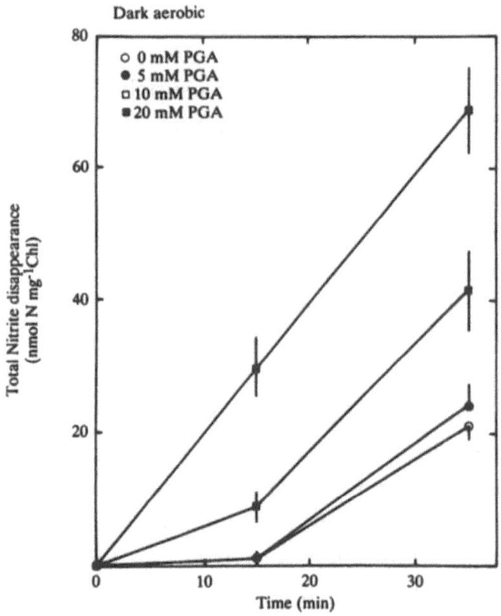

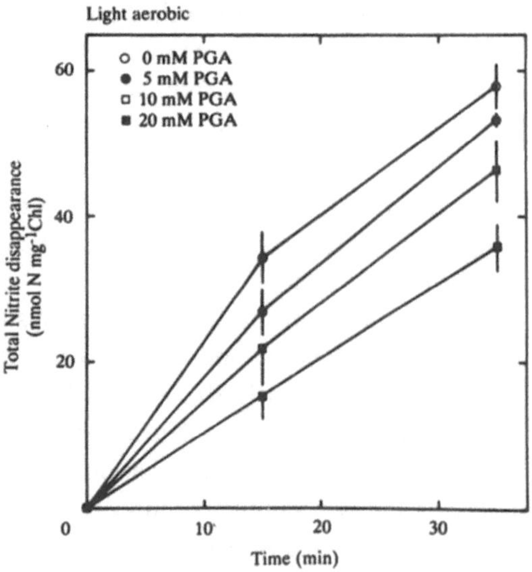

Fig. 1 *(Left)*. The effect of PGA supply on the loss of nitrite from the assay system (chloroplasts + incubation medium) under dark aerobic conditions. Bars = ±SD

Fig. 2 *(Right)*. The effect of PGA supply on the loss of nitrite from the assay system (chloroplasts + incubation medium) under light aerobic conditions. Bars = ±SD

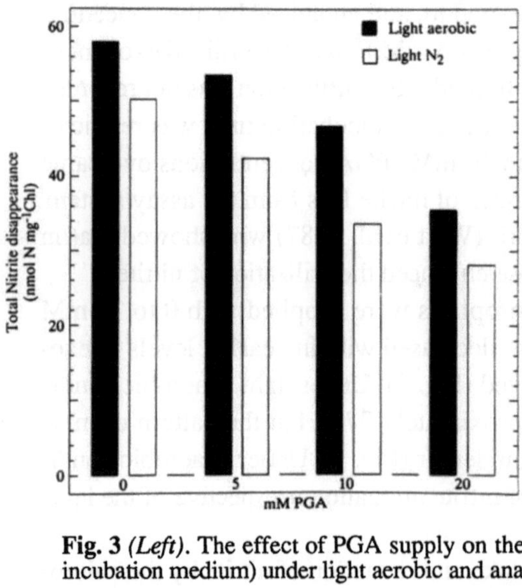

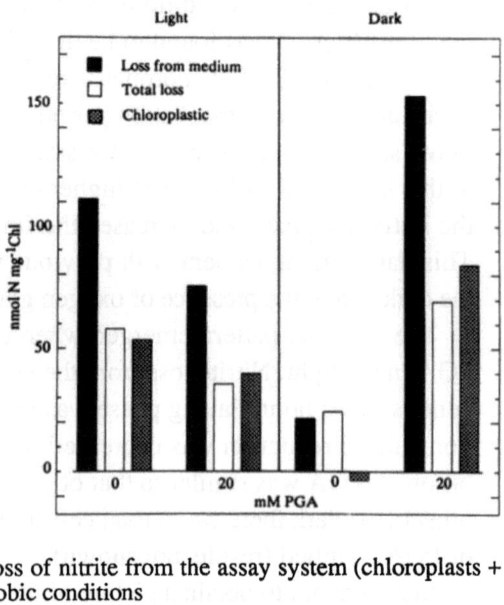

Fig. 3 *(Left)*. The effect of PGA supply on the loss of nitrite from the assay system (chloroplasts + incubation medium) under light aerobic and anaerobic conditions

Fig. 4 *(Right)*. The effect of PGA supply on the uptake and reduction of nitrite by isolated chloroplasts, under light and dark aerobic conditions

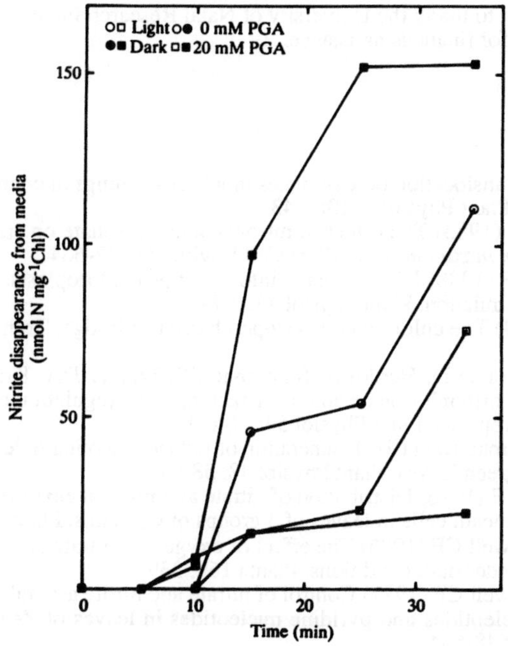

Fig. 5. The effect of PGA on the loss of nitrite from the incubation medium (nitrite uptake by chloroplasts), under light and dark aerobic conditions

were also determined (Fig. 4). From this data it was found that in the light, the presence of PGA at 20 mM reduced the total utilization of nitrite by the chloroplasts. The percentage of the loss from the medium due to a nitrite pool within the chloroplasts was about the same size in each case (± 42 %). However, the presence of 20 mM PGA reduced both the chloroplast nitrite pool size and the level of nitrite assimilation.

In the dark, a very different picture was observed: in the absence of PGA the nitrite pool size within the chloroplasts was very small, approximately 5.8% of the total nitrite loss from the medium was due to the nitrite in the chloroplastic pool. This nitrite pool size would appear not to have any negative effect on the metabolism associated with nitrite reduction. In the dark, nitrite uptake by the chloroplasts was dependent on the supply of PGA (Fig. 5).

In conclusion, this data indicates that, at least in the dark, the nitrite uptake and reduction by chloroplasts is controlled by the availability of carbohydrates, possibly trioses. It seems possible that a nitrite/triose translocator may be responsible for nitrite uptake and that the system is highly sensitive to anaerobic conditions. The permeation of nitrite into the chloroplast may play a key regulatory role in the regulation of nitrite assimilation, and possibly even carbohydrate metabolism. Experiments to test this hypothesis are presently in progress.

Acknowledgments. We wish to thank the University of Natal Research Fund and the Foundation for Research and Development for financial assistance.

References

Anisimov AA (1959) The translocation of assimilates in wheat seedlings in connection with root nutrient conditions. Soviet Plant Physiol 6: 138-143

Ariovich DI, Cresswell CF (1983) The effect of nitrogen and phosphate on starch accumulation in two varieties of *Panicum maximum* Jacq. Plant Cell Environ 6: 657-664

Brunswick P, Cresswell CF (1988) Nitrite uptake into intact pea chloroplasts. I. Kinetics and relationship with nitrate assimilation. Plant Physiol 86: 378-383

Heber U, Walker DA (1979) The chloroplast envelope - barrier or bridge. Trends in Biochem Sci 4: 252-255

Heldt HW, Chon CY, Maronde D, Herdol A, Stankovic ZS, Walker DA, Kraminer A, Kirk MR, Heber U (1977) Role of orthophosphate and other factors in the regulation of starch formation in leaves and isolated chloroplasts. Plant Physiol 59: 1146-1155

Klepper L, Flesher D, Hageman RH (1971) Generation of reduced nicotinamide adenine dinucleotide for nitrate reduction in green leaves. Plant Physiol 48: 580-590

Rathnam CKM, Edwards GE (1976) Distribution of nitrate assimilation enzymes between mesophyll protoplasts and bundle sheath cells in leaves of 3 groups of C_4 plants. Plant Physiol 57: 881-885

Watt MP, Gray VM, Cresswell CF (1986) The effect of oxygen on nitrate and nitrite assimilation in leaves of *Zea mays* L. under dark conditions. Planta 169: 339-344

Watt MP, Gray VM, Cresswell CF (1987) Control of nitrate and nitrite assimilation by carbohydrate reserves, adenosine nucleotides and pyridine nucleotides in leaves of *Zea mays* L. under dark conditions. Planta 172: 548-554

Part 4 Enzymes: Structure and Metabolic Regulation

NITRATE AND NITRITE UTILIZATION BY SUNFLOWER PLANTS

J.M. Maldonado, E. Agüera and P. de la Haba [1]

Introduction

The utilization of oxidized nitrogen forms (NO_3^- or NO_2^-) by higher plants involves: (1) uptake into the root cells; (2) reduction to NH_4^+ either in the root or in the shoot; and (3) incorporation of NH_4^+ to carbon skeletons to produce amino acids. The reduction of NO_3^- to NH_4^+ takes place by the sequential participation of the enzymes nitrate reductase (NR) and nitrite reductase (NiR) which catalyze, respectively, the reduction of NO_3^- to NO_2^- and NO_2^- to NH_4^+ (Guerrero et al. 1981; Oaks and Hirel 1985). Assimilation of NH_4^+ seems to proceed mainly via the glutamine synthetase (GS)-glutamate synthase cycle (Lea and Miflin 1979). However, under conditions of ample NH_4^+ supply, glutamate dehydrogenase (GDH) may also operate as NH_4^+-assimilating enzyme (Singh and Srivastava 1982; Cammaerts and Jacobs 1985). In most plant systems, NR and NiR are both inducible by NO_3^- (Guerrero et al. 1981; Oaks and Hirel 1985). The presence of NO_3^- also induces the development of the NO_3^--uptake system in nitrogen-starved plants (Deane-Drummond 1984; MacKown and McClure 1988). On the other hand, addition of NH_4^+ to plants growing on NO_3^- usually has a negative effect on NO_3^- utilization. The decline in NO_3^- utilization produced by ambient NH_4^+ can be the result of a short-term inhibition of net NO_3^- uptake, either by NH_4^+ *per se* (Ullrich 1987) or by a product of its assimilation (Flores et al. 1980), and/or by inactivation or repression of NR (Guerrero et al. 1981; MacKown et al. 1982). In this article, we summarize our investigations on NO_3^- and NO_2^- utilization by sunflower plants (*Helianthus annuus* L. cv Peredovic).

Induction of Nitrogen-Assimilating Enzymes during Germination

Sunflower seedlings germinated and grown in a nitrogen-free medium exhibit low levels of NR activity. However, the presence of NO_3^- in the nutrient solution promotes an increase of NR activity in the cotyledons during the course of germination. By contrast, neither NH_4^+ nor urea can elicit the appearance of NR activity (Fig. 1A). Similarly, NiR activity greatly increases when NO_3^- is present in the medium whereas, under NH_4^+ or urea, the low but significant levels of NiR activity do not appreciably increase throughout the first seven days of germination (Fig. 1B). Seedlings grown

[1] Departamento de Biología Vegetal y Ecología, División de Fisiología Vegetal, Facultad de Ciencias, Universidad de Córdoba, 14004 Córdoba, Spain

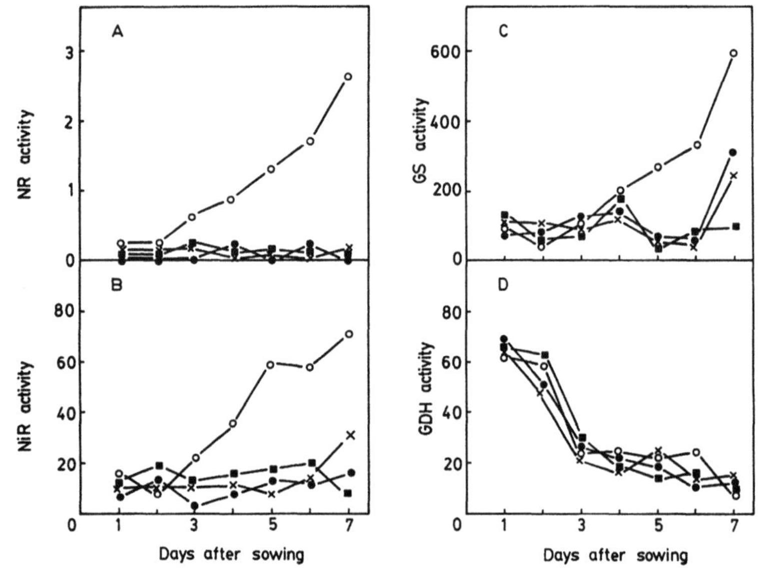

Fig. 1, A-D. Changes in the levels of NR (**A**), NiR (**B**), GS (**C**) and GDH (**D**) in cotyledons of sunflower seedlings grown either on a nitrogen-free medium (●—●) or on media with 10 mM NO_3^- (o—o), 10 mM NH_4^+ (x—x), or 5 mM urea (■—■). From de la Haba et al. 1988

with NO_3^- also develop higher levels of GS activity than either nitrogen-starved plants or plants fed exogenous NH_4^+ or urea (Fig. 1C). In contrast to NR, NiR and GS activity, GDH activity levels in sunflower cotyledons gradually decline during germination independently of the presence or absence of an exogenous nitrogen source (Fig. 1D). The concentration of soluble protein in sunflower cotyledons during germination decreases in a similar way (de la Haba et al. 1988), thus suggesting that GDH might be involved in the deamination of glutamate generated from reserve proteins rather than in the assimilation of NH_4^+.

The NO_3^--promoted appearance of NR is suppressed by both cycloheximide and chloramphenicol, which indicates that both cytoplasmic 80S and plastidic 70S ribosomes are involved in the synthesis of the NR molecule (de la Haba et al. 1988). Börner et al. (1986) also found that NR is not accumulated in chloroplast-ribosome-deficient mutants of barley and maize. The appearance of NiR activity in sunflower cotyledons during germination is inhibited by cycloheximide but not by chloramphenicol (de la Haba et al. 1988). Therefore, the NiR molecule seems to be synthesized as a whole in the cytoplasm on cycloheximide-sensitive 80S ribosomes, which implies the further transport of the enzyme molecule into the plastids, where NiR appears to be localized (Guerrero et al. 1981). Although the involvement of 70S ribosomes has been proposed by some authors (Sawhney and Naik 1973), recent work on heat-treated 70S ribosome-deficient rye leaves (Feierabend 1986) and chloroplast-ribosome-deficient mutants of barley and maize (Börner et al. 1986) has demonstrated that NiR is synthesized in the cytoplasm.

Tissue Distribution of Nitrogen-Assimilating Enzymes

Table 1 shows that, in young sunflower seedlings, the activity levels of NR and GS are substantially higher in cotyledons than in embryonic axes, whereas the opposite occurs for GDH. Likewise, in adult plants, the levels of NR and GS are higher in leaves than in roots, GDH activity showing an opposite distribution. The results on the responses of nitrogen-assimilating enzymes to the nitrogen nutrition (Fig. 1) and the enzyme distribution within the plant (Table 1) suggest that, in NO_3^- fed sunflower plants, (i) GS rather than GDH is involved in the assimilation of the NH_4^+ derived from NO_3^- reduction; and (ii) cotyledons and leaves are major sites, in contrast to embryonic axes and roots, of NO_3^- reduction and NH_4^+ assimilation (Agüera et al. 1987).

Table 1. NR, GS and GDH activities in different tissues of sunflower plants (cv Peredovic). Cotyledons and embryonic axes were from 7-day-old seedlings, and leaves and roots from 15-day-old plants. Enzyme activities are expressed as μmol substrate transformed or product formed h^{-1} g^{-1} fresh weight. (Data from Agüera et al. 1987)

Tissue	NR	GS	GDH
Cotyledons	3.79	166	3.3
Embryonic axes	1.21	81	17.5
Leaves	4.43	146	2.6
Roots	0.54	10	12.0

Nitrate and Nitrite Uptake

Nitrogen-starved sunflower plants cannot take up NO_3^- or NO_2^- initially at a reasonable rate when exposed to these anions. The ability of the plants to take up NO_3^- or NO_2^- at high rates from the beginning is induced by pretreatment with NO_3^-. The presence of cycloheximide during NO_3^--pretreatment prevents the subsequent uptake of NO_3^- and NO_2^-, indicating that NO_3^- induces *de novo* synthesis of both uptake systems. Proteins in the range of 30 to 45 kDa have been identified in the plasma membrane of corn roots (Dhugga et al. 1988) and *Synechococcus* cells (Madueño et al. 1988) grown on NO_3^-. These proteins seem to be involved in the transport of NO_3^- through the plasma membrane. The uptake of NO_3^- and NO_2^- by sunflower plants is drastically inhibited by a short pretreatment of the plants with the sulfhydryl-group reagent N-ethylmaleimide, which suggests that SH-groups of the carrier molecules are involved in the binding and transport of both anions.

Ullrich (1987) has proposed that NO_3^- is taken up by a $2H^+/NO_3^-$ symport coupled to an ATP-dependent proton-extrusion pump which generates the proton electrochemical gradient required for NO_3^- transport. We have found that the uptake both of NO_3^- and NO_2^- by sunflower plants is largely abolished when the external solution is made anaerobic, thus demonstrating the requirement of metabolic energy. Also, the uptake of both anions proceeds faster at mildly acid pH, decreasing drastically at external pH values above 5.5. Hence, it appears that not only NO_3^- uptake but also NO_2^- uptake are both stimulated by a high external proton concentration, presumably because, under those conditions, the proton electrochemical gradient is higher.

The rates of NO_3^- and NO_2^- uptake as a function of external anion concentration exhibit saturation kinetics. This observation corroborates that the uptake of both anions is mediated by carrier molecules. The K_m values of the uptake systems for NO_3^- and NO_2^-, calculated over external concentrations up to 0.2 mM, are 45 and 23 μM, respectively. The rates of NO_3^- uptake by intact sunflower plants are four to six times higher than the rates of NO_3^- reduction in roots. On the contrary, NO_2^- uptake rates, which are very similar to NO_3^- uptake rates, are much lower (about 30 times) than NO_2^- reduction rates in the roots. These observations indicate that a substantial fraction of total NO_3^- taken up by the plant is accumulated in the root and/or translocated to the shoot; in contrast, all the NO_2^- taken up would be reduced in the root.

As shown in Figure 2, the presence of NH_4^+ in the external solution greatly restricts NO_3^- uptake by intact sunflower plants without affecting NO_2^- uptake. The simultaneous addition of the GS-inhibitor L-methionine D,L-sulphoximine (MSO) does not relieve the inhibitory effect of NH_4^+. Even more, the sole addition of MSO produces a great inhibition of NO_3^- uptake. Hence, neither glutamine nor another product of NH_4^+ assimilation appears to be involved in the inhibitory effect of NH_4^+ on NO_3^- uptake by sunflower plants. We have also found that the inhibition of NO_3^- uptake is correlated with a decrease of NR activity in the roots. Moreover, NH_4^+ and/or MSO additions result in accumulation of NH_4^+ and NO_3^- in the root tissue. Hence, it appears that, under steady-state conditions of NO_3^- utilization, NO_3^- uptake is controlled by the rate of NO_3^- reduction, and not vice versa. Accumulation of NO_3^- due to decreased NR activity would restrict net NO_3^- uptake either by inhibiting NO_3^- influx or by stimulating NO_3^- efflux. In other words, internal NO_3^- would exert a negative feed back control on its own uptake system.

To further investigate the relationship between NO_3^- uptake and NO_3^- reduction, tungsten was used in order to inhibit NR activity. Pretreatment of sunflower plants

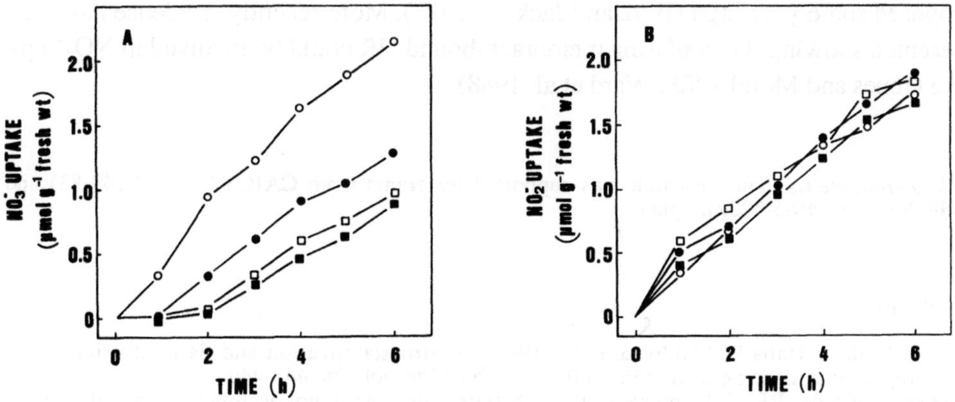

Fig. 2. Effects of NH_4^+ and MSO on NO_3^- uptake (**A**) and on NO_2^- uptake (**B**) by intact sunflower plants. NO_3^- uptake or NO_2^- uptake solutions were either supplemented or not (o——o) with 0.5 mM NH_4^+ (●——●), 2mM MSO (□——□) or 0.5 mM NH_4^+ plus 2 mM MSO (■——■)

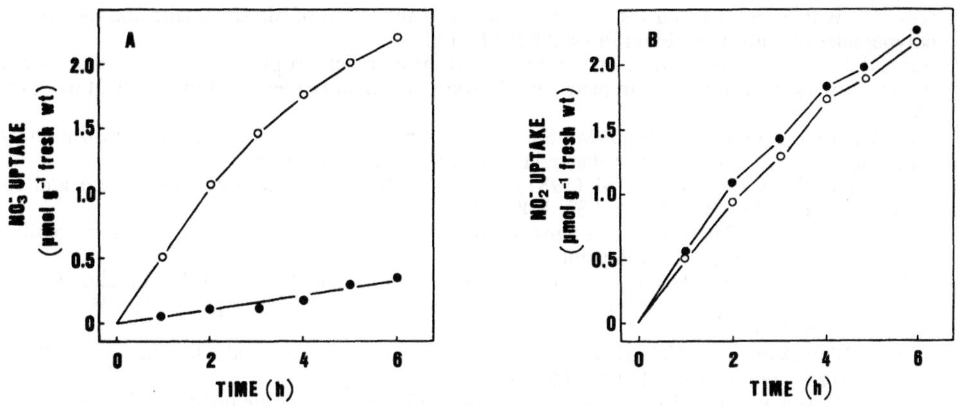

Fig. 3. Effect of tungsten on NO_3^- uptake (**A**) and on NO_2^- uptake (**B**) by intact sunflower plants. Plants were pretreated for seven days either with molybdenum (o——o) or tungsten (●——●), and subsequently assayed for NO_3^- and NO_2^- uptake

with tungsten fully inactivated root NR activity without affecting NiR activity (data not shown). As shown in Figure 3, tungsten-treated plants can hardly take up NO_3^- while NO_2^- uptake is completely unaffected. It was also found that the internal levels of NO_3^- were much lower in tungsten-treated than in molybdenum-treated plants. Hence, it appears that tungsten, besides inactivating NR activity, inhibits *per se* the NO_3^- uptake system. A model of a transmembrane NR complex acting as NO_3^- carrier molecule was proposed some years ago (Butz and Jackson 1977). More recently, evidence has been presented showing that a plasma membrane-bound NR could be involved in NO_3^- uptake (Jones and Morel 1988; Ward et al. 1988)

Acknowledgments. This research was supported by grants from CAICYT (No. 2149-83) and DGICYT (No. PB86-0232), Spain

References

Agüera E, de la Haba P, Maldonado JM (1987) *In vitro* stabilization and tissue distribution of nitrogen-assimilating enzymes in sunflower. J Plant Physiol 128: 443-449

Börner T, Mendel RR, Schiemann J (1986) Nitrate reductase is not accumulated in chloroplast-ribosome-deficient mutants of higher plants. Planta 169: 202-207

Butz RG, Jackson WA (1977) A mechanism for nitrate transport and reduction. Phytochemistry 16: 409-417

Cammaerts D, Jacobs M (1985) A study of the role of glutamate dehydrogenase in the nitrogen metabolism of *Arabidopsis thaliana*. Planta 163: 517-526

De la Haba P, Agüera E, Maldonado JM (1988) Development of nitrogen-assimilating enzymes in sunflower cotyledons during germination as affected by the exogenous nitrogen source. Planta 173: 52-57

Deane-Drummond CE (1984) The apparent induction of nitrate uptake by *Chara corallina* cells following pretreatment with or without nitrate and chlorate. J Exp Bot 35: 1182-1193

Dhugga KS, Waines JG, Leonard RT (1988) Correlated induction of nitrate uptake and membrane polypeptides in corn roots. Plant Physiol 87: 120-125

Feierabend J (1986) Investigation of the site of synthesis of chloroplastic enzymes of nitrogen metabolism by the use of heat-treated 70S ribosome-deficient rye leaves. Physiol Plant 67: 145-150

Flores E, Guerrero MG, Losada M (1980) Short-term ammonium inhibition of nitrate utilization by *Anacystis nidulans* and other cyanobacteria. Arch Microbiol 128: 137-144

Guerrero MG, Vega JM, Losada M (1981) The assimilatory nitrate-reducing system and its regulation. Annu Rev Plant Physiol 32: 168-204

Jones GJ, Morel FMM (1988) Plasmalemma redox activity in the diatom *Thalassiosira*. A possible role for nitrate reductase. Plant Physiol 87: 143-147

Lea PJ, Miflin BJ (1979) Photosynthetic ammonia assimilation. In: Gibbs M, Latzko E (eds) Encyclopedia of Plant Physiology, NS vol 6, Photosynthesis II. Springer, Berlin Heidelberg New York, pp 445-456

MacKown CT, Jackson WA, Volk RJ (1982) Restricted nitrate influx and reduction in corn seedlings exposed to ammonium. Plant Physiol 69: 353-359

MacKown CT, McClure PR (1988) Development of accelerated net nitrate uptake. Effects of nitrate concentration and exposure time. Plant Physiol 87: 162-166

Madueño F, Vega-Palas MA, Flores E, Herrero A (1988) A cytoplasmic-membrane protein repressible by ammonium in *Synechococcus* R2: Altered expression in nitrate-assimilation mutants. FEBS Lett 239: 289-291

Oaks A, Hirel B (1985) Nitrogen metabolism in roots. Annu Rev Plant Physiol 36: 345-365

Sawhney SK, Naik MS (1973) Effect of chloramphenicol and cycloheximide on the synthesis of nitrate reductase and nitrite reductase in rice leaves. Biochem Biophys Res Commun 51: 67-73

Singh RP, Srivastava HS (1982) Glutamate dehydrogenase activity and assimilation of inorganic nitrogen in maize seedlings. Biochem Physiol Pflanzen 177: 633-642

Ullrich WR (1987) Nitrate and ammonium uptake in green algae and higher plants: Mechanism and relationship with nitrate metabolism. In: Ullrich WR, Aparicio PJ, Syrett PJ, Castillo F (eds) Inorganic Nitrogen Metabolism. Springer, Berlin Heidelberg New York, pp 32-38

Ward MR, Tischner R, Huffaker RC (1988) Inhibition of nitrate transport by anti-nitrate reductase IgG fragments and the identification of plasma membrane associated nitrate reductase in roots of barley seedlings. Plant Physiol 88: 1141-1145

THE ROLE OF NO_3^- AND NH_4^+ IONS IN THE REGULATION OF NITRATE REDUCTASE IN HIGHER PLANTS

A. Oaks [1], D.M. Long, M. Zoumadakis, X.-Z. Li and C. Hertig [2]

After its identification by Evans and Nason (1953), nitrate reductase (NR) was shown to be a substrate inducible enzyme by Tang and Wu (1957). In 1960 Hageman and Flesher demonstrated further that both light and NO_3^- were required for the appearance of nitrate reductase activity (Hageman and Flesher 1960). Since that time there have been many reports confirming the initial observations with light and NO_3^- and conflicting reports concerning the repression of the induction of NR by NH_4^+ (Beevers and Hageman 1972; Ullrich 1987).

Two factors could be important in the appearance of NR activity. The activation of NR, for example, has been elegantly demonstrated in the green alga *Chlorella* by Losada et al. (1970) and by Pistorious et al. (1976). A *de novo* synthesis of the NR protein in response to NO_3^- has been demonstrated by Amy and Garrett (1979) in *Neurospora crassa* and this phenomenon may also be important in algae and higher plants. Research in higher plants has emphasized *de novo* synthesis (Remmler and Campbell 1986). However, there have also been several preliminary reports of a reversible activation of NR in higher plants (Aryan and Wallace 1983; Palacián et al. 1974; Somers et al. 1983). An activation/inactivation phenomenon may, for example, be of importance in light/dark transitions (Campbell 1987) or in the response of NR to drought (Aryan and Wallace 1983).

The ability to purify NR (Solomonson 1985), and to prepare antibodies to that protein allows us to quantitate the protein either by the ELISA method (Campbell and Remmler 1986) or by SDS-PAGE followed by immunoblotting (Oaks et al. 1988). Thus with an improved system we were able to confirm the early results with light and NO_3^- on the appearance of NR activity. A 116 kD protein which reacts with polyclonal antibody raised against a pure NR from maize leaves is also present in crude extracts prepared from light-grown but not in dark-grown seedling shoots (Oaks et al. 1988). At this stage of the investigation we can conclude either that this polypeptide is not made in the dark or that it is very unstable in dark-grown seedlings. The induction of the NR protein was apparent in the absence of NO_3^- in the light. The explanation for this unexpected observation turned out to be an induction system in maize that is extremely sensitive to low levels of external NO_3^-. Sufficient levels of NO_3^- are present in sand, vermiculite and Kimpack papers (Oaks et al. 1988). In a hydroponic system,

[1] Department of Biology, McMaster University, 1280 Main St. W., Hamilton, Ontario, Canada, L8S 4K1. *Current address*: Department of Botany, University of Guelph, Guelph, Ontario, Canada, N1G 2W1.

[2] *Current address*: Institut für Pflanzenbiologie, Zolliker Strasse 107, CH-8008 Zürich, Switzerland

Table 1. Effect of NO_3^- and NH_4^+ ions on levels of NR activity (NRA) and NR protein (NRP).

	A		B	
Nitrogen Source	NRP	NRA	NRP	NRA
No N	8.0	*n d*	faint	0.16
KNO_3 (5 mM)	100.0	12.7	100	1.46
NH_4Cl (5 mM)	49.2	0.98	faint	0.19
KNO_3 + NH_4Cl	55.4	13.2	114	1.51

Seedlings were planted on a nutrient agar with no nitrogen source and were allowed to grow in the dark for 48 h. They were then transferred to a hydroponic system which contained from the start the appropriate nitrogen source (**A**) or no nitrogen (**B**). In B the appropriate nitrogen source was added to the system 6 h before the termination of the experiment. NRP as % of 5 mM KNO_3 treatment; NRA in μmol NO_2^- h^{-1} g $frwt^{-1}$; *n d*, no activity detected. For experimental details see Oaks et al. (1988)

in the absence of NO_3^-, there was no NR protein and additions of NO_3^- resulted in the appearance of both NR protein and NR activity (Oaks et al. 1988, and Table 1). The addition of NH_4Cl led to the appearance of NR protein but not to NR activity in a 6-day growth period.

In a 6 hour induction neither NR protein nor NR activity was apparent with NH_4Cl additions. Since in non-sterile conditions NH_4^+ can be converted to NO_3^-, we think that the results of the 6 day treatment are an artifact caused by this conversion. The results of the 6 hour treatment are probably a true reflection of the influence of NH_4^+ in maize seedlings. Thus we would conclude that NH_4^+ itself does not result in the induction of the NR protein. When both NO_3^- and NH_4^+ were added to the system there was a normal induction of NR activity and appearance of NR protein. According to S.J. Rothstein a normal induction of the NR-mRNA is also seen when NH_4^+ and NO_3^- are both present in the medium (personal communication). These experiments indicate that there is an inactive form of the NR protein in maize seedling leaves that is synthesized in the presence of low levels of added NO_3^-. With higher levels of exogenous NO_3^- this inactive protein is activated. We are currently studying the properties of this protein and of its activation.

In time-course experiments both mRNA and the NR protein reached maximum levels 2 - 4 hours after the addition of 10 mM KNO_3 (Fig. 1) in our hydroponic system. Levels of mRNA then declined whereas levels of the NR-protein remained high throughout the duration of the experiment. NR activity increased over the whole 48 h induction period. Internal NO_3^- levels peaked at 24 h and then remained stable. Once again the evidence suggests that the appearance of the NR protein and the NR activity represent 2 phases in the induction process, and that these 2 phases, transcriptional and translational events, do not occur in parallel in our system. Somers et al. (1983) using barley and Remmler and Campbell (1986) using maize found that NR activity and NR protein did appear in parallel. This suggests that if an inactive protein is the first prod-

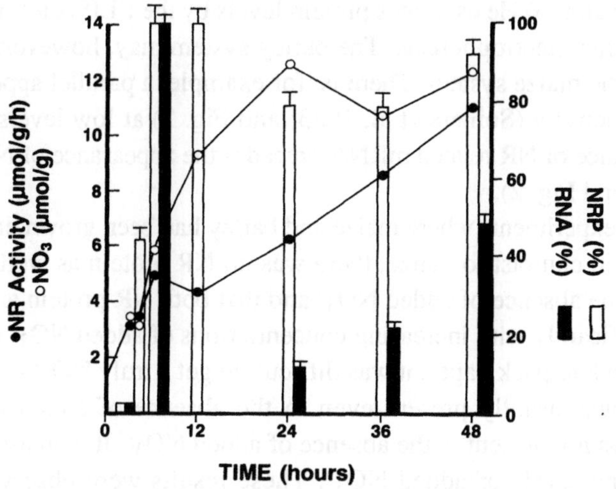

Fig. 1. Time course for the appearance of NR activity (NRA), NR protein (NRP), NR-mRNA and the accumulation of NO_3^- in maize shoots. 10 mM KNO_3 was used for the induction. Experimental set-up was as in Table 1B and in Oaks et al. (1988)

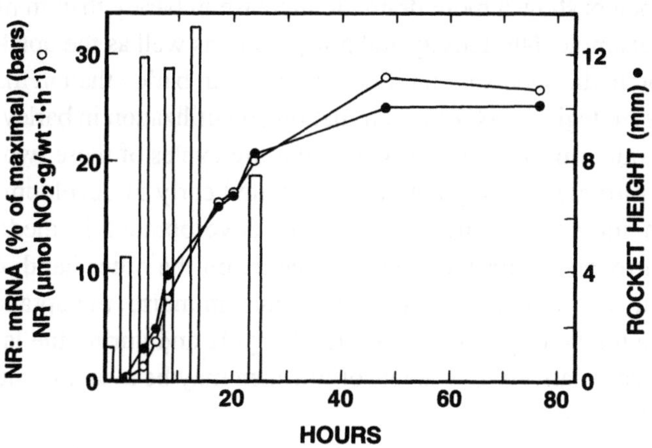

Fig. 2. Time course for the appearance of NR activity (NRA), NR protein (NRP) and NR-mRNA in barley shoots. Adapted from Somers et al. (1983) and Melzer et al. (1989)

uct of transcription it is activated immediately in their experiments. There are 3 major differences between Campbell's system and ours which may account for differences in time of appearance of NRP and NRA: (1) Campbell uses vermiculite rather than a hydroponic system; (2) He uses excessively high levels of NO_3^- in the induction system (37 - 50 mM); and (3) He estimates protein levels by the ELISA test rather than by immunoblotting after electrophoresis. The barley system may, however, be genuinely different from the maize system. There is, for example, a parallel appearance of NR protein and NR activity (Somers et al. 1983; and Fig. 2) at low levels of added NO_3^- and the appearance of NR related mRNA precedes the appearance of NR activity (Melzer et al. 1989, and Fig. 2).

From preliminary experiments where maize and barley had been grown on Kimpack paper we knew that, in contrast to maize, there was no NR protein as well as no NR activity in barley in the absence of added NO_3^- and that both NR protein and NR activity increased significantly with increasing concentrations of added NO_3^- (from 1 to 20 mM). In maize, on Kimpack paper, it was difficult to get a truly NO_3^- -free system in that NR protein was usually present even in the absence of added NO_3^-. NR activity, however, was not present in the absence of added NO_3^-. It increased to some extent with increasing levels of added NO_3^-. These results were observed with 4 different barley varieties and 3 different cultivars of maize (unpublished observations). When we examined these differences in more detail we found that not only the induction of NR activity and NR protein but also NO_3^- uptake from the medium and the accumulation of NO_3^- in the shoot tissue were more efficient in maize when low levels of NO_3^- were used (Fig. 3). However, with increased levels of NO_3^- in the medium all 4 parameters show a more dramatic increase in barley than in maize. With 20 mM KNO_3, for example, NR activity and NR protein as well as the uptake of NO_3^- and its accumulation in the shoots is significantly higher in barley than in maize. These apparently different strategies of NO_3^- accumulation and utilization in barley and maize may reflect the overall utilization of nitrogen in the life cycles of these two cereals. In barley most of the nitrogen in the plant is accumulated early in development, before anthesis. Nitrogen increase in the grain during development is mediated by a redistribution of nitrogen within the plant. In maize, on the other hand, only about 50% of the nitrogen is accumulated by anthesis. Thus, in maize, the uptake system is important through a much longer period of the life cycle and during the latter part of that life cycle, at least, relatively low levels of nitrogen may be available since nitrates are readily leached from the soil.

It is apparent from this study of NO_3^- uptake and utilization in maize and barley, and from comparisons with other well characterized systems such as *Arabidopsis* (Crawford et al. 1988) and *Nicotiana* (Galangau et al. 1988) that nitrate reductase protein and activity are indeed induced by NO_3^- but that there are important differences in the timing and the responses of these events to environmental cues. Studies at the genetic and molecular levels will certainly help in the understanding of the details of these interactions. Such studies may, however, be less successful in developing plants that use NO_3^- more efficiently, if the emphasis of research continues to be on nitrate reductase and not on other aspects of NO_3^- assimilation.

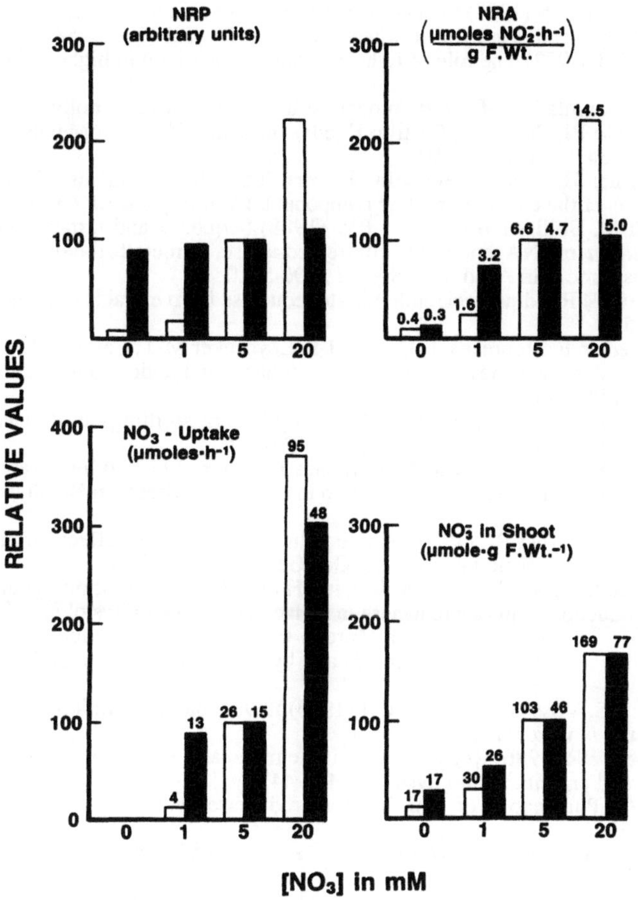

Fig. 3. The effect of NO_3^- concentration on the induction of NRA and NRP, on the uptake of NO_3^- from the medium and on the accumulation of NO_3^- in the shoots. Barley cv. Bonanza (clear bars) and maize cv. W64A x W182E (solid bars) are compared. Values for 5 mM KNO_3 were set at 100 and the bar heights are relative to this. Numbers above the bars are the actual values. Grains were sown in Kimpack paper, watered with a 1/10 strength Hoagland's solution which contained the appropriate concentration of KNO_3 and seedling shoots were harvested after 7 days growth under conditions described by Oaks et al. (1988)

Acknowledgments. We thank Andris Kleinhofs and R.L. Warner for antibody prepared against a pure NR from barley which was used in the barley experiments (Fig. 3) and for numerous very piquant discussions, Steven J. Rothstein for a cDNA clone for maize NR and N.S.E.R.C. (Canada) for financial support.

References

Amy NR, Garrett RG (1979) Immunoelectrophoretic determination of nitrate reductase in *Neurospora crassa.* Anal Biochem 95: 97-107

Aryan AP, Batt RG, Wallace W (1983) Reversible inactivation of nitrate reductase by NADH and the occurrence of partially inactive enzyme in wheat leaf. Plant Physiol 71: 582-587

Beevers L, Hageman RH (1972) The role of light in nitrate metabolism in higher plants. Photophysiology 7: 85-113

Campbell WH (1987) Regulation of nitrate reductase in maize: An immunological approach. In: Ullrich WR, Aparicio PJ, Syrett PJ, Castillo F (eds) Inorganic Nitrogen Metabolism. Springer, Berlin Heidelberg New York, pp 99-103

Campbell WH, Remmler JL (1986) Regulation of corn leaf nitrate reductase. I. Immunological methods for analysis of the enzyme's protein component. Plant Physiol 80: 435-441

Crawford NM, Smith M, Bellissimo D, Davis RW (1988) Sequence and nitrate regulation of the *Arabidopsis thaliana* mRNA encoding nitrate reductase, a metalloflavoprotein with three functional domains. Proc Natl Acad Sci USA 85: 5006-5010

Evans HJ, Nason A (1953) Pyridine nucleotide-nitrate reductase from extracts of higher plants. Plant Physiol 28: 233-254

Galangau F, Daniel-Vedele F, Moureaux T, Dorbe M-F, Leydecker M-T, Caboche M (1988) Expression of leaf nitrate reductase genes from tomato and tobacco in relation to light dark regimes and nitrate supply. Plant Physiol 88: 383-388

Hageman RH, Flesher D (1960) Nitrate reductase activity in corn seedlings as affected by light and nitrate contents of nutrient media. Plant Physiol 35: 700-708

Losada M, Paneque A, Aparicio PJ, Vega JM, Cárdenas J, Herrera J (1970) Inactivation and repression by ammonium of the nitrate reducing system in *Chlorella*. Biochem Biophys Res Commun 38: 1009-1015

Melzer JM, Kleinhofs A, Warner RL (1989) Nitrate reductase regulation: Effects of nitrate and light on nitrate reductase mRNA accumulation. Mol Gen Genet 217: 341-346

Oaks A, Poulle M, Goodfellow VJ, Cass LA, Deising H (1988) The role of nitrate and ammonium ions and light on induction of nitrate reductase in maize leaves. Plant Physiol 88: 1067-1072

Palacián E, de la Rosa F, Castillo F, Gómez-Moreno C (1974) Nitrate reductase from *Spinacia oleracea*: Reversible inactivation by NAD(P)H and by thiols. Arch Biochem Biophys 161: 441-447

Pistorius EK, Gewitz HS, Voss H, Vennesland B (1976) Reversible inactivation of nitrate reductase in *Chlorella vulgaris in vivo*. Planta 128: 73-80

Remmler JL, Campbell WR (1986) Regulation of nitrate reductase. II. Synthesis and turnover of the enzyme's activity and protein. Plant Physiol 80: 442-447

Solomonson LP (1975) Purification of NADH-nitrate reductase by affinity chromatography. Plant Physiol 56: 853-855

Somers DA, Kuo T-M, Kleinhofs A, Warner RL, Oaks A (1983) Synthesis and degradation of barley nitrate reductase. Plant Physiol 72: 949-952

Tang PS, Wu HY (1957) Adaptive formation of nitrate reductase in rice seedlings. Nature 179: 1355-1356

Ullrich WR (1987) Nitrate and ammonium uptake in green algae and higher plants: Mechanism and relationship with nitrate metabolism. In: Ullrich WR, Aparicio PJ, Syrett PJ, Castillo F (eds), Inorganic Nitrogen Metabolism. Springer, Berlin Heidelberg New York, pp 32-38

BLUE LIGHT ACTIVATION OF NITRATE REDUCTASE AND BLUE LIGHT PROMOTION OF THE BIOSYNTHESIS OF NITRITE REDUCTASE IN *MONORAPHIDIUM BRAUNII*

M.A. Quiñones and P.J. Aparicio [1]

Introduction

In green microalgae, the reduction of nitrate to nitrite and the further reduction of nitrite to ammonia have two main purposes: one assimilatory, to provide the ammonia required for building the cell material, and the other modulatory of the internal reductants by providing electron sinks in circumstances of high irradiance and low CO_2 tensions (Azuara and Aparicio 1983). Under these stressing conditions the cells release NO_2^- and/or NH_4^+ to the outer medium according to the presence of NO_3^- and/or NO_2^- in the nutrient solution. Under low CO_2 tensions, the reduction of NO_3^- to NO_2^- requires blue or UV radiation to activate nitrate reductase (Aparicio and Quiñones, in preparation).

Nevertheless, it was interesting to extend these investigations with monochromatic light (Balandín et al. 1986; Balandín and Aparicio 1987) to the biosynthetic processes of both nitrate and nitrite reductases. Blue light appears to be absolutely required for the biosynthesis of nitrite reductase. However, the activating effect of blue light on nitrate reductase appears to be exerted on the preexisting enzyme, rather than on the biosynthetic process itself.

Materials and Methods

Monoraphidium braunii cells were grown on a mineral medium with NH_4^+ as nitrogen source under a photoperiod of 14 h light to 10 h dark. The cell cultures that were sparged with regular air, were maintained inside dialysis bags in order to keep constant the composition of the nutrient solution. However, the NH_4^+ cultures that contained W and lacked Mo were grown in batches (for more details, see Balandín et al. 1986). Unless otherwise indicated, the NH_4^+-grown cells were harvested at the end of the dark period, washed, and, according to each particular experiment, resuspended in either N-free, NO_3^- or NO_2^- media with a cell density corresponding to 40 µg Chl ml^{-1}. To have an estimation of the *in vivo* activities of nitrate and nitrite reductases, the cell suspensions were sparged with CO_2-free air, and the NO_3^--dependent releases of NO_2^- and NH_4^+ were determined as described elsewhere (Azuara and Aparicio 1985).

[1] Centro de Investigaciones Biológicas, CSIC; Velázquez, 144, 28006-Madrid, Spain

In the monochromatic light experiments, red light of high irradiance was obtained from a 150 W tungsten-halogen lamp fitted with a Baird-Atomic red filter of 650-700 nm and two additional red plates of Plexiglas in a housing with three glass-fiber light conductors. Blue light was obtained from a regular slide projector with a similar tungsten-halogen lamp fitted with an Oriel 425-475 nm interference filter. Irradiance was determined with a LI-COR radiometer.

Results and Discussion

Figure 1 shows that when the NH_4^+-grown cells were irradiated with strong white light in a NO_3^- medium it took them 1 hour to start releasing NO_2^-. The same cell suspension began to release NH_4^+ also after 1 hour of light exposure. However, when a similar cell suspension was illuminated for 3 hours with strong red light, neither NO_2^- nor NH_4^+ were released to the medium. If, subsequently, red light was substituted by white light, the cells immediately began to release NO_2^-. This result indicates that during the red light exposure the biosynthesis of nitrate reductase took place, but not its photoactivation. On the contrary, the release of NH_4^+ did not start when white light substituted red light in this cell suspension but 1 hour later; this suggests that nitrite reductase was not synthesized during the 3 hours of exposure to red light. The dark cells that had been in the presence of NO_3^- as the only nitrogen source for 3 hours, did not release either NO_2^- or NH_4^+ in the first hour of white light irradiation. Hence, it seems reasonable to interpret that NO_3^- *per se* cannot promote the biosynthesis of either of these two enzymes.

In spite of the fact that the NH_4^+-grown cells were washed several times, some NH_4^+ was always present in the medium at the beginning of the experiments and, moreover, the cells that remained in darkness usually released some additional NH_4^+ to the outer medium. However, when the cell suspensions were first irradiated, this NH_4^+ became preferentially metabolized until exhaustion (see Figs. 1 and 2).

Figure 2 shows that indeed blue light was required for the biosynthesis of nitrite reductase, while it had little or no effect on the biosynthesis of nitrate reductase. In this experiment, the NH_4^+-grown cells were resuspended in a N-free medium. Samples of this suspension were kept in the dark or illuminated either with strong red light or bilaterally with strong red plus weak blue light. After 3 hours, NO_3^- was added to every cell suspension and all of them were illuminated with strong white light with the exception of the dark control. Again, NO_2^- release started immediately after the addition of NO_3^- in the cell suspensions preilluminated either with red or red plus blue light. However, the release of NH_4^+ began immediately only in the cell suspension that had been preilluminated with both red and blue light. In this respect, the cell suspension preilluminated with only red light behaved similarly to the cell suspension kept in the dark, showing a lag of more than 1 hour before releasing NH_4^+. Therefore, blue

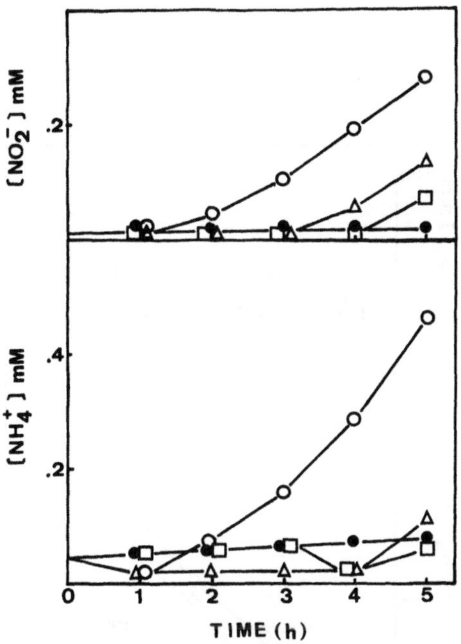

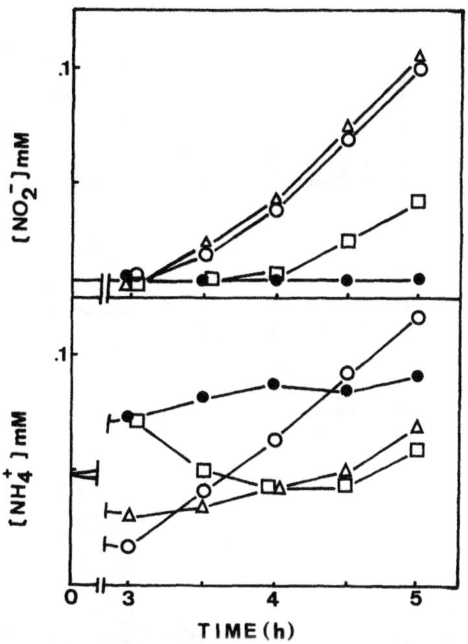

Fig. 1 *(Left)*. Light requirement for nitrate and nitrite reductase biosynthesis in *Monoraphidium braunii*. The NH_4^+-grown cells were resuspended in a 5 mM NO_3^- medium. At zero time, one sample of this suspension was irradiated with white light of 230 µE s^{-1} m^{-2} (o), another with red light of the same irradiance (Δ) and two other samples were kept in darkness (□,●). After 3 h, the suspension under red light and one of those kept in darkness (□) were illuminated with white light of the above indicated irradiance. Aliquots were taken out of the suspensions for the determinations of NO_2^- and NH_4^+ present in the outer medium

Fig. 2 *(Right)*. Blue light requirement for nitrite reductase biosynthesis and nitrate reductase activation in *M. braunii*. Cell suspensions free of nitrogen were illuminated with 150 µE s^{-1} m^{-2} red light (Δ), or bilaterally with 100 µE s^{-1} m^{-2} red light plus 50 µE s^{-1} m^{-2} blue light (o) or kept in darkness (□ , ●). After 3 h, 5 mM NO_3^- was added to all cell suspensions that were illuminated with white light, see Fig. 1, with the exception of the dark control (●)

light seems to be an environmental signal essential for the biosynthesis of nitrite reductase. Nevertheless, although NO_3^- has no promoting effect in the biosynthesis of these two reductases, the NH_4^+ concentration available in the medium might be crucial to allow these biosynthetic processes to proceed. For many years, the repressive effect of NH_4^+ on the biosynthesis of these two enzymes has been extensively documented in green algae (Losada et al. 1970).

To further test the lack of an NO_3^- effect on the biosynthesis of both nitrate and nitrite reductases (see Fig. 3), two cell suspensions were illuminated with red light, one containing NO_3^- and the other free of nitrogen. As controls, there were two N-free cell suspensions, one preilluminated bilaterally with strong red and weak blue light,

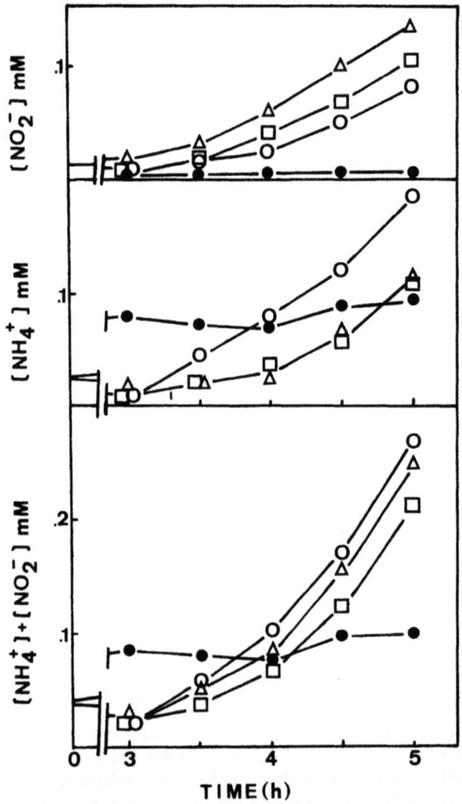

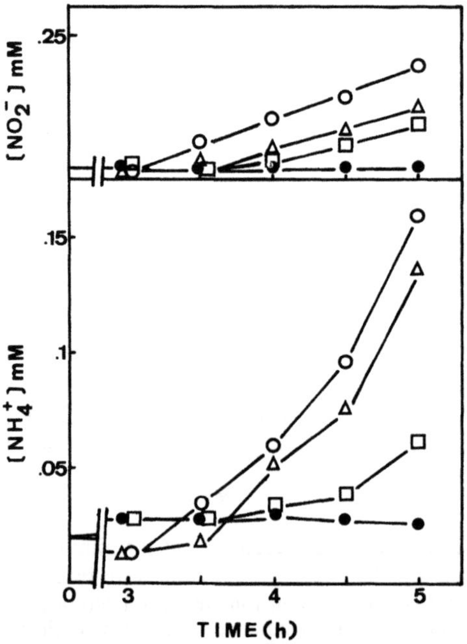

Fig. 3 *(Left)*. Lack of NO_3^- effect on the red-light promoted biosynthesis of nitrate reductase in *M. braunii*. Cells were resuspended in a N-free medium and irradiated with red plus blue (O) or red light (□) or kept in dark (●). 5 mM nitrate was added from the beginning to one of the cell suspensions and this was illuminated with red light (Δ). After 3 h, NO_3^- was added to the N-free suspensions and all of them, except the dark control, were illuminated with white light. Other experimental conditions as in Fig. 1.

Fig. 4 *(Right)*. Requirement of molybdenum for the biosynthesis of nitrate reductase in *M. braunii*. The NH_4^+-W-grown cells were resuspended in a W-N- free medium and samples of this suspension were illuminated with white light: in the presence of 5 mM NO_3^- but in the absence of Mo (Δ), or in the presence of only 100 μM Na Mo O_4 (O). The dark suspensions contained both NO_3^- and Mo at the above indicated concentrations (□, ●). After 3 h, the media were complemented with the missing components, and one of the dark suspensions (□) was also illuminated with white light. Other conditions as in Fig. 1.

and the other kept in the dark throughout the experiment. After 3 hours of pretreatment, NO_3^- was added to the N-free suspensions and all the samples were illuminated with strong white light. The release of NO_2^- started in all suspensions immediately after illumination with white light. These results show that NO_3^- is not needed for the biosynthesis of nitrate reductase, but, in its presence, the quantity of this enzyme might be higher since this anion can supply nitrogen for amino acid synthesis especially in

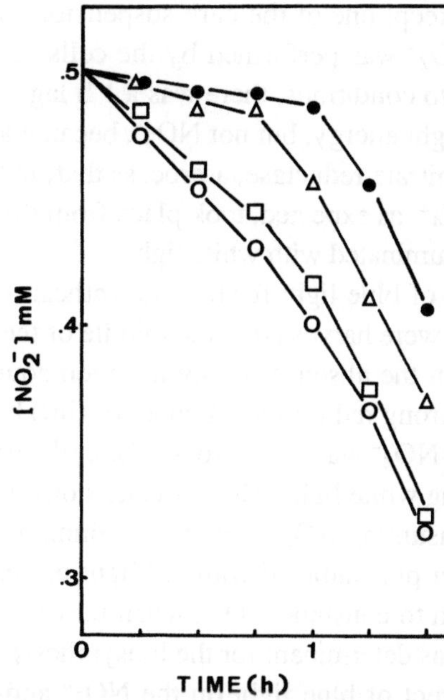

Fig. 5. Blue light dependence of nitrite reductase biosynthesis in *M. braunii* assayed as NO_2^- disappearance in the medium. The NH_4^+-grown cells were resuspended in a N-free medium, and illuminated with red light of 60 μE s^{-1} m^{-2} (△), blue light of the same irradiance (○), red light of 40 μE s^{-1} m^{-2} plus blue light of 20 μE s^{-1} m^{-2} (□) or placed in darkness (●). After 1 h, 0.5 mM NO_2^- was added to all cell suspensions and all of them were illuminated with 230 μE s^{-1} m^{-2} white light (zero time)

white light when most nitrate reductase is in its active form. The cells that were illuminated bilaterally with red and blue light initially release less NO_2^- when compared with the cells irradiated with red light, but these cells also release NH_4^+ from the beginning of white light illumination and the simultaneous NO_3^- addition. In fact, nitrate reductase activities, according to the NO_2^- and NH_4^+ release rates shown in Figure 3, were very similar.

To ensure that nitrate reductase was not present in the cells grown in NH_4^+, in the following experiment (see Fig. 4), the growth solution of the cells contained 200 μM W and was free of Mo. At zero time, 4 days old cells were harvested and resuspended in W-free media, one containing Mo but no NO_3^-; another containing NO_3^- but not Mo; and two containing both NO_3^- and Mo. The first two cell suspensions were pre-illuminated with white light for 3 hours and, as expected, neither NO_2^- nor NH_4^+ were released to the medium. Those containing NO_3^- and Mo were kept in the dark for these 3 hours. After this pretreatment time, the different nutrient solutions were complemented with the missing components (NO_3^- or Mo) and the suspensions were illuminated with white light, except one of the dark suspensions that remained so as control. A strong release of NO_2^- was performed by the cells preilluminated in the presence of Mo. In the other two conditions, there was a 1 h lag before the release of NO_2^- started. Hence, Mo and light energy, but not NO_3^-, became absolutely required factors for the biosynthesis of nitrate reductase, a process that, in this organism, only takes 1 hour. The release of NH_4^+, as expected, took place from the beginning in those cell suspensions that were preilluminated with white light.

To confirm the requirement of blue light for the biosynthesis of nitrite reductase, cells ordinarily grown on NH_4^+ were harvested in the middle of the dark period (5 h in darkness) and preilluminated in the absence of any nitrogen source with: (i) strong blue; (ii) strong red; and (iii) strong red plus weak blue; or (iiii) kept in the dark (Fig. 5). One hour later (zero time), NO_2^- was added to all the cell suspensions and all of them were irradiated with strong white light. Those suspensions that were preilluminated with blue light started consuming NO_2^- from the beginning. On the other hand, the suspensions that were either preirradiated with red light or kept in the dark had a lag of 1 hour before they began to consume NO_2^-, when they were illuminated with white light. Again, blue light was determinant for the biosynthesis of this protein.

Nevertheless, a positive effect of blue light on the NO_3^- and/or NO_2^- transport proteins cannot be ruled out, either at the plasma membrane or at the chloroplast envelope, particularly since the light effects are indeed very quick (Calero et al. 1980; Krämer et al. 1988).

In the light of the above results, the fact that *M. braunii* cells regularly grown on NO_3^- as the unique nitrogen source started to release NO_2^- only at least 1 hour after being illuminated at the beginning of the light period of their ordinary photoperiod regime, suggests a daily *de novo* synthesis of nitrite reductase. This observation is in agreement with those of *Chlorella* nitrate reductase (Velasco et al. 1989).

From these data, it can be concluded that photosynthetically active radiation seems to be required for the biosynthesis of both nitrate and nitrite reductases in autotrophically grown *M. braunii*. However, while red light is sufficient for nitrate reductase biosynthesis, nitrite reductase biosynthesis depends, in addition, on blue light as an environmental signal. As previously reported and now confirmed in this work, the cells illuminated with red light never reduced NO_3^- under these CO_2-limiting conditions and, to do so, the actinic light had to include blue or UV light for nitrate reductase activation. The above results clearly indicate that NO_3^- does not stimulate the biosynthesis of these two enzymes. As reported for *A. braunii* (Syrett and Hipkin 1973), their biosynthesis may take place when the NH_4^+ concentration in the nutrient solution becomes low enough.

References

Azuara MP, Aparicio PJ (1983) *In vivo* blue-light activation of *Chlamydomonas reinhardii* nitrate reductase. Plant Physiol 71: 286-290

Azuara MP, Aparicio PJ (1985) Spectral dependence of photoregulation of inorganic nitrogen metabolism in *Chlamydomonas reinhardii.* Plant Physiol 77: 95-98

Balandín T, Fernández VM, Aparicio PJ (1986) Characterization of the reversible inactivation of *Ankistrodesmus braunii* nitrate reductase by hydroxylamine. Plant Physiol 82: 65-70

Balandín T, Aparicio PJ (1987) Hydroxylamine metabolism in *Monoraphidium braunii.* II. Its interference with the utilization of NO_3^-. New Phytol 107: 523 -530

Calero F, Ullrich WR, Aparicio PJ (1980) Regulation by monochromatic light of nitrate uptake in *Chlorella fusca.* In: Senger H (ed) The Blue Light Syndrome. Springer, Berlin Heidelberg New York, pp 411-421

Krämer E, Tischner R, Schmidt A (1988) Regulation of assimilatory nitrate reduction at the level of nitrite in *Chlorella fusca.* Planta 176: 28-35

Losada M, Paneque A, Aparicio PJ, Vega JM, Cárdenas J, Herrera J (1970) Inactivation and repression by ammonium of the nitrate reducing system in *Chlorella.* Biochem Biophys Res Commun 38: 1009-1015

Syrett PJ, Hipkin CR (1973) The appearance of nitrate reductase activity in nitrogen-starved cells of *Ankistrodesmus braunii.* Planta 111: 57-64

Velasco PJ, Tischner R, Huffaker RC, Whitaker JR (1989) Synthesis and degradation of nitrate reductase during the cell cycle of *Chlorella sorokiniana.* Plant Physiol 89: 220-224

CONTROLLED PROTEOLYSIS OF NITRATE REDUCTASE: IDENTIFICATION AND PROPERTIES OF PRODUCTS

B.A. Notton [1], R.J. Fido [1] and M.J. Barber [2]

Physical and Catalytic Properties of Nitrate Reductase

Assimilatory nitrate reductase (NR) (E.C. 1.6.6.1) from eukaryotic sources appears to be similar in major structural features, containing one mole of FAD, haem (cytochrome b_{557}) and molybdenum per subunit. All higher plant NRs so far examined consist of a homodimer with subunit molecular mass varying from 100 to 145 kD (e.g. Kuo et al. 1982; Fido and Notton 1984; Redinbaugh and Campbell 1985). The reason for this variation in size is not known. *Chlorella vulgaris* NR is a homotetramer except at low protein concentrations when it behaves as an active homodimer (Howard and Solomonson 1982). Physiologically, NR catalyses the rate limiting step in inorganic nitrogen assimilation, i.e. the reduction of NO_3^- to NO_2^-. NADH initially donates electrons to the FAD, a process which involves sulphydryl groups. This is followed by electron transfer to the haem and egress to NO_3^- via the Mo site. Additionally the enzyme catalyses a number of partial reactions which involve one or more of the prosthetic groups. These reactions utilize various artificial electron donors (reduced flavin, $FADH_2$; methyl viologen, MV; and bromophenol blue, BPB) and acceptors (ferricyanide, F; cytochrome *c*, C; and dichlorophenol indophenol, DCPIP), and are classified as NO_3^--reducing and NADH-dehydrogenase reactions respectively (Wray and Filner 1970; Solomonson et al. 1975; Hewitt and Notton 1980; Guerrero et al. 1981; Dunn-Coleman et al. 1984; Campbell 1985).

Effect of Proteases on Enzyme Activities

Incubation of pure NR from *C. vulgaris* with either V8 protease from *Staphylococcus aureus*, which is specific for acidic amino acid residues, or a naturally occurring protease from corn roots (CIP), resulted in the loss of NADH-NR and NADH-CR (cytochrome reducing) activities with no loss of either NADH-FR (ferricyanide reducing) or MV-NR activities (Solomonson et al. 1986). The effect of proteolysis by CIP of corn (*Zea mays*) was a loss of both NADH-CR and MV-NR activities, the former being initially much more sensitive to the protease without completely disap-

[1] Department of Agricultural Sciences, University of Bristol, AFRC Institute of Arable Crops Research, Long Ashton Research Station, Bristol, BS18 9AF, UK
[2] Department of Biochemistry and Molecular Biology, College of Medicine, University of South Florida, Tampa, Florida, 33612, USA.

pearing (Poulle et al. 1987). This residual NADH-CR activity was thought to be due to the presence of cytochrome *c* reductase which was not associated with NR. The comparative effect of V8 protease and CIP on the overall and six partial activities of pure spinach and *C. vulgaris* NRs were examined (Notton et al. 1989). Brief exposure of the enzymes to V8 proteolysis caused an initial and slight increase in NADH-NR activity, followed by a loss of NADH-NR and only dehydrogenase activities associated with the haem, i.e. NADH-CR and NADH-DCPIP (Fig. 1). CIP caused loss of NADH-NR, NADH-CR and NADH-DCPIP as observed with V8. CIP also caused a small but significant loss of $FMNH_2$-NR of both *C. vulgaris* and spinach as well as MV-NR of spinach. The NADH-FR and BPB-NR activities of NR from both sources were stable. Limited digestion of spinach NR with trypsin, which is specific for basic amino acid residues, resulted in the loss of the overall and terminal NO_3^--reducing activities but retention of both NADH-FR and NADH-CR activities (Kubo et al. 1988). Subsequent digestion with V8 protease resulted in loss of NADH-CR activity. In contrast, digestion of *C. vulgaris* NR with trypsin resulted in loss of all dehydrogenase activities and retention of MV-NR activity. Cyanide protected the *C. vulgaris* NR from trypsin proteolysis (Solomonson et al. 1986).

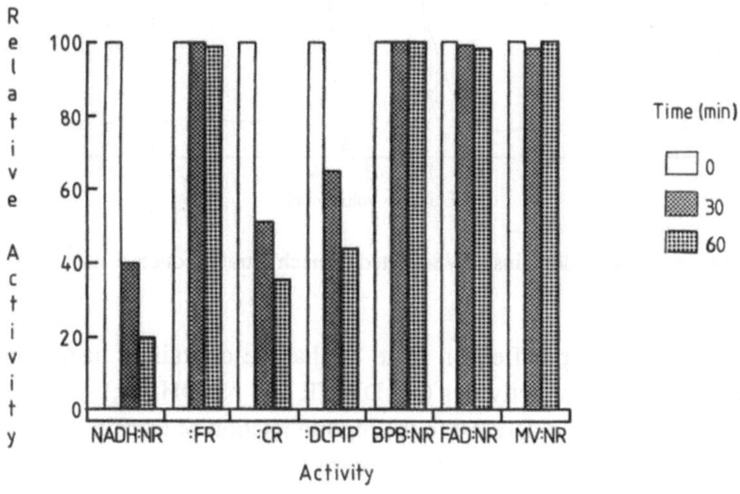

Fig. 1. Effect of V8 protease on the overall and partial activities of spinach nitrate reductase

Physical Properties of Domains Resulting from Proteolysis

The ability to obtain pure NR, initially from algae such as *C. vulgaris* (Solomonson 1975) and latterly from higher plants, especially spinach (Nakagawa et al. 1985; Fido 1987), has enabled the effect of proteases on the physical properties of the enzyme to be correlated with the loss or retention of partial activities. Both CIP and V8 protease digested *C. vulgaris* NR into two active fragments (Solomonson et al. 1986). The CIP-generated fragments were separated by column chromatography using Sephacryl S-300 and shown respectively to (a) contain FAD, retain NADH-FR activity and, by SDS-PAGE, to be a monomer of ca. 30 kDa and (b) contain haem and Mo, retain MV-NR activity and, by SDS-PAGE, to be a tetramer of ca. 260 kDa. Proteolysis by CIP of corn NR caused an increase in the mobility, on native PAGE, of a MV-NR active protein and on SDS-PAGE yielded fragments of 84 and 80 kDa (Poulle et al. 1987). V8 proteolysis of spinach followed by separation of active fragments by FPLC on Superose 12 (Notton et al. 1989) (Fig. 2) or HPLC on TSK G3000SW (Kubo et al. 1988) yielded a monomer of ca. 30 kDa and a dimer of ca. 160 kDa. Trypsin digestion of spinach NR yielded two fragments of ca. 59 kDa and 45 kDa (Kubo et al. 1988).

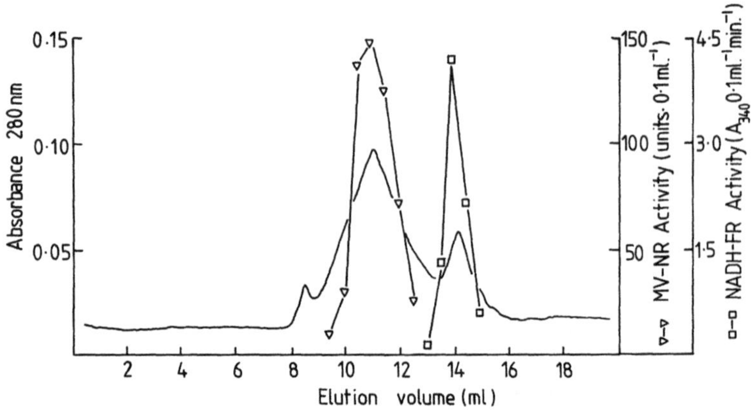

Fig. 2. Separation of active domains of V8-treated spinach nitrate reductase using Superose 12 (FPLC)

The 45 kDa fragment, purified on Blue Sepharose contained haem and FAD and associated dehydrogenase activities (NADH-FR and NADH-CR). No NO_3^--reducing activities were detected, in contrast to the reported action of trypsin on *C. vulgaris* NR (Solomonson et al. 1987). Further digestion of the purified 45 kDa fragment with V8 protease yielded two fragments of 28 kDa and 14 kDa, the former retaining only NADH-FR activity. Monoclonal antibodies MAC 74 and MAC 77 produced against the spinach NR had been shown to inhibit NO_3^--reducing activities of NR (Notton et al. 1985). After V8 proteolysis of spinach NR, the monoclonals were shown to bind to the large fragment. Chromatography of the monoclonal/digested NR mixture on Superose 12, revealed that the small ca. 30 kDa fragment retained its NADH-FR activity and original size while the large fragment was found in the excluded fraction, devoid of activity (Notton et al. 1989).

Amino Acid Sequence of Fragments

The 30 kDa fragment obtained by V8 digestion of spinach NR was isolated by chromatography on Superose 12 and either dialysed and lyophilized or further purified by Blue Sepharose affinity chromatography. A preliminary amino acid sequence of the product (Fig. 3A) was identical to the deduced sequence obtained from the cDNA clone of spinach NR (I.M. Prosser, personal communication) and showed a 60% identity to the deduced sequences obtained from cDNA of tobacco (Calza et al. 1987) and *Arabidopsis* (Crawford et al. 1988) NR. The similarity increased to 80% when conserved amino acid changes were allowed in the comparison. In *Arabidopsis* this sequence has been shown to be part of the flavin-domain of NR. A minor contaminant of this polypeptide was also sequenced (Fig. 3B) and was found to be part of the haem binding domain, being 66% and 72% identical with the deduced haem binding domain of tobacco and *Arabidopsis*. A small amount of haem (ca 10%) was also reported to be associated with the V8 proteolysis 30 kDa product of *C. vulgaris* (Solomonson et al. 1986).

```
A.                         Flavin  region
Spinach         A V P L . R N V A L N P R V K I P C N L I I K V V L M H D V R R F R F K L P F
                  | |     |   | | |   | |   | | | |   | |   |     :   | | | | : | | |   | |
Tobacco         L V P A Q R S V A L I P R E K I P C K L I D K Q S I S H D V R K F R F A L P S
Spinach         A V P L R N V A L . N P R V K I P C N L I I K V V L M H D V R R F R F K L P F
                |   | : | | : | |   | | |   | : |   : | :   |     :   | | | | : | | |   | |
Arabidopsis     A T P V R N L A L V N P R A K V P V Q L V E K T S I S H D V R K F R F A L P V

B.                         Haem  region
Spinach           . A D S G P T L K R T A S T P F M N T T V K M Y S L D I V V K H F D
                    | :       | | | : :   | | | | | | |     | | | | :     |   | |
Tobacco           S A E A P Q T L K K S I S T P F M N T A S K M Y S M S E V R K H S S
Spinach           . A D S G P T L K R T A S T P F M N T T V K M Y S L D I V V K H F D
                    | |     | : | | :     | | | | | | | |   | | | | :     |   | |
Arabidopsis       S A D A P P S L K K S V S T P F M N T T A K M Y S M S E V K K H N S
```

Fig. 3. Comparison of actual partial amino acid sequence of (**A**) NADH-FR active fragment of spinach nitrate reductase with deduced sequences from cDNA of flavin region of tobacco and *Arabidopsis* nitrate reductases. (**B**) Polypeptide association with NADH-FR fragment with deduced sequences from cDNA of haem region of tobacco and *Arabidopsis* nitrate reductases

Conclusions

Limited proteolysis of pure NRs with either V8 protease or CIP (which appear to act at similar sites) or with trypsin, has enabled the identification, separation and characterization of distinct domains, retaining one or more of the partial activities of the enzyme dependent on the presence of particular prosthetic groups.

Acknowledgments. To Professor A. Oaks for the gift of CIP, to Miss A. Robbins for purifying *C. vulgaris* NR and to Dr J. Findlay for amino acid sequencing facilities. Long Ashton Research Station is funded by the Agricultural and Food Research Council, UK. Some of the work reported was also supported by USDA, NIH and a collaborative NATO research grant to BAN and MJB.

References

Calza R, Huttner E, Vincentz M, Rouzé P, Galangau F, Vaucheret H, Chérel I, Meyer C, Kronenberger J, Caboche M (1987) Cloning of DNA fragments complementary to tobacco nitrate reductase mRNA and encoding epitopes common to the nitrate reductases from higher plants. Mol Gen Genet 209: 552-562

Campbell WH (1985) The biochemistry of higher plant nitrate reductase. In: Ludden PW, Burris JE (eds) Nitrogen Fixation and CO_2 Metabolism. Elsevier, Amsterdam, pp 43

Crawford NM, Smith M, Bellissimo D, Davis RW (1988) Sequence and nitrate regulation of the *Arabidopsis thaliana* mRNA encoding nitrate reductase, a metalloflavoprotein with three functional domains. Proc Natl Acad Sci 85: 5006-5010

Dunn-Coleman NS, Smarrelli Jr J, Garrett RH (1984) Nitrate assimilation in eukaryotic cells. Int Rev Cytol 92: 1-50

Fido RJ (1987) Purification of nitrate reductase from spinach (*Spinacea oleracea* L.) by immunoaffinity chromatography using a monoclonal antibody. Plant Sci 50: 111-115

Fido RJ, Notton BA (1984) Spinach nitrate reductase: Further purification and removal of 'nicked' subunits by affinity chromatography. Plant Sci Lett 37: 87-90

Guerrero MG, Vega JM, Losada M (1981) The assimilatory nitrate-reducing system and its regulation. Annu Rev Plant Physiol 32: 169-204

Hewitt EJ, Notton BA (1980) Nitrate reductase systems in eukaryotic and prokaryotic organisms. In: Coughlan M (ed) Molybdenum and Molybdenum-Containing Enzymes. Pergamon Press, Oxford New York, pp 273-375

Howard WD, Solomonson LP (1982) Quaternary structure of assimilatory NADH: nitrate reductase from *Chlorella*. J Biol Chem 257: 10243-10250

Kubo Y, Ogura N, Nakagawa H (1988) Limited proteolysis of the nitrate reductase from spinach leaves. J Biol Chem 263: 19684-19689

Kuo TM, Somers DA, Kleinhofs A, Warner RL (1982) NADH-nitrate reductase in barley leaves. Identification and amino acid composition of subunit protein. Biochim Biophys Acta 708: 75-81

Nakagawa H, Yonemura Y, Yamamoto H, Sato T, Ogura N, Sato R (1985) Spinach nitrate reductase: Purification, molecular weight and subunit composition. Plant Physiol 77: 124-128

Notton BA, Fido RJ, Galfre G (1985) Monoclonal antibodies to a higher plant nitrate reductase: Differential inhibition of enzyme activities. Planta 165: 114-119

Notton, BA, Fido RJ, Whitford PN, Barber MJ (1989) Effects of proteolysis on partial activities and immunological reactivity of spinach nitrate reductase. Phytochem 28: 2261-2266

Poulle, M, Oaks, A, Bzonek P, Goodfellow VJ, Solomonson LP (1987) Characterization of nitrate reductase from corn leaves (*Zea mays* cv. W64A x W182E) and *Chlorella vulgaris*. Plant Physiol 85: 375-378

Redinbaugh MG, Campbell WH (1985) Quaternary structure and composition of squash NADH: nitrate reductase. J Biol Chem 260: 3380-3385

Solomonson LP (1975) Purification of NADH-nitrate reductase by affinity chromatography. Plant Physiol 56: 853-855

Solomonson LP, Barber MJ, Robbins AP, Oaks A (1986) Functional domains of assimilatory NADH:nitrate reductase from *Chlorella*. J Biol Chem 261: 11290-11294

Solomonson LP, Lorimer GH, Hall RL, Borchers R, Bailey JL (1985) Reduced nicotinamide dinucleotide: nitrate reductase of *Chlorella vulgaris*: Purification, prosthetic groups and molecular properties. J Biol Chem 250: 4120-4127

Wray JL, Filner P (1970) Structural and functional relationships of enzyme activities induced by nitrate in barley. Biochem J 119: 715-725

THE REGULATION OF NITRATE ASSIMILATION IN YEAST

C.R. Hipkin [1]

Introduction

Since the earlier investigations of workers such as Silver (1957) and Pichinoty and Metenier (1967), progress in the study of the structure and regulation of the enzymes of nitrate assimilation in yeasts has been disappointing, compared with filamentous fungi and unicellular algae. However, more recent investigations with the imperfect, asporogenous yeast, *Candida nitratophila*, have yielded much useful information and this has helped to fill the gaps in our knowledge (Hipkin 1989). It is clear from all studies that the nitrate assimilation pathway in yeasts is similar to that found in other organisms that use nitrate as a nitrogen source and our current understanding of the pathway that incorporates nitrate-nitrogen into α-amino nitrogen in *C. nitratophila* is outlined in Figure 1. There are several noteworthy features here. Firstly, in yeasts, the heterotrophic reduction of nitrate to ammonium (via nitrite) can be driven entirely by reducing power in the form of NADPH. In species like *C. nitratophila*, the enzyme nitrate reductase (NR) is able to use NADH or NADPH as an electron donor whereas nitrite reductase is NADPH-specific (Hipkin 1989). Secondly, although it has been accepted widely that the NADPH-glutamate dehydrogenase pathway is the major pathway for the incorporation of ammonium nitrogen into α-amino nitrogen in nitrate-utilizing yeasts (Sims and Folkes 1964; Brown et al. 1973; Miflin and Lea 1980), it is clear now that a significant amount of ammonium may be assimilated via the glutamate syn-thase cycle in these organisms under non-nitrogen-replete conditions; e.g. when nitrate is the sole source of nitrogen being assimilated (Hipkin et al. 1990). Thirdly, although many forms of reduced nitrogen could be regarded as end products of nitrate assimilation, ammonium, glutamate and glutamine are the ones that occupy a central position in terms of the regulation of nitrate metabolism.

In order to appreciate the regulation of nitrate assimilation in yeasts, it will be necessary to characterize the enzymes and study the molecular genetics of the nitrate assimilation pathway in organisms like *C. nitratophila*. So far studies of nitrate metabolism in this organism have been concerned mostly with the structure and regulation of NR and the role of the pentose phosphate pathway enzymes in the provision of NADPH for nitrate assimilation. Recent progress in these areas will be the subject of this article.

[1] Biochemistry Research Group, School of Biological Sciences, University College of Swansea, Singleton Park, Swansea, SA2 8PP, UK

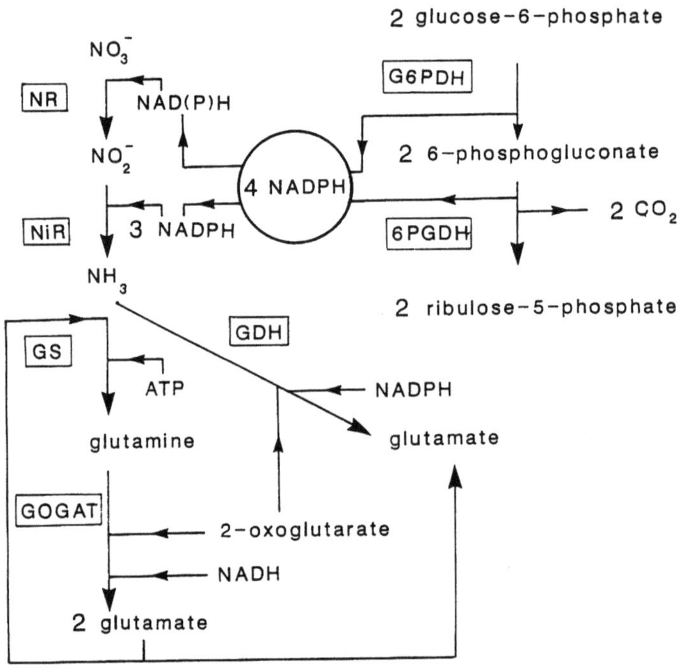

Fig. 1. The nitrate assimilation pathway in *Candida nitratophila*.
NR = nitrate reductase, NiR = nitrite reductase, GS = glutamine synthetase, GOGAT = glutamate synthase, GDH = glutamate dehydrogenase, G6PDH = glucose-6-phosphate dehydrogenase, 6PGDH = 6-phosphogluconate dehydrogenase

Physical and Biochemical Properties of Nitrate Reductase (NR)

Fortunately, *C. nitratophila* NR can be purified easily by affinity chromatography (Hipkin et al. 1986) and this has allowed detailed study of the physical and chemical properties of the enzyme (Table 1). In its subunit size and prosthetic group content, yeast NR resembles NR from *Chlorella vulgaris* (Hipkin et al. 1986, 1989; Hipkin 1989). Limited tryptic proteolysis of *C. nitratophila* NR destroys NAD(P)H-NR activity and reduces benzyl viologen-NR activity but stimulates NAD(P)H-diaphorase activity (with either cytochrome *c* or dichlorophenol indophenol as electron acceptor). Recently we (D.A. Kau and C.R. Hipkin, unpublished) have isolated and purified a 40 kD tryptic fragment of the yeast enzyme which retains diaphorase activity but has no nitrate reducing activity. The thermodynamic properties of the yeast enzyme are very similar to those of *Chlorella* (Hipkin et al. 1989).

Induction and Derepression of Nitrate Reductase (NR) Synthesis in Yeast

There is substantial evidence to indicate that the synthesis of NR is regulated at the level of transcription in plants (Cheng et al. 1986; Crawford et al. 1986; Smarrelli et al. 1987; Calza et al. 1987) and fungi (Cannons and Hipkin 1987; Kinghorn 1989; Marzluf and Fu 1989; Scazzocchio and Arst 1989). It has been shown with *C. nitratophila* that neither NR activity nor NR protein is present in ammonium-grown cultures but both appear when such cultures are transferred to nitrate medium and this correlates with the appearance of NR mRNA (Cannons et al. 1986; Cannons and Hipkin 1987; Hipkin and Cannons 1987). Small but significant amounts of NR activity, NR protein and NR mRNA also appear in ammonium-grown cultures when they are nitrogen starved (i.e. in the absence of nitrate) which indicates that NR synthesis may be derepressed in yeasts (Cannons et al. 1986; Cannons and Hipkin 1987; Hipkin 1989).

The Role of Nitrogen Metabolite Repression and Inducer Exclusion in the Regulation of Nitrate Reductase (NR) Synthesis in Yeast

In fungi, NR synthesis appears to be under the dual control of a positively acting nitrate (pathway specific) induction circuit and a positively acting, wide domain, nitrogen-metabolite repression circuit (Arst and Cove 1973; Marzluf 1981; Wiame et al. 1985). Given the advantages of assimilating reduced nitrogen sources such as ammonium, glutamate and glutamine (the end products of nitrate assimilation, Fig. 1) as opposed to nitrate, in terms of energetics and resource conservation, one can appreciate the selective advantage of a wide domain repression circuit. With *Aspergillus* and *Neurospora*, the application of both traditional and molecular genetics has provided convincing evidence for the role of nitrogen metabolite repression in the regulation of NR synthesis (Marzluf and Fu 1989; Scazzocchio and Arst 1989). With imperfect yeasts, such as *C. nitratophila*, the application of formal genetics has not been possible and although it has been demonstrated that yeast cultures grown with reduced nitrogen sources, such as ammonium, do not contain the enzymes of the nitrate assimilation pathway (see Hipkin 1989) it could be argued that this might result from inducer exclusion rather than repression. As if to confound the matter, experiments with *C. nitratophila* have shown that the addition of ammonium with nitrate does not result in the complete inhibition of nitrate assimilation (Ali and Hipkin 1986) nor a loss of NR comparable with that which occurs when cultures are resuspended in nitrate-free or carbon-free medium. Clearly, nitrate plays a positive role in sustaining high levels of active NR protein and this can be further demonstrated by the technique of *in vitro* labelling using ^{35}S-methionine (D.R. Kau, A.C. Cannons and

C.R. Hipkin, unpublished; Hipkin 1989). Making use of NR antibodies it can be shown that ^{35}S-methionine is incorporated into NR protein when either nitrate or ammonium and nitrate are added to nitrate-induced cultures of *C. nitratophila*. However, there is no incorporation of ^{35}S-methionine into NR protein if nitrate is removed, whether ammonium is present or not. The key factor here is that ammonium does not prevent the apparent induction of NR synthesis by nitrate. However, addition of glutamine with nitrate does result in an almost complete inhibition of the incorporation of ^{35}S-methionine into NR protein. Moreover, since such cultures are able to take up nitrate in the presence of glutamine (Hamoudi 1988), the glutamine effect appears to represent a repression of NR synthesis rather than a negative effect of inducer exclusion. We (Flynn and Hipkin 1990) have shown recently that glutamine is an early product of nitrate assimilation in *C. nitratophila* and cultures that express a low capacity for nitrate assimilation, such as ammonium-grown and carbon-starved cultures, exhibit relatively high ratios of glutamine:glutamate (intracellular) compared to nitrogen-starved and nitrate-grown cultures. These results indicate that in yeasts N-metabolite (glutamine) repression exists as a phenomenon distinct from inducer exclusion in the regulation of nitrate assimilation.

Table 1. Properties of *Candida nitratophila* nitrate reductase

Molecular weight (kDa)	97 (subunit)	200 (dimer)	365 (tetramer)
Pyridine nucleotide specificity			NADH, NADPH
Turnover number (s^{-1})			1200 (dimer)
Optimum pH			7.0
K_m NO_3^- (μM)			120 (NADH-NR) 110 (NADPH-NR)
K_m NADH (μM)			17
K_m NADPH (μM)			30
Spectral absorption maxima (nm)			278; 413 (ox) 278; 423; 525; 557 (red)
Haem content (mol [mol subunit]$^{-1}$)			0.9
Haem redox midpoint potential (mV)			-174
Mo content (mol [mol haem]$^{-1}$)			0.85
Mo redox midpoint potential (mV)			-3 Mo VI/Mo V -27 MoV/Mo IV

Nitrate Assimilation in Yeasts and the Regulation of Glucose-6-Phosphate Dehydrogenase and 6-Phosphogluconate Dehydrogenase

The reductive assimilation of one molecule of nitrate to one molecule of ammonium in yeasts requires 4 equivalents of NADPH. Further assimilation of this ammonium involves another molecule of NADPH if it is assimilated via the glutamate dehydrogenase pathway or NADH if it is assimilated via the glutamate synthase cycle (Fig. 1).

Bruinenberg et al. (1983 a) have calculated that the production of 100 g of *Candida* cells on ammonium medium requires 931 equivalents of NADPH whereas on nitrate medium the same biomass production requires 3651 equivalents of NADPH. Furthermore, these workers have argued convincingly that most of the extra NADPH must be generated via the initial reactions of the oxidative pentose phosphate pathway catalysed by the enzymes glucose-6-phosphate dehydrogenase (G6PDH) and 6-phosphogluconate dehydrogenase (6PGDH). Moreover, if xylose is the source of carbon in these cultures extra NADPH is needed to furnish the reduction of xylose to xylitol by the enzyme xylose reductase. It is well established that higher levels of the oxidative pentose phosphate pathway enzymes are found in cultures of fungi and algae when they are grown in nitrate medium rather than ammonium medium (Osmond and ap Rees 1969; Hankinson and Cove 1974; Bruinenberg et al. 1983 b; Hipkin and Cannons 1985), and results obtained with mutants of *Aspergillus* (Hankinson and Cove 1974) suggest that the enzymes of the nitrate assimilation pathway and the oxidative pentose phosphate pathway are subject to a coordinated control at the genetic level. Another possibility, however, is that increases in the activities of enzymes like G6PDH and 6PGDH in nitrate cultures result from post-translational regulation. This has been tested with *C. nitratophila* using rocket immunoelectrophoresis to measure levels of G6PDH protein in cultures incubated under various conditions of nitrogen source and carbon source (A. Al-Kubisi and C.R. Hipkin, unpublished). Initially, G6PDH was purified from *C. nitratophila* and some of its properties studied. The native enzyme has a molecular mass of 116 kDa and a subunit size of 56 kDa. Polyclonal antibodies raised against this protein inactivated G6PDH activity in extracts of *C. nitratophila* (and other yeasts) but did not inactivate 6PGDH activity. Immunoelectrophoretic experiments demonstrated that the higher levels of G6PDH activity in cultures transferred to nitrate medium, com-pared to cultures adapted to ammonium or glutamine medium, correlated with higher levels of G6PDH protein in these cultures. Highest levels of G6PDH protein were found in cultures transferred to nitrate medium with xylose (instead of glucose) as the carbon source. Thus an increased demand for NADPH in these cultures was reflected by a corresponding increase in G6PDH protein indicating that the synthesis of the en-zyme is regulated under these conditions. Of course, this does not rule out the additio-nal possibility of post-translational regulation, fine tuned to the availability of NADPH, for instance, operating alongside a transcriptional (or translational) control circuit. That the latter may be linked to the nitrate induction circuit and nitrogen metabolite repres-sion is intriguing indeed, providing the fungal cell with another bridge between nitro-gen and carbon metabolism.

References

Ali AH, Hipkin CR (1986) Nitrate assimilation in *Candida nitratophila* and other yeasts. Arch Microbiol 144: 263-267

Arst HN jr, Cove DJ (1973) Nitrogen metabolite repression in *Aspergillus nidulans*. Mol Gen Genet 126: 111-141

Brown CM, Burn VJ, Johnson B (1973) Presence of glutamate synthase in fission yeast and its possible role in ammonia asssimilation. Nature 246: 115-116

Bruinenberg PM, Van Dijken JP, Scheffers AW (1983 a) A theoretical analysis of NADPH production and consumption in *Candida utilis*. J Gen Microbiol 129: 953-964

Bruinenberg PM, Van Dijken JP, Scheffers AW (1983 b) An enzymatic analysis of NADPH production and consumption in *Candida utilis*. J Gen Microbiol 129: 965-971

Calza R, Huttner E, Vincentz M, Rouzé P, Galangau F, Vaucheret H, Chérel I, Meyer C, Kronenberger J, Caboche M (1987) Cloning of DNA fragments complementary to tobacco nitrate reductase mRNA and encoding epitopes common to the nitrate reductases from higher plants. Mol Gen Genet 209: 552-562

Cannons AC, Ali AH, Hipkin CR (1986) Regulation of nitrate reductase in the yeast *Candida nitratophila*. J Gen Microbiol 132: 2005-2011

Cannons AC, Hipkin CR (1987) Evidence for the transcriptional control of nitrate reductase in *Candida nitratophila* from *in vitro* translation studies. Eur J Biochem 164: 383-387

Cheng CL, Dewdney J, Kleinhofs A, Goodman HM (1986) Cloning and nitrate induction of nitrate reductase mRNA. Proc Natl Acad Sci USA 83: 6825-6828

Crawford NM, Campbell WH, Davis RW (1986) Nitrate reductase from squash: cDNA cloning and regulation. Proc Natl Acad Sci USA 83: 8073-8076

Flynn K, Hipkin CR (1990) Changes in intracellular amino acids and glutamine:glutamate during N-deprivation and feeding in *Candida nitratophila*. New Phytol 114: 435-440

Hamoudi ZS (1988) Ammonium assimilation in yeasts. M Sc Thesis, University of Wales, Swansea

Hankinson O, Cove DJ (1974) Regulation of the pentose phosphate pathway in the fungus *Aspergillus nidulans*. J Biol Chem 249: 2344-2353

Hipkin CR (1989) Nitrate assimilation in yeasts. In: Wray JL, Kinghorn JR (eds) Molecular and Genetic Aspects of Nitrate Assimilation. Oxford Sci Publications, Oxford New York Tokyo, pp 51-68

Hipkin CR, Ali AH, Cannons AC (1986) Structure and properties of assimilatory nitrate reductase from the yeast *Candida nitratophila*. J Gen Microbiol 132: 1997-2003

Hipkin CR, Cannons AC (1985) Inorganic nitrogen assimilation and the regulation of glucose-6-phosphate dehydrogenase in unicellular algae. Plant Sci 41: 155-160

Hipkin CR, Cannons AC (1987) The regulation of nitrate reductase in yeasts. In: Ullrich WR, Aparicio PJ, Syrett PJ, Castillo F (eds) Inorganic Nitrogen Metabolism. Springer, Berlin Heidelberg New York, pp 89-93

Hipkin CR, Flynn KJ, Marjot E, Hamoudi ZS, Cannons AC (1990) Ammonium assimilation in nitrate utilising yeast, *Candida nitratophila*. New Phytol 114: 429-434

Hipkin CR, Kau DA, Cannons AC, Jones DH, Gallon JR, Kay CJ, Barber MJ, Solomonson LP (1989) Physical and chemical properties of assimilatory nitrate reductase from *Candida nitratophila*. Trans Biochem Soc Lond 17: 928-929

Kinghorn JR (1989) Genetic, biochemical and structural organisation of the *Aspergillus nidulans cnrA-niiA-niaD* gene cluster. In: Wray JL, Kinghorn JR (eds) Molecular and Genetic Aspects of Nitrate Assimilation. Oxford Sci Publications, Oxford New York Tokyo, pp 69

Marzluf GA (1981) Regulation of nitrogen metabolism and gene expression in fungi. Microb Rev 45: 437-461

Marzluf GA, Fu Y-H (1989) Genetic regulation and molecular studies of nitrate assimilation in *Neurospora crassa*. In: Wray JL, Kinghorn JR (eds) Molecular and Genetic Aspects of Nitrate Assimilation. Oxford Sci Publications, Oxford New York Tokyo, pp 314

Miflin BJ, Lea PJ (1980) Ammonium assimilation. In: Miflin BJ (ed) The Biochemistry of Plants, vol 5 , Amino Acids and Derivatives. Acad Press, New York London, pp 159-202

Osmond CB, ap Rees T (1969) Control of the pentose phosphate pathway in yeast. Biochim Biophys Acta 184: 35-42

Pichinoty F, Metenier G (1966) Contribution à l'étude de la nitrate-réductase assimilatrice d'une levure. Ann Inst Pasteur 111: 282-313

Scazzocchio C, Arst HN jr (1989) Regulation of nitrate assimilation in *Aspergillus nidulans*. In: Wray JL, Kinghorn JR (eds) Molecular and Genetic Aspects of Nitrate Assimilation. Oxford Sci Publications, Oxford New York Tokyo, pp 299

Silver WS (1957) Pyridine nucleotide-nitrate reductase from *Hansenula anomala*, a nitrate reducing yeast. J Bacteriol 73: 241-246

Sims AP, Folkes BF (1964) A kinetic study of the assimilation of [^{15}N] ammonia and the synthesis of amino acids in an exponentially growing culture of *Candida utilis*. Proc Roy Soc (London) 159: 479-502
Smarrelli J jr, Malone MJ, Walters MT, Curtis LT (1987) Transcriptional control of the inducible nitrate reductase isoform from soybean. Biochem Biophys Res Commun 146: 1160-1165
Wiame JM, Grenson M, Arst HN jr (1985) Nitrogen catabolite repression in yeasts and filamentous fungi. Adv Microb Physiol 26: 1-88

PARTICIPATION OF PTERIN-MOLYBDENUM COFACTOR IN THE REGULATION OF NITRATE REDUCTASE ACTIVITY AND SIGNAL TRANSDUCTION DURING LIGHT-STIMULATED CONIDIATION OF *NEUROSPORA CRASSA*

H. Ninnemann [1] and S. Frosch [2]

The molybdenum cofactor (Moco) of *Neurospora* nitrate reductase (NR) plays a determining role in regulating the activity of the enzyme *in vitro* and *in vivo*. Roldán and Butler (1980) showed that partially purified NR of the *Neurospora* strain albino-band could be inactivated by reducing the Moco with NADPH in the presence of cyanide. The inactive enzyme could be re-activated by re-oxidizing the Moco by ferricyanide or by flavin-mediated light. The flavin participates in this reaction in its triplet state (Fritz and Ninnemann 1985). Quantitative comparison of quantum requirements and the amount of bound FAD present in partially purified NR allow the interpretation that enzyme-bound flavin could act as a photoreceptor for the photo-reactivation (Pottiez et al. 1988).

Neurospora strain al-2, bd conidiates profusely in darkness on NH_4^+-medium and much less so on NO_3^--medium. Light can stimulate this conidiation in mycelia on nitrate as sole N-source (but not in mycelia grown on ammonium), especially if they have been starved by lowering the nitrate concentration or omitting nitrate for a few hours (Table 1).

Table 1. Conidiation response of various *Neurospora crassa* strains as a function of nitrogen source and light

Strain	Nitrogen source	NADPH → NO_3^- NR activity	Conidiation in darkness	Additional photoconidiation
al-2, bd	NH_4^+	–	++	–
al-2, bd	NO_3^-	+++	+	++
nit-1	NH_4^+	diaphorase	+	–
nit-3	NH_4^+	terminal	+(+)	(+)
WC-1	NO_3^-	+++	++	++
WC-2	NO_3^-	+++	–	+

We have been trying to prove that NR can act as a photoreceptor for this light-stimulated conidiation (Klemm and Ninnemann 1979; Ninnemann 1984). WC-strains

[1] Universität Tübingen, Corrensstr. 41, D-7400 Tübingen
[2] Universität Freiburg, Schänzlestr. 1, D-7800 Freiburg, FRG

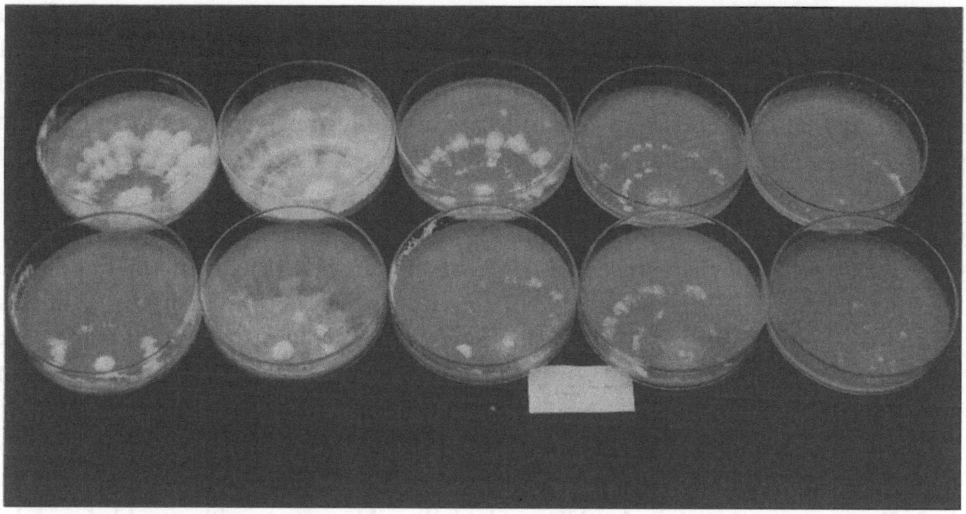

Fig. 1. Photoconidiation of mycelia of *Neurospora* al-2,bd grown on NO_3^--medium without (left) or with increasing vanadate concentrations (20 to 620 μM). Top row: irradiated mycelia (2 h, 5 W · m^{-2}), bottom row: dark controls. All dishes without molybdate

Table 2. Specific nitrate reductase activities (units per mg protein) and photoconidiation of mycelia of the experiment shown in Fig. 1.
Unit NADPH → NO_3^- and MVH → NO_3^- activity: nmol/10 min (30°C); unit diaphorase activity: nmol cyt *c* reduced/min (23°C). "0" control grown without molybdate and without vanadate. "100x" etc. refers to 100-fold etc. concentration of the normally applied 0.207 μM molybdate

Vanadate conc. μM	0	20.7 (100 x)	62 (300 x)	207 (1000 x)	620 (3000 x)
Photoconidiation	+++	+(+)	(+)	–	–
NADPH → NO_3^-activity	223	125	71	24	33
(%)	*(100)*	*(56.1)*	*(31.8)*	*(10.8)*	*(14.8)*
MVH → NO_3^-activity	575	163	83	44	33
(%)	*(100)*	*(28.3)*	*(14.4)*	*(7.6)*	*(5.7)*
Diaphorase activity	23	23	10	–	–
(%)	*(100)*	*(100)*	*(44.8)*	–	–

of *Neurospora* that are considered "blind" for blue light signals have high NR activities and their conidiation could be stimulated by light when the mycelia were grown on NO_3^-- medium (Table 1; Ninnemann, unpublished). Previously reported experiments with *Neurospora* strain al-2,bd indicated that substitution of the molybdenum in the Moco by vanadate or tungstate not only inactivated NR, but also prevented the signal transduction for photoconidiation (Ninnemann 1987). In controls in which molybdate plus vanadate were supplied simultaneously, photo-induced conidiation did occur.

A detailed analysis of the effect of vanadate substitution in the molybdenum cofactor on photoconidiation in strain al-2,bd and on NR activities is presented in Fig. 1 and Table 2. Mycelia were grown with nitrate in small (ø 5 cm) Petri dishes placed eccentrically into larger ones (ø 20 cm) where the molybdenum of the nitrate medium was substituted by vanadate (100- to 3000-fold the normal Mo concentration, i.e. 20.7 to 620 μM) so that the mycelia grew continuously from normal nitrate medium (in this experiment without 0.207 μM Mo) onto the poisonous vanadate-containing medium. Once the growth front had proceeded 4 - 5 cm onto the vanadate medium (43 - 48 h), the front 0.5 cm were scraped off for NR activity tests or, in parallel dishes, light was given for 2 hours (5 Wm^{-2}). These dishes plus the dark controls were kept in the dark for another 24 hours before they were evaluated photographically. Total NR activity of myce-lia from solid medium is always much lower than from mycelia grown in liquid me-dium. Also the ratio of total:terminal (MVH = reduced methyl viologen) activity is significantly different from that in mycelium from liquid culture where it is typically 3 - 4 : 1. We cannot yet explain that NR can be induced by nitrate without added molybdate. Possibly trace amounts are introduced with the macroelements of the medium.

A sharp decline of NADPH → NO_3^- and especially of MVH → NO_3^- activities is accompanied by decreasing efficiency in photoconidiation, whereas the NADPH → cytochrome *c* diaphorase activity remains distinctly less affected. (In the experiments of Table 2 the *per se* small absorption signal fell below the threshold of detectability at vanadate concentrations ≥ 207 μM. In many others diaphorase activity was much less affected by high vanadate concentrations). The lack of transduction of the flavin-mediated photosignal for light-promoted conidiation therefore seems to be caused by the inactivation of the molybdenum cofactor of *Neurospora* nitrate reductase.

Johnson et al. (1980) discovered a pterin derivative attached to the molybdenum of Moco's of xanthine oxidase, sulfite oxidase and *Chlorella* NR. They analysed the pterin after oxidative degradation of their samples by boiling them at low pH with or without KI + I_2. The structure of the molybdenumpterin (Mopterin) was confirmed recently as an urothione-related 6-alkylpterin (Kramer et al. 1987).

Our interest in the photobiochemistry of NR initiated us to run fluorescence spectra of partially purified NR of *Neurospora*, where, following the method of Johnson et al. (1982), we found a pterin signal with maximal excitation and emission at 410 and 480 nm ("F410/480", Siefermann et al. 1985).

In many fungi, algae and higher plants pteridines have been described (Table 3), some of which serve as chemotactant and prerequisite for actin polymerization, signal for carboxymethylation of a cell aggregating protein, inhibitor of tRNA transglycosylases (Kersten et al. 1986), photoreceptor or photosensitizer, e.g. for photophosphorylation or cytochrome oxidation (MacLean 1966).

In toto, biochemistry and function of plant and fungal pterins, especially of *Neurospora* pterins and Mopterin have been little worked on. It is conceivable that the Mo-

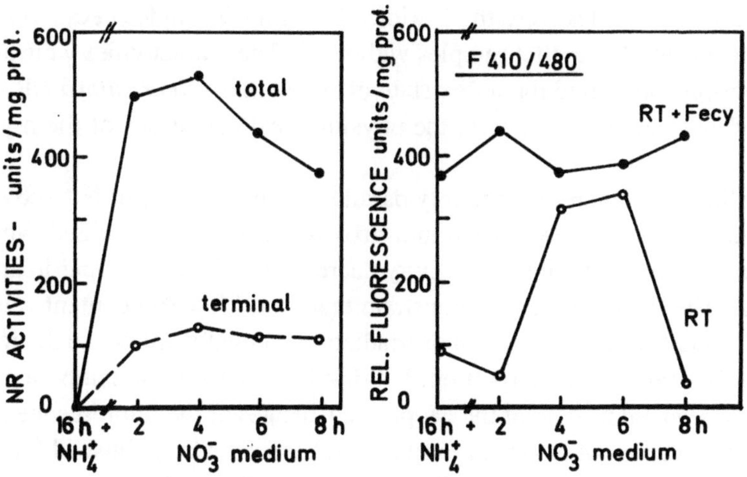

Fig. 2. Change of redox state of Mopterin in uninduced and induced mycelia as indicated by NR activities (*left*) and fluorescence (*right*)

pterin of *Neurospora* NR acts as a photoreceptor or photosensitizer, or undergoes physiological redox changes parallel to or within the electron transport chain from NADPH to nitrate, in addition to its function as a binding partner for the redox-active molybdenum.

To learn more about the redox state and function of Mopterin *in vivo*, we avoided the harsh boiling conditions for oxidative pterin degradation. We rather treated *Neurospora* supernatants (after 20 min at 13,000 x g) at room temperature (RT) for 20 min with pH 2 in the presence of 2% cholate and ran fluorescence spectra at pH 10 and at

Table 3. Presence of pteridine compounds in fungi, algae and higher plants

Dictyostelium discoideum, D. lacteum	Gerisch et al.(1982); Van Haastert et al.(1982); Van Waarde (1983)
Physarum nudum	Rakoczy (1973)
Physarum polycephalum	Wolf (1959); Wormington and Weaver (1976)
Saccharomyces cerevisiae (photolyase)	Sancar and Sancar (1987)
Phycomyces blakesleanus	Berns and Vaughn (1970); Ninnemann (1983)
Aspergillus repens	Wolf (1957)
Neurospora crassa	Siefermann et al. (1985)
Russula	Eugster et al. (1970)
Euglena	Boehme et al. (1980); Juknat et al. (1988)
Chlorella (MocoPterin NR)	Johnson et al. (1980)
Chondrus crispus and other algae	Ziegler et al. (1962)
Spinacia oleracea (chloroplasts)	MacLean et al. (1965)
Pisum sativum	Fuller et al. (1958); Wolf (1960)

pH 5 (for baseline). Samples were then oxidized with ferricyanide (Fecy) to monitor the total pterin content. In parallel samples without cholate NR activities were assayed. With this procedure we found the redox state of Mopterin in *Neurospora* NR (F 410/480 at RT without Fecy) to vary with the physiological conditions of the mycelium (Fig. 2).

The F410/480 signal increased roughly parallel to the induction of NR activity, i.e. Mopterin was reduced in the ammonium-grown culture and oxidized when NR activity became maximal. When NR activity decreased with time, Mopterin appeared reduced again. The Fecy samples demonstrate that total Mopterin content was about constant in all samples. Also starvation or irradiation seemed to influence the recorded redox state of the pterin. We interpret our F 410/480 signal at room temperature to be related to the *in vivo* redox state of the Mopterin. Changes of the fluorescence signal F 340/440 nm as a function of environmental conditions were also published for extracts of *Phycomyces* (Berns and Vaughn 1970): excitation and emission maxima decreased after irradiation and regenerated to a more oxidized form in darkness. Pteridines from dark-grown Alaska peas were more reduced than those isolated from light-grown plants (Wolf 1960), a tendency which we could confirm for Mopterin of *Neurospora* NR. More (= more oxidized?) pterins present were reported in light-grown *Rhodospirillum* and *Rhodopseudomonas* (MacLean et al. 1966). Physiological redox changes of pterins also influence their efficiency as competitive inhibitors of tRNA transglycosylases (Kersten et al. 1986).

We are not yet sure whether the differences in size of our F410/480 signal directly mirror different redox states of the Mopterin in different physiological conditions or whether they might be due to a changing stability of a protein-Mopterin association so that an increasing lability would allow the Mopterin to become more oxidized.

Acknowledgment. The financial support of the Deutsche Forschungsgemeinschaft is gratefully acknowledged.

References

Berns D, Vaughn J (1970) Studies on the photopigment system of *Phycomyces*. Biochem Biophys Res Comm 39: 1094-1103

Boehme M, Pfleiderer W, Elstner EF, Richter W (1980) Euglenapterin - ein neues, ungewöhnliches natürliches Pterin-Derivat aus *Euglena gracilis*. Angew Chemie 92: 474-475

Eugster C, Frauenfelder F, Koch H (1970) *Russula*-Farbstoffe. Helvet Chim Acta 53: 131-138

Fritz B, Ninnemann H (1985) Photoreactivation by triplet flavin and photoinactivation by singlet oxygen of *Neurospora crassa* nitrate reductase. Photochem Photobiol 41: 39-45

Fuller R, Anderson J, Nathan H (1958) Pteridines in photosynthesis. Proc Natl Acad Sci USA 44: 239-244

Gerisch G, Blank G, Schweiger M, Fuchs D, Hausen A, Reibnegger G, Wachter H (1982) Pteridines released from *Dictyostelium discoideum* cells into the extracellular medium. In: Wachter H, Curtis HC, Pfleiderer W (eds) Biochemical and Clinical Aspects of Pteridines, vol 1. W de Gruyter, Berlin, pp 253-256

Johnson J, Hainline B, Rajagopalan K (1980) Characterization of the molybdenum cofactor of sulfite oxydase, xanthine oxydase and nitrate reductase. J Biol Chem 255: 1783-1786
Johnson J, Rajagopalan K (1982) Structural and metabolic relationship between the molybdenum cofactor and urothione. Proc Natl Acad Sci USA 79: 6856-6860
Juknat A, Dörnemann D, Senger H (1988) Identification of a regulatory pteridine. Z Naturforsch 43 c: 357-362
Kersten W, Zubrod E, Emmerich B, Maubach P (1986) Queuine-lacking tRNAs in relation to the grade of malignancy of specific tumors; possible role of pteridines. In: Cooper B, Whitehead V (eds) Chemistry and Biology of Pteridines. W de Gruyter, Berlin, pp 205-208
Klemm E, Ninnemann H (1979) Nitrate reductase - a key enzyme in blue light-promoted conidiation and absorbance change of *Neurospora crassa.* Photochem Photobiol 29: 629-632
Kramer S, Johnson J, Ribeiro A, Millington D, Rajagopalan K (1987) The structure of the molybdenum cofactor. J Biol Chem 262: 16357-16363
MacLean F (1966) Stimulation of photophosphorylation and cytochrome *c* photooxydation by pteridines. Plant Physiol 41: 774-779
MacLean F, Forrest H, Hoare D (1966) Pteridine content of some photosynthetic bacteria. Arch Biochem Biophys 117: 54-58
Ninnemann H (1983) Reversible absorbance changes and modulation of biological activities by blue light. In: Montagnoli G, Erlanger B (eds) Molecular Models of Photoresponsiveness. Nato ASI Ser 68, pp 133-152
Ninnemann H (1984) The nitrate reductase system. In: Senger H (ed) Blue Light Effects in Biological Systems. Springer, Berlin Heidelberg New York, pp 95-109
Ninnemann H (1987) Regulation and signal transduction in nitrate reductase of *Neurospora crassa.* In: Ullrich WR, Aparicio PJ, Syrett PJ, Castillo F (eds) Inorganic Nitrogen Metabolism. Springer, Berlin Heidelberg New York, pp 82-88
Pottiez B, Krippahl G, Ninnemann H (1988) Quantum requirement for photoreactivation of inactivated nitrate reductase of *Neurospora crassa* strain al-2, bd. Bot Acta 101: 301-305
Rakoczy L (1973) The myxomycete *Physarum nudum* as a model organism for photobiological studies. Ber Dt Bot Ges 86: 141-164
Roldán J, Butler WL (1980) Photoactivation of nitrate reductase from *Neurospora crassa.* Photochem Photobiol 32: 375-381
Sancar G, Sancar A (1987) Structure and function of DNA photolyases. Trends in Biochem Sci 12: 259-261
Siefermann-Harms D, Fritz B, Ninnemann H (1985) Evidence for a pterin-derivative associated with the molybdenum cofactor of *Neurospora crassa* nitrate reductase. Photochem Photobiol 41: 39-45
Van Haastert P, De Wit R, Grijpma Y, Konijn T (1982) Identification of a pterin as the acrasin of the cellular slime mold *Dictyostelium lacteum.* Proc Natl Acad Sci USA 79: 6270-6274
Van Waarde A (1983) Cyclic AMP, folic acid and pterin-mediated protein carboxymethylation in cellular slime molds. FEBS Lett 161: 45-50
Wolf F (1957) The fluorescent pigment of *Aspergillus repens.* Physiol Plant 10: 825-831
Wolf F (1959) Chemical nature of the photoreceptor pigment inducing fruiting of plasmodia of *Physarum polycephalum.* In: Withrow R (ed) Photoperiodism and Related Phenomena in Plants and Animals. Publ no 55, Amer Ass Adv Sci, Washington, pp 321-326
Wolf F (1960) Pteridine pigment in microorganisms and higher plants. In: Allen M (ed) Comparative Biochemistry of Photoreactive Systems, vol 1. Acad Press, New York London, pp 53-67
Wormington W, Weaver R (1976) Photoreceptor pigment that induces differentiation in the slime mold *Physarum polycephalum.* Proc Natl Acad Sci USA 73: 3896-3899
Ziegler I, Ziegler H, Schmidt H (1962) Über Pterine bei Algen. Arch Mikrobiol 42: 80-89

NITRATE ASSIMILATION IN THE NON-N_2-FIXING CYANOBACTERIUM *PHORMIDIUM LAMINOSUM*

J.L. Serra, J.M. Arizmendi, F. Blanco, M. Martínez-Bilbao, A. Alaña, O.Fresnedo, I. Urkijo and M.J. Llama [1]

Introduction

Phormidium laminosum (strain OH-1-p.Cl_1) is a thermophilic, filamentous non-N_2-fixing cyanobacterium (blue-green alga) which can utilize either nitrate, nitrite or ammonium as the sole nitrogen source for growth.

The assimilatory utilization of inorganic nitrogen by cyanobacteria (see for a review: Guerrero and Lara 1987) comprises three main steps (Fig.1). In the first, the nitrogenous form enters the cell probably mediated by specific permeases. Secondly, the intracellular nitrate is rapidly reduced to ammonium by the sequential action of the two ferredoxin-dependent enzymes nitrate reductase (NR) and nitrite reductase (NiR). Finally, the resulting ammonium can be incorporated into 2-oxoglutarate to provide L-glutamate in reactions mainly catalyzed by the glutamine synthetase (GS)/glutamate synthase (GOGAT) cycle. In *P. laminosum* as well as in other cyanobacteria glutamate dehydrogenase (GDH) may play a role in ammonium assimilation (Fig. 1).

When all these inorganic nitrogen sources are simultaneously present ammonium is preferentially utilized because no further reduction is required. Both nitrate and nitrite must be reduced to ammonium prior to the incorporation into amino acids.

At the molecular level, the available information on the inorganic nitrogen uptake system is very scarce. Recently a possible component of the nitrate transport system (a protein of 47 - 48 kD) has been identified in the cytoplasmic membranes of nitrate-grown cells of *Synechococcus* R2 (Madueño et al. 1988) and *Anacystis nidulans* (Sivac et al. 1989). The presence of this adaptive protein is correlated with the existence of nitrate transport, suggesting that the polypeptide could be a component of the nitrate transporter (permease) (Sivac et al. 1989). The protein is absent in ammonium-grown cells (Madueño et al. 1988; Sivac et al. 1989). The possible existence of a common active uptake system for both nitrate and nitrite could be supported by the available kinetical data from *A. nidulans* (Flores et al. 1987; Madueño et al. 1987).

Nitrate assimilation in cyanobacteria is affected by a number of nutritional and environmental factors. Under autotrophic conditions, the utilization of any form of inorganic nitrogen is strictly dependent on the availability of light and CO_2 (Guerrero and Lara 1987).

[1] Departamento de Bioquímica y Biología Molecular, Facultad de Ciencias, Universidad del País Vasco/Euskal Herriko Unibertsitatea, Apartado 644, E-48080 Bilbao, Spain

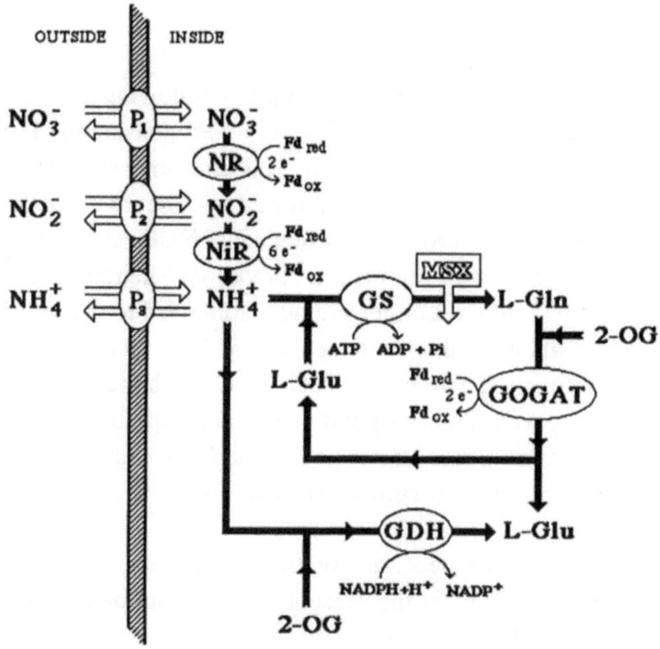

Fig. 1. Main steps in the assimilatory utilization of inorganic nitrogen in *P. laminosum:* uptake, reduction to ammonium and incorporation into amino acids. NR, nitrate reductase; NiR, nitrite reductase; GS glutamine synthetase; GOGAT, glutamate synthase; GDH, glutamate dehydrogenase; P_1, P_2, P_3, permeases; Fd, ferredoxin; 2-OG, 2-oxoglutarate; MSX, L-methionine-D,L-sulfoximine. P_1 and P_2 could represent the same permease

Cyanobacterial NR has been purified to homogeneity only from *A. nidulans* (Candau 1979) and *Plectonema boryanum* (Mikami and Ida 1984). The *Plectonema* enzyme consists of a single polypeptide chain of 85 kDa containing one atom of Mo and probably two [2Fe-2S] clusters (Mikami and Ida 1984). Only data from partially purified preparations of NiR are available in the literature (see Guerrero and Lara 1987). The native enzyme consists of a single polypeptide chain of 52 - 54 kDa. The absorption spectrum of the *Spirulina platensis* enzyme indicated that NiR is a heme protein (Yabuki et al. 1985). GS has been purified to homogeneity from a number of cyanobacterial sources (see Guerrero and Lara 1987) including two species of *Phormidium* (Serra et al. 1985; Sawa et al. 1988; Blanco et al. 1989). The native enzyme (M_r about 600,000) is a dodecameric protein composed of identical subunits of about 50 kDa. The existence of GDH activity in cyanobacteria has been questioned for many years because the enzyme seemed absent or present only in low amounts in these prokaryotes. The presence of GDH in *P. laminosum* was undoubtedly demonstrated (Martínez-Bilbao et al. 1987) and later the enzyme was purified from this cyanobacterium (Martínez-Bilbao et al. 1988) and from *Synechocystis* (Florencio et al. 1987).

Nitrate, Nitrite and Ammonium Uptake

Inorganic nitrogen uptake by intact cells was measured by following the disappearance of these ions from the extracellular medium at 45°C under continuous illumination (photon flux density at the surface of vessels ca. 100 μmol m^{-2} s^{-1}). When nitrate-grown cells of *P. laminosum* were preincubated for 24 - 72 hours in fresh nitrogen-free culture medium, the resulting nitrogen-deficient cells (NDC) showed higher inorganic nitrogen uptake rates than prior to nitrogen starvation. The rates increased with the starvation period up to 72 hours. Longer preincubations led to yellowish cultures (N chlorosis) (Fresnedo et al., unpublished data) showing lower nitrogen uptake rates.

The kinetics of inorganic nitrogen uptake of such NDC was hyperbolic denoting the possible existence of transport systems. Apparent K_s values of about 20 μM, 10 μM and <10 μM were calculated for nitrate, nitrite and ammonium ions, respectively (Martínez-Bilbao 1987). These values are similar to those found in *Anacystis nidulans* (Flores et al. 1983).

The rates of nitrogen uptake also depended on the availability of a carbon source in the medium (CO_2 or HCO_3^-). Thus, in the presence of both nitrate and nitrite, NDC preferentially took up nitrate whereas nitrogen and carbon deficient cells (NCDC) (preincubation in a medium with neither nitrogen nor carbon source) took up nitrite rather than nitrate. When nitrate and ammonium were simultaneously present in the external medium, NDC utilized exclusively ammonium whereas NCDC took up ammonium during the first 30 min but after this period nitrate was taken up, reduced to ammonium by the cells and this ion released to the external medium. When nitrite and ammonium were simultaneously present, NDC utilized only ammonium but NCDC preferentially took up nitrite. When the ability to take up nitrogen was assayed in a medium containing the three inorganic ions (Fig. 2), NDC used ammonium almost exclusively, until its exhaustion, whereas NCDC utilized nitrite in preference to nitrate. The latter ions were completely reduced to ammonium which was, finally, released to the external medium. From these data it is clear that the regulatory influence of ammonium on the uptake of nitrate or nitrite in non-NDC or NDC of *P. laminosum* was apparently absent in carbon-starved cells. Also MSX-treated *P. laminosum* cells were able to take up either nitrate or nitrite, reduce them to ammonium and release this ion to the medium (Martínez et al. 1989).

Cellular Levels of Enzymes

The cellular activity of the *P. laminosum* enzymes involved in the assimilation of nitrate strongly depended on the nature of the nitrogen source available in the medium as well as on the physiological status of the cells. In a nitrate-containing medium, max-

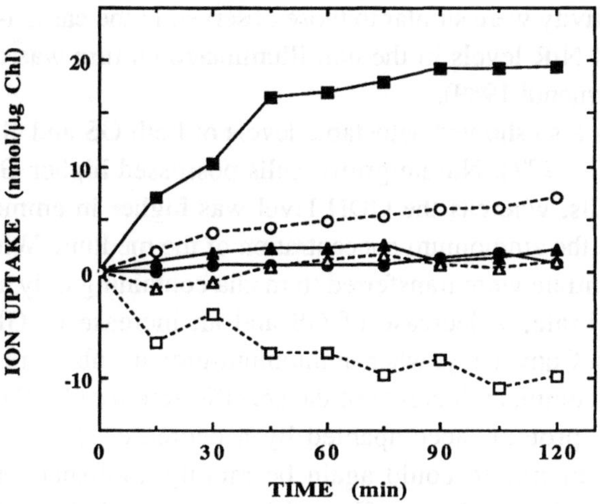

Fig. 2. Effect of nitrogen and carbon starvation on the uptake of nitrate (triangles), nitrite (circles) and ammonium (squares) by *P. laminosum*. After the indicated starvation treatment, cells were incubated with the three nitrogenous forms simultaneously (at 100 µM each). Nitrogen-deficient cells: *filled symbols*; Nitrogen- and carbon-deficient cells: *open symbols* (from Martínez-Bilbao 1987)

imum specific activity of NR was attained at the beginning of the exponential growth phase and later it decreased progressively. In contrast, similar levels of NiR activity were detected during all the growth phases.

When cells growing in media with either nitrate or nitrite were transferred to a medium containing ammonium alone or supplemented with nitrate or nitrite, a rapid decrease of NiR occurred and no activity was detected 4 h after the transfer. Similar results were obtained when nitrate-grown cells were transferred to the same media with chloramphenicol (Arizmendi et al. 1987). Cells utilizing ammonium for growth showed undetectable levels of both NR and NiR activities. The removal of ammonium from the medium was sufficient for the appearance of these activities (Arizmendi et al. 1987; Blanco and Serra, unpublished data). These results indicated that NR and NiR were both ammonium-repressible rather than nitrate- or nitrite-inducible.

When cells grown with nitrate were transferred to a carbon-free medium (without CO_2 or HCO_3^-) the growth suddenly ceased and NiR activity decreased. On the other hand, when ammonium-grown cells were transferred to a nitrate-containing and carbon-free medium, no NiR appeared suggesting that a carbon source was required to eliminate the repression exerted by ammonium (Arizmendi 1989). When nitrate-grown cells were incubated for 24 h in a carbon-free medium and then transferred to a carbon-containing medium, the observed increase of NiR depended on *de novo* protein synthesis since no increase in the enzyme level was detected in the presence of chloramphenicol. When nitrate-grown cells were transferred to darkness the observed effects on growth and NiR activity were similar to those observed in the carbon-free medium. Again the recovery of NiR levels in the non-illuminated culture was dependent on protein synthesis (Arizmendi 1989).

Nitrate-grown cells also showed detectable levels of both GS and GDH activities (Martínez-Bilbao et al. 1987). Nitrate-grown cells possessed higher GS levels than ammonium-grown cells, whereas the GDH level was higher in ammonium-grown cells and depended on the ammonium concentration of the medium. When cells of *P. laminosum* grown in nitrate were transferred to media containing only ammonium or supplemented with nitrate, a decrease of GS and an increase of GDH activities occurred in both cases. Conversely, when ammonium-grown cells were transferred to a nitrate-containing medium, an increase of the specific activity of GS (dependent on *de novo* synthesis of protein) accompanied by a decrease of that of GDH was observed. The effect of nitrate could again be rapidly abolished by addition of ammonium. Once more the regulatory effect of this ion prevailed over that of nitrate (Martínez-Bilbao et al. 1987). Addition of MSX, a specific inhibitor of GS, caused a similar effect on GDH levels as when transferred to an ammonium-containing medium. Although the cells maintained a healthy appearance for only 3 - 4 days after addition of MSX, this fact could be explained by lack of L-glutamine synthesis (Martínez-Bilbao et al. 1987). These data suggest the existence of an alternative pathway to the GS/GOGAT cycle for ammonium assimilation in *P. laminosum*.

When a culture of cells grown with nitrate or ammonium was transferred to a fresh nitrogen-free medium, an important increase of the specific activities of NR, NiR, GS and (less apparently) GDH occurred during the first days of nitrogen starvation. The level of NR reached its maximum after 3 to 4 days of nitrogen starvation when the cultures already showed a yellowish tinge. The observed increase in specific NR activity was due to a net increase in cellular enzyme content (2-fold with respect to nitrate-grown controls) in combination with a significant decrease (to about 1/10) of the cytosolic protein concentration (Fresnedo et al. 1987). However, the increase of NiR specific activity (about 6-fold) was due to a decrease in protein content, because the enzyme was synthesized similarly to non-NDC (Arizmendi et al. 1987). The specific activity of GS in NDC increased during the first 80 h of starvation. After 96 h a rapid decrease occurred, probably due to irreversible cell damage by the prolonged nitrogen deficiency. The specific activity changes of GDH were less apparent and a maximum was attained after 50 h of nitrogen starvation (Martínez-Bilbao et al. 1987).

Some Properties of the Purified Enzymes

NR (ferredoxin:nitrate oxidoreductase) was purified from nitrogen-deficient cultures. The purified enzyme, a polypeptide of about 80 kDa, was stable mainly at alkaline pH (10.5) (Blanco and Serra, unpublished data).

NiR (ferredoxin:nitrite oxidoreductase, EC 1.7.7.1) was purified from nitrate-grown cells. The enzyme also consisted of a single polypeptide chain of 54 kDa and was inhibited by cyanide and the thiol reagents p-hydroxymercuribenzoate and N-ethylmaleimide (Arizmendi and Serra, unpublished data).

GS (L-glutamate:ammonia ligase, ADP forming, EC 6.3.1.2) was purified to electrophoretic homogeneity in the absence of stabilizing ligands from nitrate-grown cells (Blanco et al. 1989). The native enzyme (602 kDa) was composed of 12 identical subunits of 52 kDa. The biosynthetic GS activity could be significantly inhibited *in vitro* by some amino acids (Gly, L-Asp, L-Ser and L-Ala), nucleotides (ADP and AMP) and by MSX. The enzyme was not activated by chemically reduced *Anabaena* thioredoxin and, in contrast to other cyanobycterial GS, neither a cumulative nor a cooperative in-hibition effect was caused by mixtures of inhibitory amino acids.

NADPH-dependent GDH (L-glutamate:NADP oxidoreductase, EC 1.4.1.4) was purified from ammonium-grown cells (Martínez-Bilbao et al. 1988). The purified enzyme (native M_r 280,000) showed no amino transferase activity and catalyzed the amination of 2-oxoglutarate preferentially to the reverse catabolic reaction. The enzyme showed hyperbolic kinetics with its substrates 2-oxoglutarate and NADPH, but negative cooperativity ($h = 0.89$ and $[S]_{0.5} = 18$ mM) was detected for ammonium.

Concluding Remarks

In *P. laminosum* the contribution of GDH in ammonium assimilation could be particularly relevant under certain conditions in which GS activity is decreased (i.e. high ammonium concentration or presence of MSX). Thus, GDH may provide an alternative pathway to the GS/GOGAT cycle for ammonium assimilation.

From data obtained in *A. nidulans*, Lara et al. (1987) proposed that nitrate utilization in cyanobacteria is regulated at the level of nitrate transport through the concerted action of ammonium assimilation and CO_2 fixation. Nitrate and nitrite uptake in *P. laminosum* depended on the presence of ammonium in the external medium and on the availability of intracellular carbon skeletons as acceptors. Ammonium also completely repressed NR and NiR activities, and increased the level of GDH and decreased that of GS. The effect of ammonium always prevailed over that of nitrate or nitrite. However, the regulatory effect of ammonium on the uptake of these ions was apparently lacking in NCDC.

It is also remarkable that the specific activities of NR, NiR, GS and, to a minor extent, GDH are higher in NDC than in non-NDC. The increase in NR specific activity of NDC is more drastic because NR is the only enzyme whose cellular levels are higher in nitrogen starvation conditions.

Acknowledgments. This work was supported by grants from Comisión Asesora de Investigación Científica y Técnica (928/84) and the Basque country Government (X-86.049). FB, MMB, OF and AA were recipients of scholarships from the Spanish Ministry of Education and Science or the Basque Country Government.

References

Arizmendi JM (1989) Regulation, purification and properties of nitrite reductase from the cyanobacterium *Phormidium laminosum* (in Spanish). PhD Thesis, Universidad del País Vasco, Bilbao

Arizmendi JM, Fresnedo O, Martínez-Bilbao M, Alaña A, Serra JL (1987) Inorganic nitrogen assimilation in the non N_2-fixing cyanobacterium *Phormidium laminosum*. II. Effect of the nitrogen source on the nitrite reductase levels. Physiol Plant 70: 703-707

Blanco F, Alaña A, Llama MJ, Serra JL (1989) Purification and properties of the glutamine synthetase from the non N_2-fixing cyanobacterium *Phormidium laminosum*. J Bacteriol 171: 1158-1165

Candau P (1979) Purification and properties of ferredoxin-nitrate reductase from the cyanobacterium *Anacystis nidulans* (in Spanish). PhD Thesis, Universidad de Sevilla

Florencio FJ, Marqués S, Candau P (1987) Identification and characterization of a glutamate dehydrogenase in the unicellular cyanobacterium *Synechocystis* PCC 6803. FEBS Lett 223: 37-41

Flores E, Guerrero MG, Losada M (1983) Photosynthetic nature of nitrate uptake and reduction in the cyanobacterium *Anacystis nidulans*. Biochim Biophys Acta 722: 408-416

Flores E, Herrero A, Guerrero MG (1987) Nitrite uptake and its regulation in the cyanobacterium *Anacystis nidulans*. Biochim Biophys Acta 896: 103-108

Fresnedo O, Blanco F, Serra JL (1987) Enhancement of the cellular level of nitrate reductase and changes of photosynthetic pigment composition associated with nitrogen starvation in the cyanobacterium *Phormidium laminosum*. Abstracts VIth Symposium on Phototrophic Prokaryotes, Noordwijkerhout, p 213

Guerrero MG, Lara C, (1987) Assimilation of inorganic nitrogen. In: Fay P, Van Baalen C (eds) The Cyanobacteria. Elsevier, Amsterdam, p 163

Lara C, Romero JM, Guerrero MG (1987) Regulated nitrate transport in the cyanobacterium *Anacystis nidulans*. J Bacteriol 169: 4376-4378

Madueño F, Flores E, Guerrero MG (1987) Competition between nitrate and nitrite uptake in the cyanobacterium *Anacystis nidulans*. Biochim Biophys Acta 896: 109-112

Madueño F, Vega-Palas MA, Flores E, Herrero A (1988) A cytoplasmic-membrane protein repressible by ammonium in *Synechococcus* R2: Altered expression in nitrate-assimilation mutants. FEBS Lett 239: 289-291

Martínez A, Llama MJ, Alaña A, Serra JL (1989) Sustained photoproduction of ammonia from nitrate or nitrite by permeabilized cells of the cyanobacterium *Phormidium laminosum*. J Photochem Photobiol 3: 269-279

Martínez-Bilbao M (1987) Purification, properties and regulation of glutamate dehydrogenase from the cyanobacterium *Phormidium laminosum* (in Spanish). PhD Thesis, Universidad del País Vasco, Bilbao

Martínez-Bilbao M, Alaña A, Arizmendi JM, Serra JL (1987) Inorganic nitrogen assimilation in the non N_2-fixing cyanobacterium *Phormidium laminosum*. I. Cellular levels of glutamine synthetase and NADPH-dependent glutamate dehydrogenase. Physiol Plant 70: 697-702

Martínez-Bilbao M, Martínez A, Urkijo I, Llama MJ, Serra JL (1988) Induction, isolation and some properties of the NADPH-dependent glutamate dehydrogenase from the non-heterocystous cyanobacterium *Phormidium laminosum*. J Bacteriol 170: 4897-4902

Mikami B, Ida S (1984) Purification and properties of ferredoxin-nitrate reductase from the cyanobacterium *Plectonema boryanum*. Biochim Biophys Acta 791: 294-304

Sawa Y, Ochiai H, Yoshida K, Tanizawa K, Tanaka H, Soda K (1988) Glutamine synthetase from a cyanobacterium, *Phormidium lapideum*: Purification, characterization and comparison with other cyanobacterial enzymes. J Biochem 104: 917-923

Serra JL, Alaña A, Blanco F, Llama MJ (1985) Purification of glutamine synthetase from the non N_2-fixing cyanobacterium *Phormidium laminosum*. Biochem Soc Transact 13: 384-385

Sivac MN, Lara C, Romero JM, Rodríguez R, Guerrero MG (1989) Relationship between a 47-kDa cytoplasmic membrane polypeptide and nitrate transport in *Anacystis nidulans*. Biochem Biophys Res Commun 158: 257-262

Yabuki Y, Mori E, Tamura G (1985) Nitrite reductase in the cyanobacterium *Spirulina platensis*. Agric Biol Chem 49: 3061-3062

BIOCHEMISTRY, REGULATION AND GENETICS OF NITRITE REDUCTION IN BARLEY

J.L. Wray [1], S.-M. Ip [2], E. Duncanson [1], A. F. Gilkes [1] and D.W. Kirk [1]

Introduction

The nitrite reductase (NiR) enzymes catalyse the six-electron reduction of nitrite to ammonium in the third step of the nitrate assimilation pathway (reviewed in Guerrero et al. 1981; Wray 1989). In higher plants the leaf enzyme (ferredoxin:nitrite oxidoreductase, EC 1.7.7.1) is located in the chloroplast and uses reduced ferredoxin as electron donor for nitrite reduction *in vivo*. The root enzyme is located in plastids (Dalling et al. 1972) and like the leaf enzyme can use the non-physiological electron donor, reduced methyl viologen. The nature of the physiological electron donor is obscure but may be a ferredoxin-like protein (Ninomiya and Sato 1984; Suzuki et al. 1985) which derives electrons from the pentose phosphate pathway (Bowsher et al. 1989). Both nitrate and light have been implicated in the regulation of NiR synthesis (reviewed by Rajasekhar and Oelmüller 1987) with the light requirement being mediated by phytochrome (Rajasekhar and Mohr 1986 a) and/or blue light (Rajasekhar and Sopory 1985). Recently, cloning of the NiR structural gene from spinach (Back et al. 1988) and maize (Lahners et al. 1988) has been accomplished and preliminary studies on their regulation have been reported.

Although in barley the nitrate reduction step is particularly well characterized in terms of biochemistry (Wray and Filner 1970; Kuo et al. 1980; Campbell and Wray 1983), regulation (Somers et al. 1983) and genetics and molecular biology (reviewed in Wray 1986; Kleinhofs et al. 1989) very little is known about these aspects of the nitrite reduction step. Here we review our studies on the biochemistry, regulation and genetics of nitrite reduction in barley.

Purification and Characterization of Nitrite Reductase

Ferredoxin:nitrite oxidoreductase (EC 1.7.7.1) was purified 1200 fold to electrophoretic homogeneity with a specific activity of 102 units/mg protein and a yield of 15.5% from leaves of barley with a procedure involving ammonium sulphate fractionation, Biogel A1.5m gel filtration and chromatography on DEAE-cellulose, DEAE-Sephadex,

---

[1] Plant Molecular Genetics Unit, Sir Harold Mitchell Building, University of St. Andrews, St. Andrews, Fife, KY16 9TH, UK

[2] Present address: Department of Applied Biology and Chemical Technology, Hong Kong Polytechnic, Hung Hom, Kowloon, Hong Kong

ferredoxin-Sepharose and hydroxylapatite (S.M. Ip and J.L. Wray, unpublished). Sedimentation equilibrium analysis, and electrophoresis in denaturing polyacrylamide gels, showed that the enzyme was a monomer with a M_r = ca. 63,000, like most nitrite reductases examined (Hucklesby et al. 1976; Vega and Kamin 1977; Serra et al. 1982; Bowsher et al. 1988).

The amino acid composition was shown to be similar to that reported for the enzyme from spinach (Ida and Mikami 1986) and marrow (Hucklesby et al. 1976). The isoelectric point of the enzyme is 4.7, similar to the value of 4.3 reported for the barley root enzyme (Ida et al. 1974) and consistent with the observation that the protein contains more acidic, than basic, amino acids. A total of 10 half-cystine residues/mol was determined and presumably four of these participate in the formation of the iron-sulphur cluster revealed by EPR spectroscopic analysis (below).

The visible absorption spectrum was similar to that reported for other NiR proteins and indicative of the presence of sirohaem (Murphy et al. 1974; Vega and Kamin 1977) with absorption maxima at 280, 392 (Soret), 574 (α) and 693 nm with shoulders at 291, 530 and 638 nm. Absorbance ratios at A_{280}/A_{392} and A_{392}/A_{574} were 2.24 and 3.72 respectively. Extinction coefficients at 280, 392 and 574 nm were 7.48 x 10^4 M cm^{-1}, 3.35 x 10^4 M cm^{-1} and 0.9 x 10^4 M cm^{-1} respectively.

The availability of purified protein in mg quantities allowed us to perform the first EPR and redox potentiometric studies of a monocotyledonous NiR. EPR spectroscopy of the oxidized enzyme showed a rhombic high-spin signal with $g_z = 6.8$ and $g_y = 5.2$. A low-spin ferric-haem signal was also observed with resonances at $g_z = 2.65$ and $g_y = 2.30$. When nitrite is added the high-spin signal disappears and the low-spin signal broadens. Both the high- and low-spin ferric-haem signals are attributed to sirohaem (Aparicio et al. 1975). On reduction of the enzyme a ferredoxin-type 4Fe-4S centre EPR signal is observed and upon reduction in the presence of nitrite an axial signal of the ferrous haem. NO-type is observed as with other NiR proteins (Aparicio et al. 1975; Cammack et al. 1978). Mid-point oxidation-reduction potential titrations of the sirohaem and the 4Fe-4S centre were performed. A best fit (n = 1) value of +80 mV at pH 7.5 was obtained for the high-spin sirohaem signal, rather higher than the published values of -50 mV and -120 mV for the spinach (Stoller et al. 1977) and marrow (Cammack et al. 1978) enzymes respectively. For the 4Fe-4S centre a value of -517 mV was determined, similar to the values of -550 (Stoller et al. 1978) and -570 mV (Cammack et al. 1978) reported for the mid-point potential of the 4Fe-4S centre of the spinach and marrow enzymes respectively.

Regulation of Nitrite Reductase Synthesis

Polyclonal antibodies, raised in rabbits against the purified barley-leaf NiR by standard procedures, specifically recognized the barley-leaf enzyme as determined by a variety of immunological techniques including tandem-crossed immunoelectrophoresis and Western blotting analysis (S.-M. Ip and J.L. Wray, unpublished).

Activity staining after non-denaturing polyacrylamide gel electrophoresis showed clearly that both leaf and roots possessed a single functional NiR species. The development of NiR activity and of NiR-cross-reacting material (NiR-CRM) was followed in tissue in response to availability of nitrate, light, ammonium and glutamine. Experiments were performed on fully-greened 6-day-old plants that had been grown in the light (115 $\mu M\ m^{-2}\ s^{-1}$) in the rigorous absence of exogenous nitrate (E. Duncanson and J.L. Wray, unpublished).

Leaves of plants grown in the absence of added nitrate in either light or dark and which contained no detectable tissue nitrate, have low levels of *in vitro* NiR activity and NiR-CRM. This suggests low-level expression of the barley NiR structural gene under these conditions, consistent with the results of Northern-blotting experiments in spinach (Back et al. 1988) and maize (Lahners et al. 1988). Addition of nitrate (25 mM) in the light or dark causes an increase in tissue nitrate to the same extent but NiR-activity and NiR-CRM increase above the low basal level only in the leaves of plants maintained in the light. These results suggest that the increase in NiR activity is due to the *de novo* synthesis of enzyme molecules and that this synthesis requires the simultaneous presence of nitrate and light. This requirement has also been reported for NiR synthesis in leaves of wheat (Small and Gray 1984), pea (Gupta and Beevers 1985) and rice (Ogawa and Ida 1987).

Unlike previous investigations we also measured tissue nitrate levels. We are thus able to say that inability to synthesize NiR molecules in leaves of plants treated with nitrate in the dark is not due to exclusion of nitrate. In the absence of light accumulated nitrate is not able to act as an inducer of NiR synthesis. The reason for this is unclear. A light signal may be required for nitrate to act as a transcriptional inducer. Alternatively light may regulate movement of nitrate to the site of NiR gene expression (the nucleus). Most of the leaf nitrate is stored in the vacuole in barley (Granstedt and Huffaker 1982) and movement of nitrate out of the vacuole is regulated by light (Aslam et al. 1976).

We looked at the possible involvement of ammonium and glutamine (as end products of nitrate assimilation) in regulation of NiR synthesis in barley leaves. Ammonium ions (25 mM), given together with nitrate, have no effect on either tissue nitrate content or on the development of NiR-activity or NiR-CRM. This is in agreement with observations in maize, where use of an NiR cDNA probe in Northern-blotting experiments demonstrates that expression of the NiR structural gene in response to nitrate is unaffected by the presence of ammonium ions (Rothstein et al. 1989), but contrasts with reports on the development of NiR activity in *Lemna* (Joy 1966) and dark-grown mustard cotyledons (Rajasekhar and Mohr 1986 b).

In contrast to ammonium, when glutamine (25 mM) was given together with nitrate the development of NiR activity and NiR-CRM was decreased, although tissue nitrate content was unchanged, suggesting a regulatory role for glutamine in the synthesis of NiR. No previous studies on the role of glutamine in NiR synthesis in higher plants seems to have been performed. However glutamine levels are known to regulate expression of the nitrate reductase (NR) structural gene in *Neurospora crassa* (Marzluf and Fu 1989) and the elevated levels of NR activity which develop in tungstate-treated plants (Wray and Filner 1970) have been ascribed, in tobacco, to the decreased production of end-products as a consequence of the inhibition of nitrate assimilation (Deng et al. 1989).

Regulation of NiR synthesis in the roots of higher plants has not been studied in any detail. Antibody to the leaf enzyme cross-reacted on Western blots with a single, nitrate-inducible, immunoreactive species of 63 kD in extracts of root tissue. The leaf and root enzymes thus appear to have antigenic determinants in common, as reported for the spinach enzymes (Hirasawa et al. 1984).

In barley we find that roots of plants grown in the absence of added nitrate in either light or dark contain no detectable nitrate and extremely low to zero levels of NiR-activity and NiR-CRM. This is in agreement with results for maize where expression of the NiR structural gene in roots of light-grown, nitrate-untreated plants was at a very low level as determined by Northern blot analysis (Lahners et al. 1988). Of some significance is the observation that, unlike the leaf, development of NiR-activity and NiR-CRM occurs in barley roots in response to nitrate, even in the absence of light suggesting perhaps that the system of *cis*-acting sequence(s) on the NiR structural gene and *trans*-acting protein(s) which mediate their interaction with nitrate and light operates differently in leaf and root.

Genetics of Nitrite Reduction

Little information is available on the genetics of nitrite reduction in higher plants. Inheritance studies performed by Heath-Pagliuso et al. (1984) suggest that the enzyme is nuclear-encoded but mutant analysis of the type carried out on the nitrate reduction step (reviewed in Wray 1986) has not been explored in any detail so far. Variant cell lines defective in nitrite reduction have been described by Gilissen et al. (1985) in *Haplopappus gracilis* but attempts at plant regeneration were unsuccessful and the molecular basis of the defect is unknown. Whole-plant mutants defective in nitrite reduction would allow analysis of the number and types of gene which determine this step in nitrate assimilation, allow studies of the consequences of such mutations in the intact, differentiated organism and complement studies performed on the nitrate reduction step.

A wide variety of mutations might be expected, *a priori*, to block nitrite reduction, e.g., within the NiR structural gene affecting either catalytic function or the transit sequence; in processing of the precursor protein; in recognition of the chloroplast envelope; in electron donation to nitrite reductase; in nitrite transport into the chloroplast, if carrier-mediated; and in regulation. However a positive selection scheme of the type utilized so successfully in the isolation of mutants defective in nitrate reduction (chlorate resistance) does not seem to be available.

We have devised a selection screen based on the rationale that mutants defective in nitrite reduction should be identifiable by virtue of the fact that they would likely accumulate nitrite after nitrate treatment (Duncanson et al. 1987).

Using this procedure we have screened nearly 150,000 M_2 seedlings (over 32,500 M_1 ears) from eight barley cultivars and have identified fifteen whole-plant nitrate accumulators (E. Duncanson, A. Gilkes, D.W. Kirk and J.L. Wray, unpublished). Some had low levels of accumulated nitrite (ca. 0.3 μmol/g) whilst others had extremely high levels (ca. 200 μmol/g). Wild-type plants do not accumulate nitrite under screen conditions. The selected plants define nine independent M_1 ears which carry one or more nitrite-accumulating M_2 plants. All M_2 plants examined (except one which is not considered further here and may represent a false-positive) lack detectable leaf NiR-CRM as determined by Western blotting and have very low to zero leaf *in vitro* NiR activity. They retain leaf nitrate reductase activity and the ability to take up nitrate. The M_2 plants were maintained in a hydroponic culture system with glutamine (1 mM) as sole N source. Homozygous M_3 seed is available from some of the selections and M_3 plants have been examined to determine whether the variant phenotypes are inherited.

All M_3 plants examined also lack detectable leaf NiR-CRM and have a very low to zero leaf *in vitro* NiR activity but retain leaf nitrate reductase activity and the ability to take up nitrate. They also lack root NiR-CRM. However, very surprisingly, they do not accumulate nitrite (or do so very weakly) under the conditions from which their M_2 parents were selected. We currently have no explanation for this.

On the basis of the evidence that is so far available it appears that these lines carry a defect in the formation of the NiR protein. Loss of both leaf and root NiR-CRM in the M_3 progeny of the primary M_2 selections indicates that a common locus determines NiR formation in leaf and root. The most likely explanation is that these selections are altered in the NiR structural gene. However we cannot, at present, dicount the possibility that (some of) the selections are altered at a regulatory locus. Further biochemical and genetical analysis is in progress.

Acknowledgments. This work was supported by grants to JLW from the Agricultural and Food Research Council and the Science and Engineering Research Council of the United Kingdom. EPR and redox potentiometric studies were performed in collaboration with W.J. Ingledew.

References

Aparicio PJ, Knaff DB, Malkin R (1975) The role of an iron-sulfur center and siroheme in spinach nitrite reductase. Arch Biochem Biophys 169: 102-107
Aslam M, Oaks A, Huffaker RC (1976) Effect of light and glucose on the induction of nitrate reductase and on the distribution of nitrate in etiolated barley leaves. Plant Physiol 58: 588-591

Back E, Burkhardt W, Moyer M, Privalle L, Rothstein S (1988) Isolation of cDNA clones coding for spinach nitrite reductase: Complete sequence and nitrate induction. Mol Gen Genet 212: 20-26
Bowsher CG, Emes MJ, Cammack R, Hucklesby DP (1988) Purification and properties of nitrite reductase from roots of pea (*Pisum sativum* cv Meteor). Planta 175: 334-340
Bowsher CG, Hucklesby DP, Emes MJ (1989) Nitrite reduction and carbohydrate metabolism in plastids purified from roots of *Pisum sativum* L. Planta 177: 359-366
Cammack R, Hucklesby DP, Hewitt EJ (1978) Electron-paramagnetic-resonance studies of the mechanism of leaf nitrite reductase. Biochem J 171: 519-526
Campbell JMcA, Wray JL (1983) Purification of barley nitrate reductase and demonstration of nicked subunits. Phytochemistry 22: 2375-2382
Dalling MJ, Tolbert NE, Hageman RH (1972) Intracellular location of nitrate reductase and nitrite reductase. Biochim Biophys Acta 283: 505-512
Deng M, Chérel I, Daniel-Vedele F, Moureaux T, Pouteau S, Caboche M (1989) Regulation of the expression of the nitrate reductase structural gene in higher plants. In: Abstracts Adv Course Inorg Nitrogen Metabolism, Seiano, p 109
Duncanson E, Wilshin A, Holyoake A, Kirk DW, Wray JL (1987) Genetic analysis of nitrite reduction in barley. In: Abstracts AFRC Meeting on Plant and Soil Nitrogen Metabolism, Sussex, UK P 22
Gilissen LJW, Barneix AJ, van Staveren M, Breteler H (1985) Variant cell lines of *Haplopappus gracilis* with disturbed activities of nitrate reductase and nitrite reductase. Plant Physiol 78: 658-660
Granstedt RC, Huffaker RC (1982) Identification of the leaf vacuole as a major nitrate storage pool. Plant Physiol 70: 410-413
Guerrero MG, Vega JM, Losada M (1981) The assimilatory nitrate-reducing system and its regulation. Annu Rev Plant Physiol 32: 169-204
Gupta SC, Beevers L (1985) Regulation of synthesis of nitrite reductase in pea leaves: *In vivo* and *in vitro* studies. Planta 166: 89-95
Heath-Pagliuso S, Huffaker RC, Allard RW (1984) Inheritance of nitrite reductase and regulation of nitrate reductase, nitrite reductase and glutamine synthetase isoenzymes. Plant Physiol 76: 353-358
Hirasawa M, Fukushima K, Tamura G, Knaff DM (1984) Immunochemical characterization of nitrite reductases from spinach leaves, spinach roots and other higher plants. Biochim Biophys Acta 791: 145-154
Hucklesby DP, James PM, Banwell MG, Hewitt EJ (1976) Properties of nitrite reductase from *Cucurbita pepo*. Phytochemistry 15: 599-603
Ida S, Mikami B (1986) Spinach ferredoxin-nitrite reductase: A purification procedure and characterization of chemical properties. Biochim Biophys Acta 871: 167-176
Ida S, Mori E, Morita Y (1974) Purification, stabilisation and characterisation of nitrite reductase from barley roots. Planta 121: 213-224
Joy KW (1969) Nitrogen metabolism of *Lemna minor* II. Enzymes of nitrate assimilation and some aspects of their regulation. Plant Physiol 44: 849-853
Kleinhofs A, Warner RL, Lawrence JM, Melzer JM, Jeter JM, Kudrna DA (1989) Molecular genetics of nitrate reductase in barley. In: Wray JL, Kinghorn JR (eds) Molecular and Genetic Aspects of Nitrate Assimilation. Oxford Univ Press, Oxford, p 197
Kuo TM, Kleinhofs A, Warner RL (1980) Purification and partial characterization of nitrate reductase from barley leaves. Plant Sci Lett 17: 371-381
Lahners K, Kramer V, Back E, Privalle L, Rothstein S (1988) Molecular cloning of complementary DNA encoding maize nitrite reductase. Molecular analysis and nitrate induction. Plant Physiol 88: 741-746
Marzluf GA, Fu Y-H (1989) Genetics, regulation and molecular studies of nitrate assimilation in *Neurospora crassa*. In: Wray JL, Kinghorn JR (eds) Molecular and Genetic Aspects of Nitrate Assimilation. Oxford Univ Press, Oxford, p 314
Murphy MJ, Siegel LM, Tove SR, Kamin H (1974) Siroheme: A prosthetic group participating in six-electron reduction reactions catalyzed by both sulfite and nitrite reductase. Proc Natl Acad Sci USA 71: 612-616
Ninomiya Y, Sato S (1984) A ferredoxin-like electron carrier from non-green cultured tobacco cells. Plant Cell Physiol 25: 453-458
Ogawa M, Ida S (1987) Biosynthesis of ferredoxin-nitrite reductase in rice seedlings. Plant Cell Physiol 28: 1501-1508
Rajasekhar VK, Sopory SK (1985) The blue-light effect and its interaction with phytochrome in the control of nitrite reductase activity in *Sorghum bicolor* Wild. New Phytol 101: 251-258

Rajasekhar VK, Mohr H (1986 a) Appearance of nitrate reductase in cotyledons of the mustard (*Sinapis alba* L.) seedlings as affected by nitrate, phytochrome and photooxidative damage of plastids. Planta 168: 369-376

Rajasekhar VK, Mohr H (1986 b) Effect of ammonium and nitrate on growth and appearance of nitrate reductase and nitrite reductase in dark- and light-grown mustard seedlings. Planta 169: 594-599

Rajasekhar VK, Oelmüller R (1987) Regulation of induction of nitrate reductase and nitrite reductase in higher plants. Physiol Plant 71: 517-521

Rothstein S, Lahners K, Kramer V, Privalle L, Back E (1989) Expression of corn nitrite reductase in hydroponically grown seedlings and cell suspensions. In: Abstracts Adv Course Inorg Nitrogen Metabolism, Seiano, p 103

Serra JL, Ibarlucea JM, Arizmendi JM, Llama JM (1982) Purification and properties of the assimilatory nitrite reductase from barley (*Hordeum vulgare*) leaves. Biochem J 201: 167-170

Small IS, Gray JC (1984) Synthesis of wheat leaf nitrite reductase *de novo* following induction with nitrate and light. Europ J Biochem 145: 291-297

Somers DA, Kuo T-M, Kleinhofs A, Warner RL, Oaks A (1983) Synthesis and degradation of barley nitrate reductase. Plant Physiol 72: 949-952

Stoller ML, Malkin R, Knaff DM (1977) Oxidation-reduction properties of photosynthetic nitrite reductase. FEBS Lett 81: 271-274

Suzuki A, Oaks A, Jacquot J-P, Vidal J, Gadal P (1985) An electron transport system in maize roots for reactions of glutamate synthase and nitrite reductase. Plant Physiol 78: 374-378

Vega JM, Kamin H (1977) Spinach nitrite reductase: Purification and properties of a siroheme-containing iron-sulfur enzyme. J Biol Chem 252: 896-909

Wray JL (1986) The molecular genetics of higher plant nitrate assimilation. In: Blonstein A, King PJ (eds) A Genetic Approach to Plant Biochemistry. Springer, Vienna, Austria, p 101

Wray JL (1989) Molecular and genetic aspects of nitrite reduction. In: Wray JL, Kinghorn JR (eds) Molecular and Genetic Aspects of Nitrate Assimilation. Oxford Univ Press, Oxford, p 245

Wray JL, Filner P (1970) Structural and functional relationships of enzyme activities induced by nitrate in barley. Biochem J 119: 715-725

NITRITE REDUCTION IN THE ROOTS AND LEAVES OF *PISUM SATIVUM*

D.P. Hucklesby [1], M. J. Emes [2], C.G. Bowsher [3] and R. Cammack [4]

Introduction

The location of the nitrate/nitrite reducing mechanisms in plants is distributed between the root and the shoot in proportions which vary according to species. At one extreme, nitrate may be reduced almost entirely in the shoot, e.g. *Beta* and *Chenopodium*, while at the other extreme most reduction occurs in the root system as in some woody species. In the majority of species, however, including many important crop plants, both root and shoot participate in reduction. This subject has been discussed in some detail (Pate 1973). The mechanisms of function of both nitrate and nitrite reduction are better understood in leaves than roots. Root nitrogen assimilation has been reviewed by Oaks and Hirel (1985). Shortly after methods became available for the extraction and assay of nitrite reductase (Hageman et al. 1962), the enzyme was shown to be active in tomato roots (Sanderson and Cocking 1964). A few years previously an interesting study by Butt and Beevers (1961), approaching the subject from the direction of carbon-nitrogen relations had shown that glucose-6-phosphate could increase the rate of disappearance of nitrite in contact with preparations from maize roots. The association of both nitrite reductase and the enzymes of the pentose-phosphate pathway (PPP) with root plastids, has been developed in a number of studies in the intervening years (see, e.g. Emes and Fowler 1979 a,b). The present paper describes a dual approach to the subject of nitrite reduction by roots, in which studies were made of the purified nitrite reductase (NIR) enzyme and of carbohydrate-nitrite relationships as found in root plastids. Pea was chosen for this work as a species in which nitrate was thought to be reduced in roots, and from which well authenticated root plastids can be extracted. A large number of labelling techniques was used in this study. Details of these may be found in Bowsher et al. (1988, 1989).

Results and Discussion

Nitrite reductase from pea roots

Nitrite reductase prepared from various non-photosynthetic tissues (Dalling et al. 1973;

[1] Department of Molecular Biology, Long Ashton Research Station, Long Ashton, Bristol BS18 9AF, UK. [2] School of Biological Sciences, University of Manchester, Manchester M13 9PL, UK. [3] Dept. of Molecular Studies, University of Guelph, Guelph, Ontario, Canada. [4] Kings College, University of London, Kensington Campus, London, UK

Ida et al. 1974: Nagaoka et al. 1984) share many properties with leaf NIR. The current work with pea roots confirms and extends this conclusion, particularly in respect of EPR characteristics.

Activity of NIR in both roots and leaves of pea was dependent upon presence of nitrate in the nutrient solutions. Maximal induction of the enzyme by 8 mM nitrate in plants previously nitrate-starved was reached in 3 - 5 days. After removal of nitrate from the culture media of fully nitrate-induced plants, a steady decline to about 40% activity occurred in 5 days. The specific activity of nitrite reductase in 5 - 10 day old plants was approximately 8-fold higher in roots than in leaves. Later, the activities became more nearly equal, maximizing in roots between 20 and 30 days. Plants of this age were used for enzyme extraction after accumulation of sufficient material in storage at -18°C. The enzyme was purified to a specific activity of 372 nkats mg^{-1} protein (Bowsher et al. 1988). The purified enzyme ran as a single band on native and SDS polyacrylamide gels. The absorption spectrum showed maxima at 278, 389, 530 and 574 nm with a ratio E_{389}: E_{278} = 0.4. This spectrum shows the characteristics described for NIR enzymes from leaves of several species. The low wavelength of the Soret band, and its low absorbance ratio with reference to the absorbance maximum at 278 nm, appear to be typical of siroheme-containing enzymes. The electron paramagnetic resonance spectrum (EPR) of the root NIR in its oxidized state as prepared showed signals at g = 6.9, 5.0 and 1.95. On reduction with dithionite in the presence of KCN, a wholly different spectrum, characteristic of iron-sulphur prosthetic groups was obtained, with a principal signal at g = 1.94. During turnover of the enzyme with nitrite and dithionite, a third type of signal appeared, recognizable as a nitrosyl-siroheme signal. This originates from the nitrosyl radical (the one electron reduction product of nitrite, oxidation state +2) in association with the Fe atom of the siroheme to which the nitrosyl group is a ligand. Hyperfine structure as a triplet, representing interaction between the nuclei of the N atom of the nitrosyl ligand and the Fe of the siroheme, was visible. The three types of signal are considered to be virtually identical with those obtained from the *Cucurbita* leaf NIR (Cammack et al. 1978) and strongly resemble those described for spinach leaf (Aparicio et al. 1975). These EPR spectra appear to provide definitive proof of the siroheme / iron sulphur nature of the pea root NIR.

Immunological studies of the pea root and pea leaf nitrite reductases were made by cross-reaction tests against antibody raised in rabbits against NIR prepared from *Cucurbita* leaves. On the basis of Ouchterlony double immuno-diffusion tests, the enzymes from the two sources were indistinguishable.

Carbohydrate-nitrite relationships

Plastids prepared from the roots of nitrate-grown pea plants were 75 - 95% intact as estimated from studies of enzyme latency. Incubation of such plastids with glucose-6-

phosphate and nitrite resulted in a loss of nitrite at rates of 6 nmol min^{-1} mg^{-1} protein. Activity was negligible on rupturing the organelles. Similar rates of nitrite disappearance were obtained when glucose-6-phosphate was replaced by fructose-6-phosphate or ribose-5-phosphate. Rates of about one half of these were observed with glucose-1-phosphate, while 6-phosphogluconate, fructose-1,6-bisphosphate, glyceraldehyde-3-phosphate, dihydroxyacetone phosphate, fructose and glucose were ineffective. Ability of a substrate to support nitrite reduction may reasonably be supposed to depend upon both its ability to penetrate the plastid membrane and upon the presence in the plastid of enzymes able to convert the substrate to glucose 6-phosphate. The K_m value (0.66 mM) for glucose-6-phosphate in nitrite reduction was almost identical with the K_m (0.68 mM) for glucose-6-phosphate in glucose-6-phosphate dehydrogenation. Similarly, the K_m (0.32 mM) for fructose-6-phosphate when reducing nitrite, was close to that (0.35 mM) of hexose-phosphate isomerase.

To observe the response of the pentose-phosphate pathway (PPP) to nitrite reduction in purified organelles, studies were made of the release of $^{14}CO_2$ from glucose-6-phosphate labelled in specific carbon atoms. In the absence of nitrite, $^{14}CO_2$ was released by plastids from carbons 6, 1, 2, 3 = 4 in order of increasing yield of CO_2. Addition of nitrite caused a marked increase in $^{14}CO_2$ release only from the carbon-1 position. Chloride and nitrate were ineffective in increasing the release of $^{14}CO_2$ from carbon-1. The stimulation was not observed if the plastids were prepared from plants grown without nitrate and therefore lacking NIR. The K_m value for nitrite-dependent release of CO_2 from carbon-1 was 0.85 mM, which is of the same order as the K_m for extracted nitrite reductase assayed with dithionite and methyl viologen (1.08 mM). The data clearly support the view that it is only the PPP which is stimulated to meet the demand for reductant for nitrite reduction in the root plastids.

Electron carrier studies

Reduction of one mol of nitrite to ammonium requires 6 electrons. The oxidation of glucose-6-phosphate to ribulose-5-phosphate produces 2 mol of NADPH, one from the glucose 6-phosphate dehydrogenase (G6Pdh) and one from 6-phosphogluconate dehydrogenase. The expected stoichiometric ratio for nitrite coupled to the PPP would therefore be 1.5. Experiments conducted with glucose-6-phosphate labelled in carbon 1 gave ratios between 1 and 1.33. This lower ratio reflects the complexity of plastid metabolism, and in particular, a recycling of products of the dehydrogenase reactions to unlabelled glucose 6-phosphate with consequent underestimation of the amount of glucose-6-phosphate used.

Leaf nitrite reductase can be coupled to glucose-6-phosphate and $NADP^+$ in a reconstituted system comprising G6Pdh, NADPH-ferredoxin reductase, ferredoxin and NIR. This system is much slower than nitrite reduction *in vivo* in the light, or reduction *in vitro* using systems based on illuminated chloroplasts. A major difficulty with such a system is that electrons must be transferred from NADPH (E_0 = -0.321) to ferredoxin (E_0 = -0.421). This can only occur if the glucose-6-phosphate dehydrogenase maintains the nicotinamide nucleotide in a rather complete state of reduction. A 90% reduction of NADP can be calculated to give approximately 10% reduction of ferredoxin. Nevertheless such a system may operate as a physiological means of nitrite reduction in darkness.

Extracts from maize scutellum - an obligate non-photosynthetic tissue - can also reduce nitrite using a similar enzyme sequence. G6Pdh, NADPH-diaphorase and nitrite reductase are present in the extracts and the system appears to function in the way described for the leaf, but ferredoxin from a leaf source must be added (Hucklesby et al. 1972). This experimental system immediately suggests a possible means for the reduction of nitrite in non-photosynthetic tissues, with, moreover, a rationale for the process to be based in the plastid, where NADPH is available from the PPP. Such a hypothesis requires that the tissue contain a ferredoxin or other electron transport compound, together with an NADPH reductase for its reduction. A ferredoxin-type carrier, operating in this context, might be expected to have a less negative mid-point potential value than that of leaf ferredoxin.

The presence of a carrier, prepared from cultured tobacco cells, has been reported by Ninomiya and Sato (1984). The carrier was shown to replace methyl viologen in the reduction of nitrite by a proplastid-enriched fraction, and was active *in vitro* with dithionite and nitrite reductase. The absorption spectrum was very similar to that of leaf ferredoxin, but the molecular weight (approximately 19,500) appeared to be considerably larger. The cofactor had an EPR spectrum indicating an iron-sulphur group. Detection of a cofactor from young maize roots has been reported by Suzuki et al. (1985). This carrier was recognized by its ability to mediate the reduction of cytochrome *c* from NAD(P)H, but could not be coupled to nitrite reduction using a bispecific NADPH ferredoxin reductase prepared from roots. A ferredoxin from radish roots has also been described (Wada et al. 1986).

In our work with pea roots, extracts were examined for a ferredoxin by using as criterion their ability to support nitrite reduction in the presence of exogenous nitrite reductase with dithionite present to provide electrons. Such a technique enabled the detection of a nitrite reducing cofactor in etiolated wheat leaves. In a further set of experiments, the assumption was made that the root electron carrier might be a ferredoxin sufficiently similar to *Cucurbita* leaf ferredoxin to react with it antigenically. Extracts were therefore run on SDS electrophoresis gels, and the separated products transblotted into nitrocellulose sheets and challenged with antibody to *Cucurbita* leaf ferredoxin. Although it proved possible to detect a single band of antigenic material in etiolated pea leaves, none was found in total extracts of pea roots.

The use of pea root plastid preparations as the starting material was more successful. Extracts were examined by the same technique of PAGE, Western blotting and treatment with antibody to *Cucurbita* leaf ferredoxin. Two bands of stained material

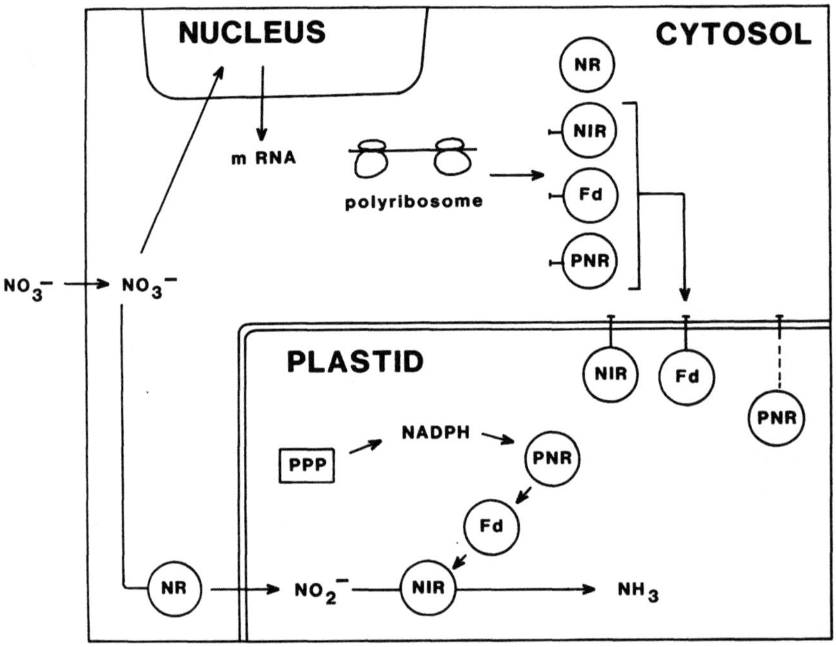

Fig. 1. Proposed scheme for the induction of the components of nitrate assimilation in root cells. NR, nitrate reductase; NIR, nitrite reductase; PNR, pyridine nucleotide reductase; PPP, oxidative pentose phosphate pathway; Fd, ferredoxin; T, transit peptide

were obtained at positions corresponding to molecular weights of 13.9 and 14.9 kDa. These bands were absent from preparations made from plastids obtained from pea plants which had not been nitrate induced.

NAD(P)H ferredoxin reductases in the root

Partially purified extracts from pea plastids were examined for NAD(P)H ferredoxin reductase by supplying exogenous pea leaf ferredoxin and measuring the rate of reduction of cytochrome *c*. The NADPH-specific activity was found to be 7 - 10 fold higher than activity with NADH. Both activities were 4 - 5 times higher in nitrate-induced compared with non-nitrate induced pea roots. This was further demonstrated by comparing the extracts by means of polyacrylamide electrophoresis, followed by specific staining for diaphorase type enzymes with tetrazolium (Bowsher et al. 1989). Our current understanding of the relationship between these components and the PPP during nitrate assimilation is illustrated in Figure 1.

References

Aparicio PJ, Knaff DB, Malkin R (1975) The role of an iron-sulphur centre and siroheme in spinach nitrite reductase. Arch Biochem Biophys 169: 102-107

Bowsher CG, Emes MJ, Cammack R, Hucklesby DP (1988) Purification and properties of nitrite reductase from roots of pea. Planta 175: 334-340

Bowsher CG, Hucklesby DP, Emes MJ (1989) Nitrite reduction and carbohydrate metabolism in plastids purified from roots of *Pisum sativum* L. Planta 177: 359-366

Butt VS, Beevers H (1961) The regulation of pathways of glucose metabolism in maize roots. Biochem J 80: 21-27

Cammack R, Hucklesby DP, Hewitt EJ (1978) Electron paramagnetic resonance studies of the mechanism of leaf nitrite reductase. Signals from the iron-sulphur centre and haem under turnover conditions. Biochem J 171: 519-536

Dalling MJ, Hucklesby DP, Hageman RH (1973) A comparison of nitrite reductase enzymes from green leaves, scutella and roots of corn (*Zea mays* L.). Plant Physiol 51: 481-484

Emes MJ, Fowler MW (1979 a) The intracellular location of the enzymes of nitrate assimilation in the apices of seedling pea roots. Planta 144: 249-253

Emes MJ, Fowler MW (1979 b) Intracellular reactions between the pathways of carbohydrate oxidation and nitrate assimilation in plant roots. Planta 145: 287-292

Hageman RH, Cresswell CF, Hewitt EJ (1962) Reduction of nitrate, nitrite and hydroxylamine to ammonia by enzymes extracted from higher plants. Nature 193: 247-250

Hucklesby DP, Dalling MJ, Hageman RH (1972) Some properties of two forms of nitrite reductase from corn (*Zea mays* L.) scutellum. Planta 104: 220-233

Ida S, Mori E, Morita Y (1974) Purification, stabilization and characterization of nitrite reductase from barley roots. Planta 121: 213-224

Nagaoka S, Hirasawa M, Fukushima K, Tamura G (1984) Methyl viologen-linked nitrite reductase from bean roots. Agricultural Biol Chem 48: 1179-1188

Ninomiya Y, Sato S (1984) A ferredoxin-like electron-carrier from non-green cultured tobacco cells. Plant Cell Physiol 25: 453-458

Oaks A, Hirel B (1985) Nitrogen metabolism in roots. Ann Rev Plant Physiol 36: 345-364

Pate JS (1973) Uptake, assimilation and transport of nitrogen compounds by plants. Soil Biol Biochem 5: 109-119

Sanderson GW, Cocking EC (1964) Enzymic assimilation of nitrate in tomato plants I. Reduction of nitrate to nitrite. Plant Physiol 39: 416-422

Suzuki A, Jacquot JP, Vidal J, Oaks A (1985) An electron-transport system in maize roots for reactions of glutamate synthase and nitrite reductase. Physiological and immunological properties of the electron carrier and pyridine nucleotide reductase. Plant Physiol 78: 374-378

Wada K, Onda M, Matsubara H (1986) Ferredoxin isolated from plant non-photosynthetic tissues. Purification and characterization. Plant Cell Physiol 27: 407-415

PRODUCTION OF NITROUS OXIDE AND NITRIC OXIDE BY SOME NITRATE-RESPIRING BACTERIA

R. Voßwinkel and H. Bothe [1]

Many systematically unrelated bacteria utilize nitrate as a terminal respiratory electron acceptor when oxygen becomes limiting. The composition of the enzymes involved in nitrate respiration as well as the end product differ from organism to organism among the relatively few species investigated extensively (Knowles 1982; Stouthamer 1988; Zumft et al. 1988). The bacteria which produce N_2O and NO besides N_2 (denitrifying bacteria *sensu stricto*) are of particular interest currently. The NO_x produced by either denitrifying or nitrifying bacteria in soils may have an impact on the destruction of the ozone layer in the stratosphere (Söderlund and Svensson 1976; Delwiche 1981). More than 50% of the terrestrial NO_x production might come from biological sources and less from human activities. The biological reactions resulting in the production or consumption of NO_x are not well understood currently.

The organism of our choice for studying nitrate respiration and denitrification has been *Azospirillum* in the past few years. *Azospirillum* lives in association with grasses and is abundant in tropical soils. A summary of our major results was already given in the proceedings of the last Advanced Course on Inorganic Nitrogen Metabolism (Zimmer et al. 1987). Therefore only some newer results will be mentioned here.

Azospirillum can grow with either nitrate, nitrite or nitrous oxide as terminal respiratory electron acceptors. Energy transduction efficiencies in nitrogenous oxide respirations were extrapolated from molar growth yield determinations and from H^+-translocation measurements (Danneberg et al. 1989). In O_2- and N_2O-respiring cells, maximal molar growth yields in continuous cultures grown with malate limiting were essentially the same and no H^+-consumptions could be detected. Both findings indicate that the energy transduction efficiencies are the same in O_2- and N_2O-respirations. Maximal molar growth yields were 1/3 or 2/3 lower with NO_3^- or NO_2^-, respectively, than with O_2 or N_2O as electron acceptor. A H^+-uptake of 2.5 - 3.0 H^+ per molecule of NO_3^- or NO_2^- was observed when each of these compounds served as sole respiratory electron acceptor. Assuming that the buffer capacity inside but not outside the cell is high enough to compensate for the H^+-consumption during reduction of any nitrogenous compounds, the data allow to conclude that the conversions of NO_3^- to NO_2^- and N_2O to N_2 proceed inside the cells, whereas the reduction of NO_2^- to N_2O is at the periplasmic face of the cytoplasmic membrane in *Azospirillum*. The experiments performed with ionophores only partly support this conclusion (Danneberg et al. 1989). It

[1] Botanisches Institut, Universität zu Köln, Gyrhofstrasse 15, D-5000 Köln 41, FRG

was also suggested that nitrate for dissimilatory reduction is taken up by an energy-independent nitrate/nitrite antiport system in *Azospirillum*.

In batch culture, *Azospirillum* can form as much as 10 mM nitrite by nitrate reduction, depending on the density of the population assayed. Unexpectedly, nitrite was found to exert phytohormonal effects on plants (Zimmer and Bothe 1988; Zimmer et al. 1988). Nitrite in the concentration range of 0.1-10 mM substituted indole-3-acetic acid in a number of plant tests for auxins. The effect of nitrite was enhanced by adding ascorbate in assays like the C_2H_4-formation by pea epicotyl sections and the increase in the wet weight of wheat root segments. Nitrite formed by bacterial nitrate respiration may interact with ascorbate in the cells, and an unidentified reaction product seems to act as phytohormone. This is an alternative to explain the positive effects of associative bacteria on the plant growth. Further experiments have to show whether such positive effects are restricted to nitrate-respiring and nitrite-excreting bacteria.

Nitrite and nitrous oxide are well established intermediates in denitrification. The role of nitric oxide is being investigated currently. Nitric oxide is not liberated by intact cells of the nitrate-respiring bacteria commonly investigated. The gas in intact cells can, however, be trapped either by adding hemoglobin (Goretski and Hollocher 1988; Carr et al. 1989), or be detected by adding low concentrations of Triton X-100 to the suspensions (Carr et al. 1989). Membrane preparations of nitrite-respiring bacteria generally reduce nitrite to N_2O, whereas the solubilized nitrite reductase produces significant amounts of NO (see Zumft et al. 1988 a). This has also been shown for the cytochrome *cd*-containing nitrite reductase of *Azospirillum brasilense* (Danneberg 1987). The product formation in cell-free systems also depends on the electron donor system used. Ascorbate and phenazine methosulfate as electron donors favour NO-formation, whereas more N_2O is formed with dithionite and methylviologen (see Zumft et al. 1988 a). Transposon Tn 5-induced mutants from *Pseudomonas stutzeri* (Zumft et al. 1988 b) lacking cytochrome *cd* were able to reduce exogenously supplied NO, indicating that cytochrome *cd* is not part of the nitric oxide reductase. The role of this enzyme remains debatable, because intact cells of *Paracoccus denitrificans* (Carr et al. 1989) and *Pseudomonas stutzeri* (Zumft et al. 1988 b) can reduce NO, but do not grow with it as sole electron acceptor. It does not appear to be resolved currently whether NO is part of a linear or branched reaction sequence in denitrification. Likewise, the conditions which result in both production and consumption of NO are not yet clear. Recently Ji and Hollocher (1988) reported, without giving detailed informations, that several species of the *Enterobacteriaceae* produce NO which is affected by nitrate and azide in intact cells.

Azospirillum brasilense Sp 7 does not evolve NO and *Azospirillum lipoferum* Sp 59 produces only small amounts of the gas under anaerobic growth conditions with NO_3^- as respiratory electron acceptor (R. Voßwinkel, unpublished). To find better NO-producers, samples were isolated from wheat straw which had been decomposed in soils of the Braunschweig area for 4 to 6 months. The straw material was incubated

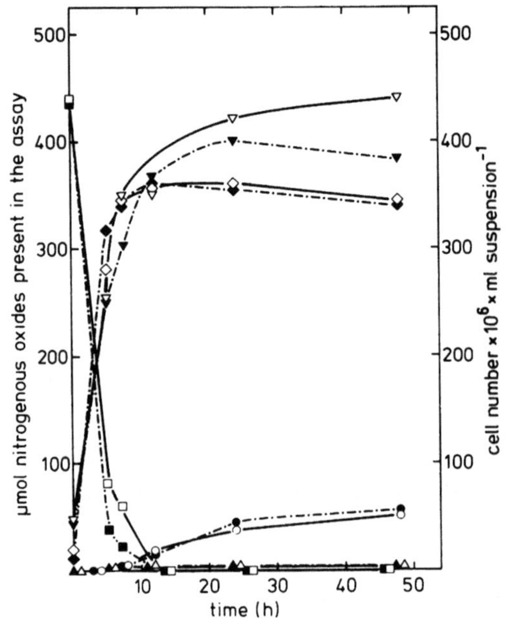

Fig. 1. Time dependence of the formation of NO and N_2O by the isolate E 8/2 (= *Pseudomonas* spp.)
The cells were grown in 40 ml peptone-yeast extract medium supplemented with nitrate under argon in 100 ml flasks sealed with rubber stoppers. For other experimental details and for the methods to determine the concentrations of nitrogenous oxides see text.
Straight lines and open symbols: experiment performed without C_2H_2; dashed lines and closed symbols: experiment performed under argon plus 30% (v/v) C_2H_2.
□,■ = nitrate; ◇,◆ = nitrite; ▽,▼ = cell number; ○,● = NO; △,▲ = N_2O

in soft agar containing mineral salts and malate. Cells were then plated on King agar B Base or Endo C agar. The authors are indebted to Prof. G. Jagnow, FAL, Braunschweig, FRG, for the help in isolating the bacteria. Twenty isolates from the agar plates were grown in peptone-yeast extract supplemented with 10 mM KNO_3 under argon at 30°C. When 4 x 10^7 cells/ml of each of the isolates were inoculated into 40 ml medium in 100 ml sealed flasks, growth proceeded for at least 3 days and the cells increased to numbers varying between 2 x 10^8 and 7 x 10^8/ml. 15 of the 20 isolates produced NO which was readily measurable after 3 days.

For the measurements, N_2O concentrations were determined in a gas chromatograph fitted with a thermal conductivity detector (Carbosieve S-II column, He as carrier gas). Nitrate and nitrite were determined colorimetrically, NO was quantified luminometrically by a NO/NO_x-Analyzer (model CLD 502, TECAN AG, CH-8634

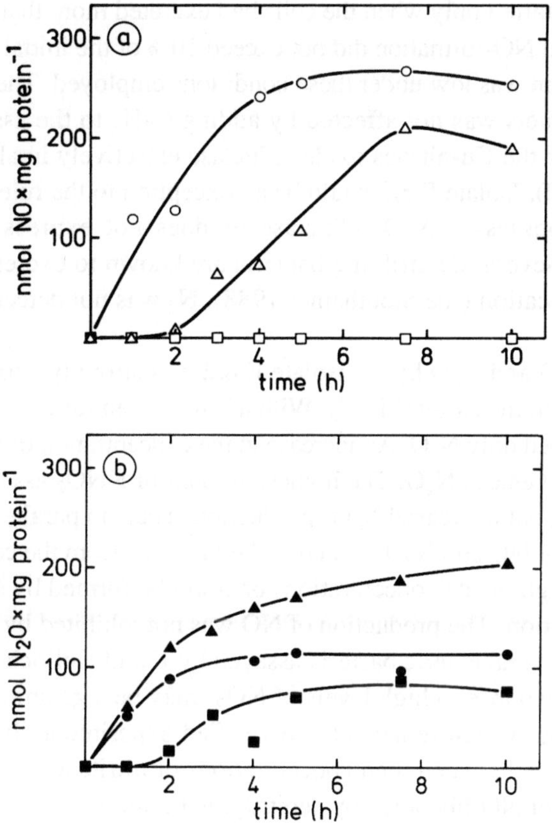

Fig. 2. The dependence of the formation of NO and N_2O by E 8/2 on the concentration of nitrite. For this experiment the cells were grown anaerobically in a medium containing 10 mM malate and 10 mM KNO_3 and mineral salts for 48 h. The suspension was centrifuged and the cells were suspended in the medium which, however, did not contain any of the nitrogenous oxides. The experiment was performed in 7.0 ml sealed Fernbach flasks under argon in a shaking waterbath at 30°C. The flasks contained 1.9 ml of the bacterial suspension with an OD at 560nm of 2.25. The experiment was started by injecting 0.1 ml anaerobic solutions with the different amounts of nitrite
a) NO-formation: □——□ 1 mM KNO_2; △——△ 5 mM KNO_2; ○——○ 10 mM KNO_2.
b) N_2O-formation: ■——■ 1 mM KNO_2; ▲——▲ 5 mM KNO_2; ●——● 10 mM KNO_2.
(always final concentration in the Fernbach flasks)

Hombrechtikon). The gas was supplied to the continuous stream of N_2 as carrier gas in the NO_x-analyzer which allowed to monitor 0.15 ppm of NO. Experimental details will be published elsewhere.

Figure 1 gives a typical experiment with the isolate E 8/2 which was tentatively identified as *Pseudomonas* spp. using the api 20 E test system (api bio Mérieux, 7440 Nürtingen). The isolate reduced more than 90% of the initial nitrate to nitrite within 10 hours. NO-production started only when the cells had excreted more than 5 mM nitrite to the medium. Maximal NO-formation did not exceed 10% of the initial concentration of nitrate. N_2O-formation was low under these conditions employed. The formation of N_2O and of other products was not affected by adding C_2H_2 to the assays (Fig. 1). C_2H_2 is known to block the Cu-nitrous oxide reductase effectively in all denitrifying bacteria (Knowles 1982). Isolate E 8/2 could be an exception to the rule. More likely, the isolate does not possess a N_2O reductase or does not express it under the conditions employed. Several denitrifying bacteria are known to excrete N_2O as the end product of denitrification (see Stouthamer 1988). N_2 was not detectable with the isolate E 8/2.

The formation of NO and N_2O by the isolate E 8/2 was strongly dependent on the concentration of nitrite in the assay (Fig. 2). With a low concentration of 1 mM KNO_2 initially, the cells liberated only N_2O. An increase in the concentration of NO_2^- resulted in the formation of NO besides N_2O. The highest amount of KNO_2 assayed (10 mM) caused a maximal NO-, but decreased N_2O- production. Thus no parallelism was seen between the formation of NO and N_2O. NO could be liberated from the cells to prevent deleterious effects of high nitrite concentrations or could be formed by severely damaged cells in the population. The production of NO was not inhibited by nitrate with E 8/2, as observed with several enteric bacteria assayed by Ji and Hollocher (1988). The liberation of NO as a response to high levels of NO_2^- may be a general phenomenon for denitrifying bacteria, which remains to be verified experimentally. Likewise, it remains to be shown whether such high concentrations of nitrite are found in soils to meet the requirements for NO-liberation by denitrifying bacteria.

Acknowledgments. This work was performed in cooperation with the International Biotechnology Laboratories GmbH, D-6900 Heidelberg, and was kindly supported by grants from the Bundesministerium für Forschung und Technologie, Bonn, FRG.

References

Carr GJ, Page MD, Ferguson SJ (1989) The energy-conserving nitric-oxide reductase system in *Paracoccus denitrificans*. Eur J Biochem 179: 683-692

Danneberg G (1987) Untersuchungen zur Denitrifikation an *Azospirillum brasilense* Sp 7. Thesis, University of Cologne FRG, pp 1-128

Danneberg G, Zimmer W, Bothe H (1989) Energy transduction efficiencies in nitrogenous oxide respirations of *Azospirillum brasilense* Sp 7. Arch Microbiol 151: 445-453

Delwiche CC (1981) Denitrification, nitrification and atmospheric nitrous oxide. Wiley and Sons, New York, pp 1-286

Goretski J, Hollocher TC (1988) Trapping of nitric oxide produced during denitrification by extracellular hemoglobin. J Biol Chem 263: 2316-2323

Ji XB, Hollocher TC (1988) Reduction of nitrite to nitric oxide by enteric bacteria. Biochem Biophys Res Commun 157: 106-108

Knowles R (1982) Denitrification. Microbiol Rev 46: 43-70

Stouthamer AH (1988) Dissimilatory reduction of oxidized nitrogen compounds. In: Zehnder AJB (ed) Biology of Anaerobic Microorganisms. Wiley and Sons, New York, pp 245-303

Söderlund R, Svensson BH (1976) The global nitrogen cycle. In: Svensson BH, Söderlund R (eds) Nitrogen, Phosphorus and Sulphur Global Cycles. Scope Report 7, Ecological Bulletins, Swedish Natural Science Research Council, Stockholm, pp 23-73

Zimmer W, Bothe H (1988) The phytohormonal interactions between *Azospirillum* and wheat. Plant and Soil 110: 239-247

Zimmer W, Roeben K, Bothe H (1988) An alternative explanation for the plant growth promotion by bacteria of the genus *Azospirillum*. Planta 176: 333-342

Zimmer W, Roeben K, Danneberg G, Bothe H (1987) The bacterial genus *Azospirillum* and its potential applications. In: Ullrich WR, Aparicio PJ, Syrett PJ, Castillo F (eds) Inorganic Nitrogen Metabolism. Springer, Berlin Heidelberg New York, pp 177-182

Zumft WG, Döhler K, Körner H, Löchelt S, Viebrock A, Frunzke K (1988 b) Defects in cytochrome cd_1-dependent nitrite respiration of transposon Tn5-induced mutants from *Pseudomonas stutzeri*. Arch Microbiol 149: 492-498

Zumft WG, Viebrock A, Körner H (1988 a) Biochemical and physiological aspects of denitrification. In: Cole JA, Ferguson S (eds) The Nitrogen and Sulphur Cycles. Cambridge Univ Press, Cambridge, UK, pp 245-279

ROLE OF AMINATION AND TRANSAMINATION IN AMMONIUM INCORPORATION IN *CHLAMYDOMONAS REINHARDTII*

J. Cárdenas, B. Lain-Guelbenzu, E. Moyano and J. Muñoz-Blanco [1]

Introduction

In *C. reinhardtii*, like in other unicellular green algae, the main route of ammonium incorporation into carbon skeletons is the so-called glutamine synthetase/glutamate synthase (GS/GOGAT) cycle (Miflin and Lea 1977).

Besides this main route, there is an alternative pathway of ammonium incorporation into 2-oxoglutarate to yield L-glutamate, through the amination reaction catalyzed by glutamate dehydrogenase (Kates and Jones 1964; Shatilov and Kretovich 1977). Once L-glutamate is formed, it constitutes a pool from which, through different transamination reactions, other L-amino acids are formed and enter the different routes of biosynthetic metabolism.

The contribution and importance of the GDH route to L-glutamate synthesis and its relationship with nitrogen and carbon metabolism in green algae are still unclear. Similarly, the contribution of transaminations to the total carbon and nitrogen balance in algal cells remains to be elucidated.

In the following paragraphs we present evidence of the role of NADH-glutamate dehydrogenase (GDH, EC 1.4.1.2) and the two most active transaminases involved in glutamate assimilation, namely L-aspartate aminotransferase (AAT, EC 2.6.1.1) and L-alanine aminotransferase (AST, EC 2.6.1.2), in the carbon and nitrogen metabolism of the unicellular green alga *C. reinhardtii* 6145.

NADH-Glutamate Dehydrogenase

In aged *Chlamydomonas* cells GDH plays an anabolic role under certain environmental and trophic stress conditions: darkness, high ammonium concentrations and GS/GOGAT inhibition (Muñoz-Blanco and Cárdenas 1989). GDH activity markedly increased in cells grown on ammonium and subjected to prolonged dark conditions. GS and GOGAT decreased significantly in the first 5 hours of treatment. Another enzyme utilizing glutamate, L-aspartate aminotransferase, also increased in the dark. GDH increase in the dark depended on protein synthesis *de novo*, since cycloheximide prevented the appearance of activity. In darkened cells a smooth and steady diminution

[1] Departamento de Bioquímica, Biología Molecular y Fisiología, Universidad de Córdoba, E-14071 Córdoba, Spain

Table 1. Changes under stress conditions in aged cells of *C. reinhardtii*

	N-starvation	Darkness	High ammonium concentration	GS/GOGAT inhibition
GDH	+	+++	++	++
GS	++	– –	– –	– –
GOGAT	++	– –	– –	– –

+ = 100%; ++ = 200%; +++ = 300%; – = -50%; – – = -100%. Changes were recorded after 30-40 h treatment

of intracellular L-glutamine was observed, accompanied by an increase of the internal L-glutamate pool and a strong decrease of intracellular ammonium (Muñoz-Blanco and Cárdenas 1989). Similar variations in GDH, GS and GOGAT were observed in cells subjected to high ammonium concentrations or when either GS or GOGAT was inhibited with L-methionine-D,L-sulfoximine (MSX) or azaserine, specific inhibitors of GS and GOGAT, respectively (Muñoz-Blanco and Cárdenas 1989) (Table 1).

The possible connection between GDH aminating activity and carbon metabolism in *Chlamydomonas* has been investigated by following, under stress conditions, the variation with time of certain representative enzymes of carbon metabolism in the alga. In C-starved and darkened cells, levels of GDH, pyruvate kinase (PK) and isocitrate dehydrogenase (ICDH) increased with time, whereas those of malate dehydrogenase (MDH) remained fairly constant after 30 hours of treatment. Under the same conditions MDH, ICDH and glucose-6-phosphate dehydrogenase (G6PDH) were high in N-starved cells, whereas GDH did not change significantly (Table 4).

All these evidences indicate that GDH is an adaptive enzyme which maintains the intracellular levels of L-glutamate when the GS/GOGAT cycle is not fully operative. The role of the GDH system is, therefore, ancillary rather than alternative, it operates when the GS/GOGAT cycle is blocked or malfunctioning, i.e. under nutritional and environmental stress conditions and, thus, it is more connected with carbon than with nitrogen metabolism.

The versatility of the GDH system under stress conditions has been corroborated by the recent finding of several isozymes differently responsive to different nitrogen sources and situations of stress metabolism (Muñoz-Blanco et al. 1989). In *Chlamydomonas* cells grown on ammonium there are, at least, three different GDH isozymes, named GDH_1, GDH_2 and GDH_3 after their increasing mobility in polyacrylamide gels (Fig. 1). These isozymes depended, in their induction, on the trophic and environmental conditions of the algal cells. Thus, in cells grown on ammonium, GDH_1 and total GDH increased in parallel with extracellular ammonium concentration, whereas GDH_2 changed slightly and GDH_3 did not change significantly at all. In contrast, in urea-grown cells GDH_1 levels decreased and those of GDH_3 were enhanced when the urea concentration was raised.

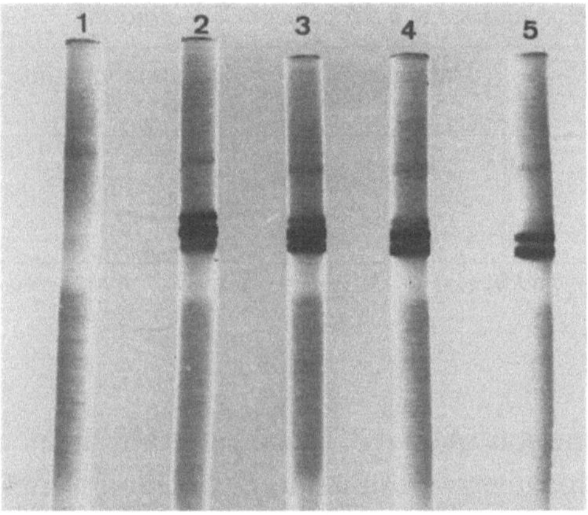

Fig. 1. GDH isozyme induction under carbon starvation conditions. Cells grown on ammonium were harvested in the light in ammonium medium lacking CO_2, and GDH activity was measured at 0 h (**2**), 4 h (**3**), 7 h (**4**) and 24 h (**5**) induction. Bands correspond to GDH_1, GDH_2 and GDH_3 from the top to the bottom of the gel rod, respectively. **1**, control without L-glutamate

Table 2. Effect of different treatments on induction of GDH isozymes of *C. reinhardtii*

Treatment (h)	Specific activity ($mU\ mg\ protein^{-1}$) Total GDH	GDH_1	GDH_2	GDH_3
A. Dark conditions				
0	39	4	12	23
7	67	28	18	21
24	114	62	34	18
B. Carbon starvation				
0	37	7	17	13
7	71	2	35	34
24	81	1	37	43
C. + 5 mM MSX				
0	33	9	15	9
2	36	17	16	3
7	44	22	20	2
D. Nitrogen starvation				
0	35	10	13	12
7	38	6	14	18
24	37	2	13	22

Cells were cultured phototrophically with 10 mM ammonium until mid-exponential phase of growth (A_{660} = 1.4), harvested and subjected to the indicated treatments. Intracellular ammonium concentration did not change significantly (0.14 µmol mg $protein^{-1}$) in A and B. In C intracellular ammonium ranged between 0.19 (0h) and 0.24 (7h) µmol mg $protein^{-1}$, and in D values of 0.15 (0h) and 0.07 (24 h) µmol mg $protein^{-1}$ were measured

The time course of isozyme induction has been studied under conditions in which total GDH activity was increased (Table 2). In darkened cells, both GDH_1 and GDH_2

were augmented after 24 h induction, whereas GDH_3 levels fell significantly. Similar results were found in cells treated with MSX. A different course of induction was observed in C-starved cells: GDH_2 and GDH_3 levels increased, whereas GDH_1 diminished strongly after 24 h treatment. Similarly, in N-starved cells GDH_1 decreased and GDH_3 was enhanced, whereas GDH_2 remained practically unchanged.

All these facts lead to the conclusion that GDH_1 is involved predominantly in the reductive amination of 2-oxoglutarate, and GDH_3 in L-glutamate deamination. GDH_2 might participate in both aminating and deaminating functions, depending on the cellular metabolic requirements, since its adaptive course seems to follow the induction pattern of either GDH_1 or GDH_3, according to the prevailing trophic or environmental conditions.

This dual response of GDH_2 might be related to a different intracellular localization of the three isozymes. In fact, in unicellular green algae GDH isozymes located in chloroplasts (Gayler and Morgan 1976; Prunkard et al. 1986; Fisher and Klein 1988) and in cytosol and mitochondria (Fayyaz-Chaudhary et al. 1984; 1985) have been reported.

Lately we have succeeded in separating these three isozymes by anion exchange chromatography. Preliminary kinetic results confirm the anabolic and catabolic roles assigned to GDH_1 and GDH_3.

L-Aspartate Aminotransferase

The L-aspartate aminotransferase of *C. reinhardtii* has been purified 400-fold to electrophoretic homogeneity, and it is the first case reported of a pure aminotransferase in green algae and higher plants.

The enzyme has an apparent relative molecular mass (M_r) of 138,000 and is a dimer consisting of two identical or similar-sized subunits of M_r 65,000 each, it contains pyridoxal-5'-phosphate (PLP), and its absorption spectrum is characteristic of a Schiff base. Its physicochemical and kinetic properties are listed in Table 3.

Aminooxyacetate, hydroxylamine and semicarbazide, all known antagonists of PLP, fully inhibited AAT activity. Significant inhibitions by p-hydroxymercuribenzoate (pHMB) (43% at 1 mM), thiosemicarbazide (50% at 2 mM) and L-ascorbate (30% at 2 mM) were also observed. By contrast, metal chelating agents showed very low or no inhibition when used at concentrations of 2 mM under the same conditions.

Purified AAT of *C. reinhardtii* exhibited substrate multispecificity to a certain extent. With 2-oxoglutarate as oxoacid acceptor, AAT catalyzed transamination of L-phenylalanine and L-tyrosine, but to a very low degree (10 and 9%, respectively), whereas with oxaloacetate, it catalyzed transamination of L-serine (27%), L-alanine (14%) and L-glutamine (14%). Kinetically, the transamination of L-aspartate in *Chla-*

mydomonas is a bisubstrate reaction with a bi-bi ping-pong mechanism, and is not inhibited by substrates .

AAT activity levels in *C. reinhardtii* were dependent on the nitrogen and carbon sources utilized by the cells in the light. With urea, urate, xanthine and hypoxanthine as nitrogen sources, a much higher specific activity was obtained than with nitrate and CO_2 as carbon source. Higher activities were found when, in addition to CO_2, carbon skeletons were present (Muñoz-Blanco et al. 1988).

Table 3. Physicochemical and kinetic properties of purified AAT from *C. reinhardtii*

Stokes' radius (nm)	4.5
M_r	138,000
Number of subunits	2
M_r of subunit	65,000
PLP/ enzyme molecule	2
Isoelectric point	8.48
Absorption spectrum maxima (nm)	412, 330
Optimum pH	7.8
Optimum temperature (^oC)	37
Activation energy ($kJmol^{-1}$)	8.9
Q_{10}	1.13
Apparent K_m (mM) for:	
2-oxoglutarate	0.55
oxaloacetate	0.04
$S_{0.5}$ (mM) for:	
L-aspartate	2.53
L-glutamate	3.88
Hill coefficient for:	
L-aspartate	1.57
L-glutamate	0.90

Other factors that increased AAT levels were nitrogen and carbon starvation, and transfer of the cells from light to dark conditions. Under these conditions, a pronounced enhancement of AAT activity, dependent on protein synthesis *de novo*, took place, possibly caused by an accelerated degradation of protein and carbon reserves in algal cells. These processes, in which AAT could play a relevant role, would provide nitrogen and/or carbon skeletons to keep the cells functioning under these circumstances. This interpretation is reinforced by the fact that, under nitrogen and carbon deprivation, *C. reinhardtii* cells have increased free amino acid pools (Muñoz-Blanco et al. 1988). Besides, levels of several enzymes related to carbon metabolism such as MDH and G6PDH were enhanced in N-starved cells, and in C-starved cells also increases in G6PDH, ICDH and PK were found. Increases in ICDH and PK were likewise detected in darkened cells (Table 4).

L-Alanine Aminotransferase

AST has been also purified and its main physicochemical and kinetic properties have been investigated. It has an M_r of 141,000 and a Stokes' radius of 4.9 nm. Its optimum pH and temperature are 7.3 and 60°C, respectively. The reaction has an activation energy of 19.1 kJ mol^{-1} and a Q_{10} of 1.29. AST activity was very sensitive to aminooxyacetate (100% inhibition) and hydroxylamine (97% inhibition), known antagonists of PLP. Also EDTA and pHMB strongly inhibited AST activity. 10 mM Mn^{2+} enhanced the activity three times when present in reaction mixtures, whereas other cations (Mg^{2+}, Ca^{2+}, K^+, Na^+) were ineffective or strongly inhibitory (Zn^{2+}, Ni^{2+}, Co^{2+}, Fe^{2+}, Cu^{2+}). The apparent K_m value for 2-oxoglutarate was 0.54 mM and the enzyme exhibited a non-hyperbolic behaviour for L-alanine (negative cooperativity) with an $S_{0.5}$ of 0.72 mM and a Hill coefficient of 0.5.

The activity of AST changed in response to the trophic conditions of algal cells. Under phototrophic conditions, the highest activity levels were found in cells grown on urea, urate, allantoin, allantoate or xanthine. Likewise, high levels of AST were detected in *Chlamydomonas* cells cultured mixotrophically with acetate and ammonium in the light. With ammonium as sole nitrogen source and phototrophic conditions, AST increased in parallel with increasing concentrations of extracellular ammonium. Like that of AAT, AST activity was sensitive to prolonged periods of carbon or nitrogen deprivation, showing increased activity after 32 h treatment. When algal cultures were subjected to alternating periods of light and darkness, no changes in AST levels were observed. A summary of these changes compared with those of GDH and other enzymes related with carbon metabolism (MDH, G6PDH, ICDH, and PK) is shown in Table 4.

Table 4. Changes in *C. reinhardtii* cultures as a consequence of nitrogen, carbon and light status

	N-starved cells	C-starved cells	Darkened cells
Free amino acid pool	++	++	+
Protein content	+	+	+
GDH	+	++	+++
AAT	+++	+++	+++
AST	+++	+++	+
MDH	++	+	+
G6PDH	++	+++	–
ICDH	++	+/	++
PK	– –	+/	++

+/ = 50%; + = 100%; ++ = 200%; +++ = 300%; – = -50%; – – = -100%. Changes were recorded after 32 h treatment

Acknowledgments. The work in the authors' laboratory has been supported by Grant No. PB86-01 67-CO3-01 and Junta de Andalucía, Spain. The secretarial help of C. Santos and I. Molina is acknowledged.

References

Fayyaz-Chaudhary M, Cannons AS, Merrett MJ (1984) Photoregulation of NADPH-glutamate dehydrogenase in regreening cultures of *Euglena gracilis*. Plant Sci Lett 34: 89-94

Fayyaz-Chaudhary M, Javed Q, Merrett MJ (1985) Effect of growth conditions on NADPH-specific glutamate dehydrogenase activity of *Euglena gracilis*. New Phytol 101: 367-376

Fischer P, Klein V (1988) Localization of nitrogen-assimilating enzymes in the chloroplast of *Chlamydomonas reinhardtii*. Plant Physiol 88: 947-952

Gayler KR, Morgan WR (1976) An NADP-dependent glutamate dehydrogenase in chloroplasts from the marine green alga *Caulerpa simpliciuscula*. Plant Physiol 58: 283-287

Kates JR, Jones RF (1964) Variation in alanine dehydrogenase and glutamate dehydrogenase during the synchronous development of *Chlamydomonas*.Biochim Biophys Acta 86: 438-447

Miflin BJ, Lea PJ (1977) Amino acid metabolism. Annu Rev Plant Physiol 28: 299-329

Muñoz-Blanco J, Cárdenas J (1989) Changes in glutamate dehydrogenase activity of *Chlamydomonas reinhardtii* under different trophic and stress conditions. Plant Cell Environ 12: 173-182

Muñoz-Blanco J, Lain-Guelbenzu B, Cárdenas J (1988) Characterization of an L-aspartate aminotransferase activity in *Chlamydomonas reinhardtii*. Physiol Plant 74: 433-439

Muñoz-Blanco J, Moyano E, Cárdenas J (1989) Glutamate dehydrogenase isozymes of *Chlamydomonas reinhardtii*. FEMS Microbiol Lett. (in press)

Prunkard DD, Bascomb NF, Robinson RW, Schmidt RR (1986) Evidence for chloroplastic localization of an ammonium-inducible glutamate dehydrogenase and synthesis of its subunits from a cytosolic precursor-protein in *Chlorella sorokiniana*. Plant Physiol 81: 349-355

Shatilov VR, Kretovich WL (1977) Glutamate dehydrogenase from *Chlorella*: Forms, regulation and properties. Mol Cell Biochem 15: 201-212

NADP-ISOCITRATE DEHYDROGENASES IN HIGHER PLANTS

R.D. Chen [1], P. Gadal [1], E. Bismuth [2] and M.L. Champigny [2]

Introduction

The keto-acid, 2-oxoglutarate, provides the carbon skeleton implicated in the biosynthesis of glutamate. For higher plants, it is widely accepted that this oxo-acid is derived from the Krebs Cycle where it is produced through the action of the mitochondrial NAD-isocitrate dehydrogenase (NAD-IDH, E.C. 1.1.1.41) (Metzler 1977). The resulting keto-acid then traverses the mitochondrial and chloroplast envelopes to reach the chloroplast stroma where it is used by glutamate synthase (EC. 1.1.7.1). However, there is increasing evidence suggesting that the reaction catalyzed by NADP-isocitrate dehydrogenase (NADP-IDH, E.C.1.1.1.42) provides an alternative pathway for the production of 2-oxoglutarate for ammonia assimilation (Elias and Givan 1977; Randall and Givan 1981; Chen et al. 1988, 1989 a,b; Chen and Gadal 1989). The significance of this scheme in ammonia assimilation is sustained by the fact that an increase in the NADP-IDH activity coincides with the onset of nitrogen fixation in nodules where a high supply for 2-oxoglutarate is required (Nautiyal and Modi 1987). In this review, we summarize the recent data obtained in our laboratory on this enzyme.

Results and Discussion

Distribution of NADP-IDH in higher plants

In higher plants, NADP-IDH is widely distributed. Its presence has been demonstrated in mosses, ferns, gymnosperms and angiosperms (Chen et al. 1989 b). The activity of NADP-IDH is detectable in all living tissues and organs of pea, including dry seeds, germinating and maturing seeds, roots, host part of nodules, cotyledons, etiolated and green leaves (Chen 1989).

Depending on the species and organs, the NADP-IDH activity varied from 10 to 300 nmoles NADPH produced min^{-1} mg protein^{-1}. As a general rule, the activity in roots is higher than in green leaves, reaching 150-250 and 30-90 nmoles NADPH produced min^{-1} mg protein^{-1}, respectively (Chen et al. 1989 b).

[1] Physiologie Végétale Moléculaire, [2] Photosynthèse et Métabolisme, Université Paris Sud, U.R.A. CNRS D 1128, Bâtiment 430, 91405 Orsay Cedex, France

Intracellular localization and occurrence of isoforms

NADP-IDH is mostly present in the plant cytosol (Randall and Givan 1981). In addition, some enzyme activity has been detected also in pea chloroplasts purified through a percoll gradient (Chen et al. 1989 a).

Table 1. Relative proportions of NADP-IDH isoenzymes, expressed as % of total activity in leaf extracts

Plant species	Cytosolic isoenzyme	Chloroplastic isoenzyme
Funaria hygrometrica	19	81
Athyrium filix-femina	67	33
Quercus pedunculata	78	22
Nicotiana tabacum	88	12
Pisum sativum	86	14
Triticum aestivum	83	17

As with many other enzymes, multiple isoforms of NADP-IDH occur in pea and other higher plants. The enzymes associated with the cytosol and the chloroplasts of pea leaves are distinguishable with respect to their physicochemical features, kinetic (Table 2) and immunological properties. Thus the NADP-IDHs in cytosol and chloroplasts correspond to two isoenzymes (see below). On the other hand, the cytosolic isoenzyme appears to be tissue-specific, the root and leaf isoenzyme exhibiting different electrophoretic mobilities and amino acid compositions (Chen et al. 1988).

The patterns of NADP-IDH isoenzymes have been investigated in some representative species of higher plants. Using chromatographic and immunological techniques combined with subcellular localization studies, it has been demonstrated that both cytosolic and chloroplastic isoenzymes reported previously for pea leaves are present in all leaf extracts (Chen 1989).

The relative importance of the isoenzymes was estimated by comparing the area of the two isoenzyme peaks eluted from a DEAE-cellulose column. Significant differences are observed (Table 1). In some plants such as mosses, the chloroplastic isoenzyme is the predominant form, while in others, it represents only 10 to 33% of total activity (Chen 1989).

Purification and properties of the isoenzymes

The isoenzymes of NADP-IDH were purified to homogeneity from pea roots and leaves by ammonium sulfate precipitation, DEAE-cellulose chromatography, gel filtration, Matrex red-A affinity chromatography and Phenyl-Superose HPLC (Chen et al. 1988; Chen 1989). Some properties of these isoenzymes from pea roots and leaves are presented in Table 2.

Table 2. Some characteristic properties of NADP-IDH isoenzymes in pea roots and green leaves

Parameter	Root	Leaf	
	Cytosolic NADP-IDH	Cytosolic NADP-IDH	Chloroplastic NADP-IDH
Electrophoretic mobility (R_f)	0.31	0.11	
Molecular weight (kD)	82	82	152
Subunit molecular weight (kD)	41	41	77
With Mg^{2+} as cofactor:			
$K_m Mg^{2+}$ (μM)	34	29	724
K_mDL-isocitrate (μM)	15.2	20.3	4.4
K_mNADP (μM)	2.6	3.1	2.9
With Mn^{2+} as cofactor:			
$K_m Mn^{2+}$ (μM)	2.1	2.2	36
K_mDL-isocitrate (μM)	5.7	5.7	7.3
K_mNADP (μM)	2.7	2.6	2.0

They are all dimeric, but their molecular weights are very different, being 82 kDa for the cytosolic enzyme and 152 kDa for the chloroplastic form. A similar subunit molecular weight of about 40 to 41 kDa was found for the cytosolic isoenzyme from other species such as sorghum, spinach and wheat.

Both cytosolic and chloroplastic isoenzymes have an absolute requirement for a divalent cation such as Mn^{2+} or Mg^{2+}. The enzymes exhibit hyperbolic kinetics with respect to the substrates isocitrate and NADP. The K_m values are presented in Table 2. Despite of these differences, all the isoenzymes display a high affinity towards isocitrate or NADP, at the micromolar level.

As far as the control of NADP-IDH is concerned, although the activity of these isoenzymes is inhibited to some extent by NADPH or metabolites such as glutamate, citrate and adenine nucleotides, no specific regulatory properties have yet emerged; no post-translational modification such as phosphorylation has been found (Chen 1989).

Comparative immunochemistry of cytosolic NADP-IDH

The degree of analogy between NADP-IDH from various plant species has been compared using polyclonal antibodies raised against the cytosolic isoenzyme of pea roots and shoots.

Table 3. Comparison of the amounts of anti-pea root NADP-IDH antibodies required for the immunoprecipitation of 50% of the activity of NADP-IDH extracted from the leaves of the listed species

Species	Antibodies necessary for precipitate 50% of NADP-IDH	
	μl	μl sample/μl pea leaves
Pisum sativum L.	23	1.0
Hypnum cupressiforme L.	58	2.6
Equisetum arvense L.	23	1.0
Araucaria imbricata Pav.	23	1.0
Spinacia oleracea L.	26	1.2
Sorghum vulgare Pers.	30	1.4

For these enzymes, complete fusion lines were formed between them, and the NADP-IDH activities were almost completely precipitated in a range of concentrations very similar to the one used for the pea enzyme. The volumes of antiserum required to immunoprecipitate 50% of the initial activity from the various extracts are very similar, being 1 to 2.5 fold that for pea leaves (Table 3). The similar immunochemical properties suggest that the cytosolic enzyme has been highly conserved in the course of evolution.

Physiological roles of NADP-IDH in higher plants

In higher plants, 2-oxoglutarate, one of the products of the IDHs serves both as intermediate in the Krebs cycle and as the donor of carbon skeletons for amino-acid synthesis. Despite of the early discovery of NADP-IDH its role has never been seriously assessed and it is generally considered that the mitochondrial NAD-IDH is the sole catalyst producing this keto-acid. However, the present studies strongly suggest that NADP-IDH is also involved in this process (Chen and Gadal 1989).

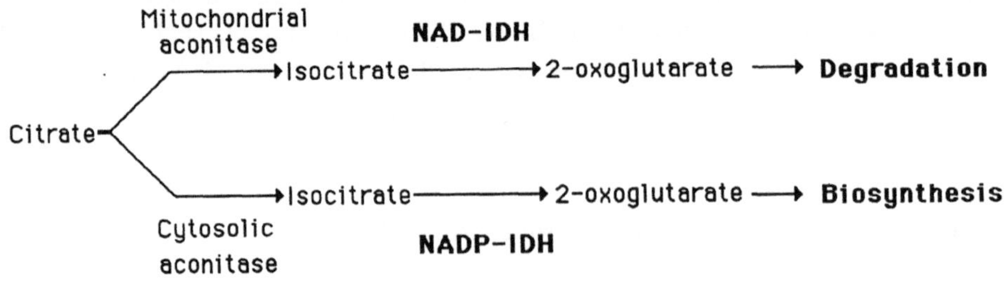

Scheme 1. "Division of labor" between NAD-IDH and NADP-IDH in higher plants

Thus, one may wonder what the biological significance for the biogenesis of this keto-acid could be both inside and outside the mitochondria. From the comparison of their distribution, subcellular localization and properties, we propose that there is a "division of labor" for 2-oxoglutarate production between the two IDHs: NAD-IDH provides 2-oxoglutarate inside the mitochondria preferentially for tricarboxylic acid degradation while NADP-IDH produces this keto-acid in the cytosol and/or in the chloroplasts mainly for biosynthetic purposes (Scheme 1). In addition, NADP-IDH has also a potential role in supplying NADPH.

Conclusion

NADP-IDH is shown to be widely present in higher plants. It is localized in both cytosol and chloroplasts. All the isoenzymes exhibit very high affinity for both NADP and isocitrate, although they are easily distinguishable in their physicochemical, kinetic or immunological properties. In most higher plants, the cytosolic enzyme is the predominant form and has been highly conserved throughout evolution.

Our studies support the hypothesis that one of the physiological roles of NADP-IDH could be to provide the major part of 2-oxoglutarate for ammonium assimilation (Chen and Gadal 1989). However, this point remains to be ascertained by making use of other techniques such as isolation of idh^{-1} mutants.

References

Chen RD (1989) Recherches sur l'origine de l'acide α–cétoglutarique associé à l'assimilation de l'azote chez les végétaux supérieurs: Etude de l'isocitrate deshydrogénase à NADP. Doctoral Thesis, Université Paris-Sud

Chen RD, Bismuth E, Champigny ML, Gadal P (1989 a) Chromatographic and immunological evidence that chloroplastic and cytosolic pea (*Pisum sativum* L.) NADP-isocitrate dehydrogenases are distinct isoenzymes. Planta 178: 157-163

Chen RD, Bismuth E, Issakidis E, Pacot C, Champigny ML, Gadal P (1989b) Comparative immunochemistry of the NADP-isocitrate dehydrogenases: A highly conserved cytosolic enzyme in higher plants. CR Acad Sci Paris, Série III 308: 459-465

Chen RD, Gadal P (1990) Do the mitochondria provide the 2-oxoglutarate needed for glutamate synthesis in higher plants? Plant Physiol Biochem (in press)

Chen RD, Le Maréchal P, Vidal J, Jacquot JP, Gadal P (1988) Purification and comparative properties of the cytosolic isocitrate dehydrogenases (NADP) from pea (*Pisum sativum*) roots and green leaves. Eur J Biochem 175: 565-572

Elias BA, Givan CV (1977) Alpha-ketoglutarate supply for amino acid synthesis in higher plant chloroplasts. Plant Physiol 59: 738-740

Metzler DE (1977) (ed) Biochemistry. Academic Press, New York London

Nautiyal CS, Modi VV (1987) Malate dehydrogenase and isocitrate dehydrogenase in root nodules of *Trigonella*. Phytochemistry 7: 1863-1865

Randall DD, Givan CV (1981) Subcellular location of NADP-isocitrate dehydrogenase in *Pisum sativum* leaves. Plant Physiol 68: 70-73

REGULATION OF NITROGEN METABOLISM IN *RHIZOBIUM*

M. Iaccarino, M. Rossi, R. Defez, M. Chiurazzi, G. Manco, G. Espin, A. Lamberti and A. Riccio.[1]

Introduction

Bacteria of the genus *Rhizobium* infect roots of leguminous plants and induce the development of root nodules. The nodules are specialized structures in which some cells contain a large number of *Rhizobium* bacteria. These, called bacteroids, are modified in their morphology and physiology as compared to the free-living forms and their multiplication is controlled.

The symbiosis should be viewed as a metabolic complementation. The plant provides energy in the form of photosynthate; bacteroids make ATP from these carbon compounds through a respiration taking place at a very low oxygen tension and synthesize NH_3, which is assimilated by the plant. Lack of photosynthate, experimentally imposed by darkening plants or by harvesting, results in decreased nitrogen fixation (Wong and Evans 1971), and may be a consequence of the decreased ATP concentration in bacteroids.

The use of fixed carbon by nodules is intensive: in *Pisum* it was found that about 30% of the carbon fixed is consumed by the root nodules, while this value was about 10% in lupin (Pate 1980). Photosynthate appears to be translocated to the nodules in the form of sucrose, but it is most probably taken up by bacteroids in the form of dicarboxylic acids (Ronson et al. 1984; O'Gara et al. 1988).

In the study of regulation of nitrogen metabolism in *Rhizobium* three different physiological states should be considered: *a*) free-living bacteria, which assimilate various nitrogen sources, depending on the soil or laboratory conditions; *b*) bacteria actively growing in the plant: these originate from a single cell to produce up to 10^9 daughter cells; they assimilate from the plant as yet unknown nitrogen sources; *c*) fully differentiated bacteroids; these have little or no protein synthesis, a high content of nitrogenase and probably of ancillary proteins; they assimilate N_2 and convert it into NH_3.

Nitrogen Metabolism

Two main pathways for NH_4^+ assimilation exist in bacteria: via glutamate dehydrogenase (GDH; EC 1.4.1.3)(reaction 1) and via glutamine synthetase (GS; EC6. 3.1.2) and

[1] International Institute of Genetics and Biophysics, CNR, via Marconi 10, 80125 Napoli, Italy

glutamate synthase (GOGAT; 1.4.1.13) (reactions 2 and 3). The end product of both pathways is glutamate, which is either directly used for protein synthesis, or acts as an amino group donor in various amino transferase reactions, thus giving rise to the other amino acids. Glutamine is used for protein synthesis and, in addition, serves as a source of nitrogen atoms in the biosynthesis of all purine and pyrimidine nucleotides, of glucosamine-6-phosphate, of p-aminobenzoic acid and of nicotinamide derivatives. Therefore glutamine synthetase activity is still needed when the GDH pathway is operative, since it links the assimilation of NH_4^+ with biosynthetic pathways leading to the formation of proteins, nucleic acids, complex polysaccharides, and various coenzymes.

$$NH_4^+ + \text{2-oxoglutarate} + NAD(P)H \xrightarrow{GDH} \text{glutamate} + NAD(P)^+ \quad (1)$$

$$NH_4^+ + \text{glutamate} + ATP \xrightarrow{GS} \text{glutamine} + ADP + Pi \quad (2)$$

$$\text{Glutamine} + \text{2-oxoglutarate} + NADPH \xrightarrow{GOGAT} 2\ \text{glutamate} + NADP^+ \quad (3)$$

GDH has a high K_m for NH_4^+ and is thus only operative at high NH_4^+ concentrations. GS/GOGAT is the major route of NH_4^+ assimilation when NH_4^+ is limiting. In enteric bacteria, the use of these two pathways is genetically controlled in response to the availability of NH_4^+. These enzymes form part of a general nitrogen control system (see Magasanik 1982; Gussin et al. 1986, for reviews) which includes a number of enzymes involved in the breakdown of nitrogenous compounds. When NH_4^+ is present at high concentrations the genes for GS and other enzymes under nitrogen control are repressed, while GDH is induced. The converse occurs when NH_4^+ is limiting, giving the cell the capability to use a variety of nitrogen sources. Additionally, GS activity is controlled biochemically in response to NH_4^+. In the presence of NH_4^+ adenyl groups are reversibly attached to GS, causing inactivation.

The Nitrogen Control System

When NH_4^+ is limiting and other specific conditions are encountered, the genes for nitrogen fixation (*nif*), or those for histidine utilization (*hut*), or for proline utilization (*put*) etc., are induced (Magasanik 1982). In most studies the level of expression of *glnA*, the structural gene for GS has been used to monitor the modulation of the two pathways for NH_4^+ assimilation. The GS level, fully induced in bacteria growing in a poor nitrogen source, decreases when NH_4^+ is used, and decreases even further in a medium containing a rich nitrogen source (like NH_4^+) and a poor carbon source. Thus, bacteria adjust the potential of expression of the nitrogen control system to the ratio of N/C, namely to the ratio of glutamine/2-oxoglutarate.

Promoters of genes regulated by the nitrogen control system have a similar structure and are characterized by the consensus sequence CTGGYAYR-N4-TTGCA extending approximately from -26 to -10 relative to the transcription-initiation site (see Gussin et al. 1986). This consensus sequence defines the *ntr* promoters and it is different from that of the other *E. coli* promoters. The *ntr* promoters require a specific sigma factor for RNA polymerase, sigma-54, the product of the *ntrA* gene (also called *glnF* or *rpoN*). They also require the product of the *ntrC* (*glnG*) gene for expression. This protein, NtrC, is an activator of gene expression at all *ntr* promoters except at those of the *nif* genes, where a more specific protein, NifA, is the activator (Dixon et al., 1980). The activity of NtrC is regulated by NtrB, the product of the *ntrB* (*glnL*) gene, which catalyzes a phosphorylation and dephosphorylation of NtrC (Ninfa and Magasanik 1986). It appears that NtrC-phosphate activates transcription from *ntr* promoters and that NtrB controls the formation and breakdown of NtrC-phosphate. The activity of NtrB is modulated by P_{II}, the product of the *glnB* gene: in the presence of P_{II} NtrB catalyzes the removal of phosphate from NtrC-phosphate thus preventing transcription at *ntr* promoters. The P_{II} protein becomes inactive when uridylylated by the product of the *glnD* gene, GlnD, a protein catalyzing two post-translational modifications of the P_{II} protein, namely uridylylation or deuridylylation. Finally, GlnD activity is regulated by the intracellular glutamine/2-oxoglutarate ratio. Thus, NH_4^+ starvation, which results in a low ratio, causes a conversion of P_{II} to P_{II}-UMP; this causes a NtrB-catalyzed phosphorylation of NtrC to NtrC-P, which activates transcription of *ntr* promoters. Growth with NH_4^+ excess or a poor carbon source causes an increase in the glutamine/2-oxoglutarate ratio thereby increasing the ratio P_{II}/P_{II}-UMP.

Regulation of GS Activity

GS of *Enterobacteriaceae* is a dodecamer composed of identical subunits and has a molecular weight of approximately 600,000. GS activity is modulated by several different mechanisms (Ginsburg and Stadtman 1973) including the interconversion of inactive to active configuration in response to variations in the concentrations of divalent cations, the cumulative feedback inhibition by a number of different end products of glutamine metabolism and a post-translational modification, namely attachment of a single adenyl residue to a specific tyrosine residue on each of the subunits.

Adenylylation and deadenylylation are two different activities of the same protein, ATase, the product of the *glnE* gene. ATase activity is regulated by the P_{II} protein described above, which activates adenylylation while P_{II}-UMP is essential for deadenylylation.

Ammonia Assimilation in *Rhizobium* spp.

In free-living *Rhizobium* spp. the level of GDH, GS and GOGAT differs with the strains and the culture media (Brown and Dilworth 1975; Ludwig 1978; Osburne and Signer 1980). It has been reported that at least some *Rhizobium* spp. are poor at assimilating NH_4^+ and have a marked preference for an organic nitrogen source such as L-glutamate (Ludwig 1978). However, several groups have been able to grow *Rhizobium* on NH_4^+ as the sole nitrogen source provided that the medium had been well buffered (Howitt and Gresshof 1985; Poole et al. 1987; Rossi et al. 1989). Mutants of different wild type isolates of *R. meliloti* (Ali et al. 1981; Osburne and Signer 1980) and of *Bradyrhizobium japonicum* (Lane et al. 1986) lacking GOGAT were isolated and their phenotype on arginine, proline etc. as a nitrogen source shows that the GS/GOGAT pathway is operational in the wild-type strain. Some of these strains failed to grow on NH_4^+, thus showing that the GDH pathway is missing in some, but not all, of the tested strains. A plasmid coding for *E. coli* GDH was introduced in one of the latter strains (Lane et al. 1986) which in this way aquired the ability to grow on NH_4^+.

Two forms of glutamine synthetase, GSI and GSII have been demonstrated in all species tested of the genera *Rhizobium* and *Agrobacterium* (Fuchs and Keister 1980). GSI has been shown to be similar to the single GS of enteric bacteria. It is a polymeric enzyme consisting of 12 identical subunits of M_r 59,000, is relatively heat stable, and can be adenylylated (Darrow et al. 1981). In contrast, GSII is made of smaller subunits, M_r 36,000, which are heat labile and not known to be modified after translation. These proteins are products of different genes (Somerville and Kahn 1983; Carlson et al. 1985; Filser et al. 1986; Colonna-Romano et al. 1987). Using extracts of *R. leguminosarum* containing both GS forms we showed (Rossi et al. 1989) that GSI is adsorbed to Blue-Sepharose, while GSII is not. Repeated adsorption/elution experiments showed that one form does not generate the other, thus confirming that they should be considered isozymes. When the enzyme is purified from *R. leguminosarum* grown in NH_4Cl it is found to be 90% adenylylated. Treatment with snake venom phosphodiesterase causes a decrease in adenylylation to a value of 20%. An assay for the biosynthetic activity on the same sample shows that adenylylation inhibits the forward reaction of GS, similarly to what happens for enteric GS and contrary to what is reported for GSI of *Rhizobium* spp ANU289 (Howitt and Gresshof 1985).

Adenylylation of GSI appears to have an *in vivo* significance since the following values were obtained when *R. leguminosarum* was grown in different nitrogen sources: 90% when grown in glutamine, 11% in glutamate, 92% in NH_4^+ and 50% in nitrate. Moreover, when glutamate-grown cells were centrifuged and resuspended for 2 hours in a medium containing either nitrate or NH_4^+, the level of adenylylation became 5% and 98% respectively.

The levels of GSI activity change very little with nitrogen sources in *R. leguminosarum* (Howitt and Gresshof 1985; Rossi et al. 1989). Instead, GSII levels change dramatically, from a specific activity of <50 (nmol/min/mg protein in the transferase assay) when glutamine or NH_4^+ are the nitrogen sources, to about 1,000 when grown on glutamate or nitrate. When glutamate-grown cells were treated with NH_4Cl there was a rapid removal of GSII activity and addition of chloroamphenicol prevented this effect (Rossi et al. 1989). Thus, it appears that the decrease in GSII activity observed after addition of NH_4Cl might be the consequence of a post-translational modification and that this requires protein synthesis. If confirmed, this complex mechanism of regulation may be related to the fact that a non-dividing bacteroid cannot achieve repression of a gene product by dilution among daughter cells.

We reported (Colonna-Romano et al. 1987) the cloning and complete nucleotide sequence of the *R. leguminosarum glnA* gene, the structural gene for GSI. The gene of *B. japonicum* coding for GSII has been cloned and sequenced (Carlson and Chelm 1986). Comparison of the deduced amino acid sequence to that of other GSs showed limited homology to the bacterial GSs but extensive homology to a GS of *Phaseolus vulgaris* roots, thus suggesting that the gene coding for GSII might be of eukaryotic origin. Upstream of the gene coding for GSII there is a consensus sequence for an *ntr* promoter (Carlson et al. 1987).

Not much is known yet about the significance of the enzymes for NH_4^+ assimilation with respect to the symbiotic phenotype of *Rhizobium* spp. As described above, GDH is missing in several strains and perhaps not very active in others. This might cause a slow growth rate in NH_4Cl versus other nitrogen sources, but it does not affect the Nod and Fix phenotype. Strains defective for GOGAT are Nod$^+$ and Fix$^+$. GSI mutants of *R. meliloti* (Somerville and Kahn 1983) and *R. leguminosarum* (our observations) are Gln$^+$ in the free-living form, probably because GSII expression substitutes for GSI; they are Nod$^+$ and Fix$^+$. Mutants of *B. japonicum* lacking either GSI or GSII are Gln$^+$, Nod$^+$, Fix$^+$; double mutants are Gln$^-$, Nod$^-$ (Carlson et al. 1987). *Azorhizobium caudolinans* ORS571 contains only one GS and a mutant devoid of it is Gln$^-$, Fix$^-$ (Donald and Ludwig 1984).

Although GSI is adenylylated, nothing is known yet on the presence of an ATase in *Rhizobium* spp. We reported (Colonna-Romano et al. 1987) of a DNA sequence located upstream *glnA* coding for a protein similar to *E. coli* P_{II}. More recently (Holtel et al. 1989) we showed that this sequence complements the Gln$^-$ phenotype of a *glnB* mutant of *K. pneumoniae*. Therefore, while *glnB* and *glnA* are unlinked in enteric bacteria, *glnB* is contiguous to *glnA* in *R. leguminosarum*. A similar sequence appears to be present upstream of *glnA* in *B. japonicum* and *A. brasilense*. No transcription-termination signal is observed between *glnB* and *glnA* and S1 mapping and Northern experiments indicate (M.C., unpublished) the presence of a *glnB-glnA* transcript as well as of a *glnA* transcript. Upstream of *glnB* we find an *ntr* promoter. Since the *glnB* sequence is expressed in a minicell experiment in *E. coli* we inserted it in an appropriate vector to make a fused protein, which was purified to homogeneity. With this protein preparation we raised an antiserum (R.D., unpublished) and assayed crude extracts of *R. leguminosarum* by immunoblot. The antiserum reacts with proteins separated on non-denaturating gels giving a doublet of two very close bands (may be the urydylylated and non-uridylylated form of the protein).

The Nitrogen Control System in *Rhizobium* spp

Not very much is known about the regulation of nitrogen metabolism in *Rhizobium* spp, but the progress is quite rapid. Consensus sequences for *ntr* promoters have been described (see Gussin et al. 1986, for a review) and an *ntrA*-like gene has been reported (Ronson et al. 1987). Szeto et al. (1987) identified a *R. meliloti ntrC* gene on the basis of cross-hybridization with the *E. coli ntrC* gene. DNA sequence analysis showed that this gene is homologous to the *ntrC* gene of *K. pneumoniae* and *Bradyrhizobium* spp (Parasponia)(Nixon et al. 1986) and that an *ntrB*-like sequence is upstream of *ntrC*. The cloned sequence was used to generate a mutant and this showed that the *ntr* gene is required for growth in a poor nitrogen source (nitrate) as it is the case for a *K. pneumoniae ntrC* mutant. The latter is also Nif$^-$ while the *R. meliloti ntrC* mutant is Nod$^+$ Fix$^+$, thus indicating the presence of separate regulatory pathways for regulation of nitrogenase and of the other genes for nitrogen metabolism. A *ntrC* gene has been isolated from *A. caulinodans* ORS571 and an insertion mutant has been constructed (Pawlowski et al. 1987). Analysis of the phenotype indicates that in *A. caulinodans ntrC* is essential for efficient derepression of the *nif* genes and proper nodule development.

Acknowledgments . Partially supported by Ministero Agricoltura e Foreste, DGPA, Italy, special grant "Tecnologie Avanzate in Agricoltura" and by EEC grant CI1-0408-I. G.E. was the recipient of EEC Fellowship CI1-0242-I (SP).

References

Ali H, Niel A, Guillame JB (1981) The pathway of ammonium assimilation in *Rhizobium meliloti*. Arch Microbiol 129: 391-394

Brown CM, Dilworth MJ (1975) Ammonia assimilation by *Rhizobium* cultures and bacteroids. J Gen Microbiol 86: 39-48

Carlson TA, Guerinot M, Chelm BK (1985) Characterisation of the gene encoding glutamine synthetase I (*glnA*) from *Bradyrhizobium japonicum*. J Bacteriol 162: 698-703

Carlson TA, Chelm BK (1986) Apparent eukaryotic origin of glutamine synthetase II from the bacterium *Bradyrhizobium japonicum*. Nature 322: 568-570

Carlson TA, Martin GB, Chelm BK (1987) Differential transcription of the two glutamine synthetase genes of *Bradyrhizobium japonicum*. J Bacteriol 169: 5861-5866

Colonna-Romano S, Riccio A, Guida M, Defez R, Lamberti A, Iaccarino M, Arnold W, Priefer U, Pühler A (1987) Tight linkage of *glnA* and a putative regulatory gene in *R. leguminosarum*. Nucleic Acids Res 15: 1951-1964

Darrow RA, Crist D, Evans WR, Jones BL, Keister DL, Knotts RR (1981) In: Gibson AH, Newton WE (eds),Current Perspectives in Nitrogen Fixation. Australian Academy of Sciences, Canberra, Australia, pp 182-185

Dixon R, Eady RR, Espin G, Hill S, Iaccarino M, Kahn D, Merrick MJ (1980) Analysis of the regulation of the *Klebsiella pneumoniae* nitrogen fixation (*nif*) gene cluster with genetic fusions. Nature 286: 128-132

Donald RGK, Ludwig RA (1984) *Rhizobium* sp. strain ORS571 ammonium assimilation and nitrogen fixation. J Bacteriol 158: 1144-1151

Filser MMK, Moscatelli C, Lamberti A, Vincze E, Guida M, Salzano G, Iacarino M (1986) Characterization and cloning of two *Rhizobium leguminosarum* genes coding for glutamine synthetase activities. J Gen Microbiol 132: 2561-2569

Fuchs RL, Keister DL (1980) Comparative properties of glutamine synthetase I and II in *Rhizobium* and *Agrobacterium* spp. J Bacteriol 144: 641-648

Ginsburg A, Stadtman ER (1973) Regulation of glutamine synthetase in *Escherichia coli*. In: Prusiner S, Stadtman ER (eds) The Enzymes of Glutamine Metabolism. Acad Press, New York, pp 9-44

Gussin GN, Ronson CW, Ausubel FM (1986) Regulation of nitrogen fixation genes. Annu Rev Genet 20: 567-591

Holtel A, Colonna-Romano S, Guida M, Riccio A, Merrick MJ, Iaccarino M (1989) The *glnB* gene of *Rhizobium leguminosarum* biovar. *viciae*. FEMS Lett 58: 203-207

Howitt SM, Gresshof PM (1985) Ammonia regulation of glutamine synthetase in *Rhizobium* sp. ANU289. J Gen Microbiol 98: 477-484

Lane M, Meade J, Manian SS, O'Gara F (1986) Expression and regulation of the *Escherichia coli* glutamate dehydrogenase gene (GDH) in *Rhizobium japonicum*. Arch Microbiol 144: 29-34

Ludwig RA (1978) Control of ammonium assimilation in *Rhizobium* 32H1. J Bacteriol 135: 114-123

Magasanik B (1982) Genetic control of nitrogen assimilation in bacteria. Annu Rev Genet 16: 135-168

Ninfa AJ, Magasanik B (1986) Covalent modification of the *glnG* product NRI by the *glnL* product NRII regulates transcription of the *glnA* 5 operon in *Escherichia coli*. Proc Natl Acad Sci USA 83: 5909-5913

Nixon BT, Ronson CW, Ausubel FM (1986) Two component regulatory systems responsive to environmental stimuli share strongly conserved domains with the nitrogen assimilation regulatory genes *ntrB* and *ntrC*. Proc Natl Acad Sci USA 83: 7850-7854

O'Gara F, Birkenhead K, Noonan B, Manian SS (1988) Dicarboxylic acid utilization and regulation of nitrogen fixation in *Rhizobium* species. J Plant Physiol 132: 439-441

Osburne MS, Signer ER (1980) Ammonium assimilation in *Rhizobium meliloti*. J Bacteriol 143: 1234-1240

Pate JS (1980) Transport and partitioning of nitrogenous solutes. Annu Rev Plant Physiol 31: 313-328

Pawlowski K, Ratet R, Schell J, de Bruijn FJ (1987) Cloning and characterization of *nifA* and *ntrC* genes of the stem nodulating bacterium ORS571, the nitrogen fixing symbiont of *Sesbania rostrata*; regulation of nitrogen fixation (*nif*) genes in the free-living versus symbiontic state. Mol Gen Genet 206: 207-219

Poole PS, Dilworth MJ, Glenn AR (1987) Ammonia is the preferred nitrogen source in several rhizobia. J Gen Microbiol 133: 1707-1712

Ronson CW, Astwood PM, Downie JA (1984) Molecular cloning and genetic organization of C4-dicarboxylate transport genes from *Rhizobium leguminosarum*. J Bacteriol 160: 903-909

Ronson CW, Nixon BT, Albright LM, Ausubel FM (1987) *Rhizobium meliloti ntrA* (*rpoN*) gene is required for diverse metabolic functions. J Bacteriol 169: 2424-2431

Rossi M, Defez R, Chiurazzi M, Lamberti A, Fuggi A, Iaccarino M (1989) Regulation of glutamine synthetase isoenzymes in *Rhizobium leguminosarum* biovar. *viciae*. J Gen Microbiol 135: 629-637

Somerville E, Kahn ML (1983) Cloning of glutamine synthetase I gene from *Rhizobium meliloti*. J Bacteriol 156: 167-176

Szeto WW, Nixon BT, Ronson CW, Ausubel FM (1987) Identification and characterization of the *Rhizobium meliloti ntrC* gene: *R. meliloti* has separate regulatory pathways for an activation of nitrogen fixation genes in free-living and symbiontic cells. J Bacteriol 169: 1423-1432

Wong PP, Evans HJ (1971) Poly-B-hydroxybutyrate utilization by soybean (*Glycine max* Merr.) nodules and assessment of its role in maintenance of nitrogenase activity. Plant Physiol. 47: 750-755

OXYGEN AND NITROGENASE IN NON-HETEROCYSTOUS CYANOBACTERIA

J.R. Gallon [1]

This year, 1989, sees the centenary of the first suggestions that cyanobacteria might fix atmospheric N_2 (Frank 1889; Prantl 1889). Nevertheless, widespread interest in N_2-fixing cyanobacteria goes back only about 30 years, to the discovery that nitrogenase was rapidly and irreversibly inhibited by O_2. N_2-fixing cyanobacteria also catalyse an oxygenic photosynthesis, and much research has therefore been directed towards understanding how these organisms maintain an active nitrogenase in such an apparently hostile environment.

Over the years, there have been many reviews concerning cyanobacterial N_2 fixation, most of which mention the interrelations between O_2 and nitrogenase in these organisms. A list is given in Gallon and Chaplin (1988 a) but it is also worth consulting Gallon and Chaplin (1988 b) and Gallon (1990) for more recent information.

Most N_2-fixing cyanobacteria possess heterocysts, differentiated cells that are the site of aerobic N_2 fixation in these organisms. Heterocysts do not photoevolve O_2 and are surrounded by a thick cell wall that impedes the ingress of atmospheric O_2. The rate of diffusion of O_2 into heterocysts is balanced by respiratory O_2 consumption so that the intracellular concentration of O_2 is close to zero, and ideal for the O_2-sensitive process of N_2 fixation.

However, some unicellular and filamentous non-heterocystous cyanobacteria can fix N_2. These show no cellular differentiation and, with the possible exception (Carpenter and Price 1976) of *Trichodesmium*, a marine *Oscillatoria*, probably locate their N_2-fixing apparatus in the same cell that photoevolves O_2. Although most non-heterocystous cyanobacteria fix N_2 only when grown under subatmospheric concentrations of O_2, a few strains capable of aerobic N_2 fixation are known. These include unicellular cyanobacteria, such as *Gloeothece* (Gallon et al. 1988) and *Synechococcus* (Mitsui et al. 1986; Huang and Chow 1988), and filamentous genera, such as *Microcoleus* (Malin and Pearson 1988) and *Oscillatoria* (Stal and Krumbein 1985).

Non-heterocystous cyanobacteria that fix N_2 only microaerobically practise an avoidance strategy, favouring environments where the prevailing O_2 concentration is low. Such environments might include O_2-depleted microzones created in aggregates of cyanobacteria and heterotrophic eubacteria (Paerl and Carlton 1988). However, the tendency of certain non-heterocystous cyanobacteria to form aggregates in pure labora-

[1] Biochemistry Research Group, School of Biological Sciences, University College of Swansea, Singleton Park, Swansea, SA2 8PP, UK

tory culture could not be shown to facilitate aerobic N_2 fixation (Malin and Pearson 1988; Ohki and Fujita 1988).

In *Trichodesmium*, a spatial separation has been suggested between nitrogenase, confined to certain lightly pigmented cells, and photosynthetic O_2 evolution (Carpenter and Price 1976). Although, more recently, Ohki and Fujita (1988) have proposed that the separation is intracellular rather than intercellular, the general principle remains the same in that N_2 fixation and photosynthesis may be physically separated in this organism. However, at present, there is no evidence for a spatial separation between N_2 fixation and photosynthesis in other non-heterocystous cyanobacteria. Similarly, there is no evidence that non-heterocystous cyanobacteria produce any external barrier that might limit the ingress of atmospheric O_2 (Gallon 1990).

The interrelations between O_2 and N_2 fixation in non-heterocystous cyanobacteria have been most extensively studied in aerotolerant strains. There are, of course, differences between individual strains, and most rely on several strategies functioning in concert in order to fix N_2 aerobically. However, some general principles have emerged and it is therefore possible to write in broad terms about these strategies. The first strategy is seen when cultures are grown under alternating 12 hours light and 12 hours darkness. With the exception of *Trichodesmium*, non-heterocystous cyanobacteria grown under these conditions, fix N_2 exclusively or mainly during the period of darkness (Mullineaux et al. 1981; Stal and Krumbein 1985; Khamees et al. 1987; Huang and Chow 1988). In this way they achieve a temporal separation between N_2 fixation in the dark and photosynthetic O_2 evolution, which occurs only in the light. On the other hand, cultures of heterocystous cyanobacteria (Khamees et al 1987) and *Trichodesmium* (Bryceson and Fay 1981) fix N_2 and photoevolve O_2 simultaneously during the period of illumination. However, it should be noted that these cyanobacteria effect a spatial separation between N_2 fixation and photosynthesis, and are therefore unlikely to benefit from any additional temporal separation of these processes.

The mechanism by which most non-heterocystous cyanobacteria confine N_2 fixation to the period of darkness involves regulation of nitrogenase synthesis, but further details appear to vary according to the organism studied. For example, in *Synechococcus*, incubation under alternating light and darkness imposes an endogenous rhythm of N_2 fixation that is retained when cultures are transferred to constant illumination (Grobbelaar et al. 1986; Mitsui et al. 1986) and, in one strain, can be correlated with synchronization of cell division (Mitsui et al. 1986). In contrast, in *Gloeothece*, the pattern of N_2 fixation under alternating 12 hours light and 12 hours darkness is imposed by metabolic changes related to accumulation and degradation of glucan reserves (Gallon et al. 1988) and is rapidly lost following transfer of cultures to continuous illumination (Gallon and Chaplin 1988 a).

By effecting this temporal separation between N_2 fixation and photosynthesis, natural populations of non-heterocystous cyanobacteria avoid contact between photosynthetically produced O_2 and nitrogenase. However, the problem of atmospheric O_2 remains. In addition, when incubated in the laboratory under continuous artificial illumination, many non-heterocystous cyanobacteria simultaneously fix N_2 and photoevolve O_2. It is clear therefore that these diazotrophs must possess additional strategies for coping with O_2.

Like all aerobes, N_2-fixing non-heterocystous cyanobacteria can prevent the formation of toxic radicals from O_2 (Tözüm and Gallon 1979). Furthermore, the ability to consume O_2 by carbon- or, less usually, H_2-supported respiration functions to maintain a low intracellular concentration of O_2, more conducive to N_2 fixation. Diazotrophic non-heterocystous cyanobacteria often show high rates of respiration. In *Gloeothece*, respiratory consumption of O_2 is stimulated in the short term by elevated concentrations of O_2 (Gallon and Hamadi 1984) and is, in part, only poorly coupled to ADP phosphorylation (Maryan et al. 1986 a), implying a significant role in consuming O_2 as well as in providing energy. There is evidence that Ca^{2+} is important in respiratory O_2 consumption, not only in *Gloeothece* (Gallon and Hamadi 1984), but also in *Synechococcus* (Chen et al. 1988). However, in *Gloeothece*, respiration does not prevent the deleterious effects of superatmospheric concentrations of O_2 for more than about 45 - 60 min (Gallon 1990). Thereafter, continued N_2 fixation appears to depend on *de novo* synthesis of nitrogenase in order to replace O_2-inactivated enzyme. Consistent with this, cultures of *Gloeothece* (Maryan et al. 1986 b) and, perhaps, also *Oscillatoria* sp. UCSB 25 (Gallon and Chaplin 1988 b) differ from other diazotrophs in that O_2 does not permanently repress nitrogenase synthesis, though a transient inhibition is observed when cultures are transferred from air to 1 atm of O_2. Incubation under O_2 abolishes nitrogenase activity in *Gloeothece*, but not in *Oscillatoria* UCSB 25. However, it is not yet known whether *Oscillatoria* nitrogenase is modified to give an O_2-stable form, as occurs in certain heterocystous cyanobacteria (Tabita et al. 1988; Reich and Böger 1989).

The inhibitory effect of O_2 on nitrogenase synthesis, found in most diazotrophs, may be exerted through regulation of the degree of supercoiling of one or more of the *nif* genes (Kranz and Haselkorn 1986; Dimri and Das 1988; Dixon et al. 1988). Facultative anaerobes carry the genetic information for growth under both aerobic and anaerobic conditions, but those genes needed for anaerobic growth are expressed only in the absence of O_2, whilst those genes needed for aerobic growth are expressed only in the presence of O_2. Incubation of *Escherichia coli* or *Salmonella typhimurium* under anaerobic conditions increases the degree of negative supercoiling of chromosomal DNA (Dorman et al. 1988) and a model has been proposed in which those genes expressed only anaerobically are transcribed preferentially from negatively supercoiled promoters, whilst aerobically expressed genes are transcribed most strongly from relaxed promoters. Several genes have been identified as candidates for this form of regulation (Yamamoto and Droffner 1985; Dorman et al. 1988). In most diazotrophs, the *nif* genes are transcribed only anaerobically so, according to the above model, would be expressed preferentially from supercoiled promoters. Support for this comes from several observations. Firstly, inhibition of the enzyme system that introduces negative supercoils into DNA decreased the expression of a regulatory *nif* gene in *Rhodobacter capsulatus* (Kranz and Haselkorn 1986). Secondly, anaerobic expression of the *Klebsiella pneumoniae nif* genes in *E. coli* was dependent upon either the ability of the host cells to catalyse DNA supercoiling or the simultaneous presence of a constitutively expressed positive regulatory *nif* A gene (Dimri and Das 1988). Thirdly, *in vitro* transcription from the promoter of the regulatory *nif*LA operon of *K. pneumoniae* was dependent on negative supercoiling at physiological salt concentrations (Dixon et al. 1988).

The degree of supercoiling of intracellular DNA is affected by several factors. These include transcription, which creates positive supercoils (overwinding of the double helix) ahead of RNA polymerase and negative supercoils (underwinding of the double helix) behind this enzyme (Wu et al. 1988), and also the action of DNA topoisomerases (Wang 1985). In bacteria, two topoisomerases predominate: DNA topoisomerase I, which relaxes negatively supercoiled DNA, and DNA topoisomerase II (DNA gyrase), which relaxes positive supercoils and also introduces negative supercoils into relaxed DNA. Together, these enzymes therefore regulate the degree of supercoiling of the bacterial chromosomes. This supercoiling may be local (related, for example, to transcription) or more general.

Aerobically grown cultures of *S. typhimurium* contain an active DNA topoisomerase I, but show no significant DNA gyrase activity. In contrast, anaerobically grown cultures show gyrase activity but no significant activity of topoisomerase I. Furthermore, mutants lacking any detectable topoisomerase I activity are strict anaerobes and have a negatively supercoiled chromosome, whilst mutants with no detectable gyrase activity are strict aerobes and contain relaxed DNA (Yamamoto and Droffner 1985). The balance between the activities of these two enzymes may therefore regulate the expression of 'aerobic' and 'anaerobic' genes, through the degree of supercoiling of all or part of the chromosome. That topoisomerases may also have a role in diazotrophs is suggested by the observations that inhibition of DNA gyrase prevents anaerobic expression of the *nif* genes in *R. capsulatus* and *K. pneumoniae* (Kranz and Haselkorn 1986; Dixon et al. 1988).

Gloeothece differs from other diazotrophs in that nitrogenase synthesis in this cyanobacterium is only transiently, not permanently, repressed by O_2 (Maryan et al. 1986 b). Furthermore, nitrogenase is synthesized in both aerobically and microaerobically incubated cultures, though the rate of synthesis may be very slightly greater under N_2 than in air (Gallon and Chaplin 1988 a). Extracts of aerobically grown *Gloeothece* contain an active DNA gyrase, as measured by their ability to convert relaxed ColE1 plasmid DNA into a supercoiled form (Fig. 1). Furthermore, the activity of enzymes that introduce negative supercoils into ColE1 DNA (presumably DNA gyrase), relative to that of DNA-relaxing enzymes (presumably DNA topoisomerase I), fluctuated in parallel with changes in the rate of nitrogenase synthesis in cultures of *Gloeothece*

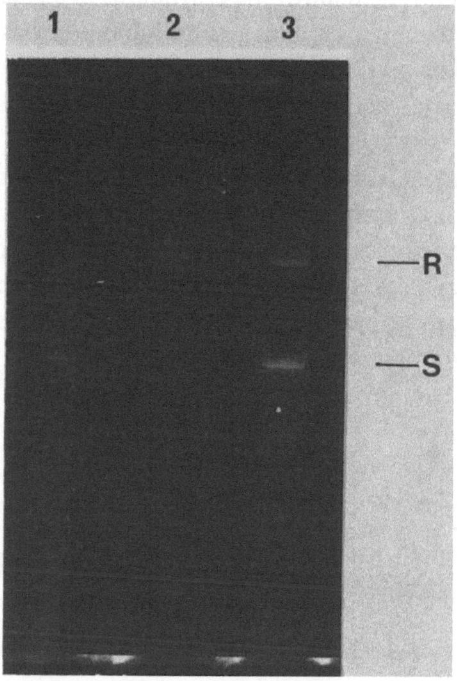

Fig. 1. Supercoiling of relaxed ColE1 plasmid DNA by a cell-free extract of aerobically grown *Gloeothece*.
Track 1: complete assay; *Track 2*: boiled extract of *Gloeothece* ; *Track 3*: standard of partially relaxed ColE1 DNA, showing relaxed (**R**) and supercoiled (**S**) bands.

Relaxed ColE1 DNA was isolated from partially relaxed plasmid by electrophoresis in 1% (w/v) agarose. After elution, approximately 15 fmol was incubated for 30 min in 50 mM tris-HCl (pH 7.6), 20 mM KCl, 10 mM $MgCl_2$, 2 mM dithiothreitol, 1.5 mM ATP, 5 mM spermidineCl_3 and 50 μg/ml of bovine serum albumin, along with an extract of *Gloeothece* containing 20 μg protein. The total assay volume was 30 μl. The reaction was terminated by addition of 10 μl of a solution containing 0.2 μg sodium lauryl sulphate, 2 μg Ficoll and 4 μg of bromophenol blue. The incubation mixture was then subjected to electrophoresis in 1% agarose in TBE buffer (89 mM tris, 89 mM boric acid, 2.5 mM EDTA). The gel was stained with ethidium bromide (0.5 μg/ml) for 15 min and photographed under UV light

transferred from air to 1 atm of O_2 (Gallon et al. 1989). In aerobically and, to a greater extent, in microaerobically grown cultures, DNA gyrase activity predominated over that of DNA topoisomerase I. However, during the first 2 hours following transfer of aerobically grown cultures to 1 atm of O_2 (the period when nitrogenase synthesis was inhibited) DNA topoisomerase I activity predominated over that of DNA gyrase. After longer incubation under O_2, nitrogenase synthesis recovered and DNA gyrase activity again predominated over that of topoisomerase I. Transient inhibition of nitrogenase synthesis may therefore correlate with transient relaxation of all or part of the *Gloeothece* chromosome.

Inhibitors of DNA gyrase prevented nitrogenase synthesis to a greater extent in aerobic cultures of *Gloeothece*, though a small degree of inhibition was observed under microaerobic conditions (Gallon and Chaplin 1988 a). This difference could be a consequence of lower topoisomerase I activity in microaerobically incubated cells, so that inhibition of DNA gyrase did not result in significant relaxation of the *nif* genes during the 6 h duration of the experiment. However, this remains to be shown directly.

It therefore appears that in at least one N_2-fixing cyanobacterium, *Gloeothece*, nitrogenase synthesis may be regulated by O_2 through the relative activities of DNA supercoiling and relaxing enzymes. However, because elevated concentrations of O_2 only transiently inhibit nitrogenase synthesis in this organism, this regulation may be more subtle than in other diazotrophs.

References

Bryceson I, Fay P (1981) Nitrogen fixation in *Oscillatoria (Trichodesmium) erythrea* in relation to bundle formation and trichome distribution. Mar Biol 61: 159-166

Carpenter EJ, Price CC (1976) Marine *Oscillatoria (Trichodesmium)*: explanation for aerobic nitrogen fixation without heterocysts. Science (Wash) 191: 1278-1280

Chen T-H, Huang T-C, Chow T-J (1988) Calcium requirement in nitrogen fixation in the cyanobacterium *Synechococcus* RF-1. Planta 173: 253-256

Dimri GP, Das HK (1988) Transcriptional regulation of nitrogen fixation genes by DNA supercoiling. Mol Gen Genet 212: 360-363

Dixon RA, Henderson NC, Austin S (1988) DNA supercoiling and aerobic regulation of transcription from the *K. pneumoniae nif* LA promoter. In: Bothe H, de Bruijn FJ, Newton WE (eds) Nitrogen Fixation: Hundred Years After. Fischer, Stuttgart New York, p 306

Dorman CJ, Barr GC, Ni Bhriain N, Higgins CF (1988) DNA supercoiling and the anaerobic and growth phase regulation of *ton B* expression. J Bacteriol 170: 2816-2826

Frank B (1889) Über den experimentellen Nachweis der Assimilation freien Stickstoffs durch erdbodenbewohnende Algen. Ber deut bot Ges 7: 34-42

Gallon JR (1990) The physiology and biochemistry of N_2 fixation by non-heterocystous cyanobacteria. Phykos (to be published)

Gallon JR, Chaplin AE (1988 a) Nitrogen fixation. In: Rogers LJ, Gallon JR (eds) Biochemistry of the Algae and Cyanobacteria. Oxford University Press, Oxford, pp 147-173

Gallon JR, Chaplin AE (1988 b) Recent studies on N_2 fixation by non-heterocystous cyanobacteria. In: Bothe H, de Bruijn FJ, Newton WE (eds) Nitrogen Fixation: Hundred Years After. Fischer, Stuttgart New York, pp 183-188

Gallon JR, Dingley KH, Le Noury PJ, Hipkin CR (1989) Topoisomerase activity in a diazotrophic cyanobacterium. Biochem Soc Trans 17: 926-927

Gallon JR, Hamadi AF (1984) Studies on the effects of oxygen on acetylene reduction (nitrogen fixation) in *Gloeothece* sp ATCC 27152. J Gen Microbiol 130: 495-503

Gallon JR, Perry SM, Rajab TMA, Flayeh KAM, Yunes JS, Chaplin AE (1988) Metabolic changes associated with the diurnal pattern of N_2 fixation in *Gloeothece*. J Gen Microbiol 134: 3079-3087

Grobbelaar N, Huang T-C, Lin H-Y, Chow T-J (1986) Dinitrogen-fixing endogenous rhythm in *Synechococcus* RF-1. FEMS Microbiol Lett 37: 173-177

Huang T-C, Chow T-J (1988) Comparative studies of some nitrogen-fixing unicellular cyanobacteria isolated from rice fields. J Gen Microbiol 134: 3089-3097

Khamees HS, Gallon JR, Chaplin AE (1987) The pattern of acetylene reduction by cyanobacteria grown under alternating light and darkness. Br Phycol J 22: 55-60

Kranz RG, Haselkorn R (1986) Anaerobic regulation of nitrogen-fixation genes in *Rhodopseudomonas capsulata*. Proc Natl Acad Sci USA 83: 6805-6809

Malin G, Pearson HW (1988) Aerobic nitrogen fixation in aggregate-forming colonies of the non-heterocystous cyanobacterium *Microcoleus chthonoplastes*. J Gen Microbiol 134: 1755-1763

Maryan PS, Eady RR, Chaplin AE, Gallon JR (1986 a) Nitrogen fixation by *Gloeothece* sp (PCC 6909): Respiration and not photosynthesis supports nitrogenase activity in the light. J Gen Microbiol 132: 789-796

Maryan PS, Eady RR, Chaplin AE, Gallon JR (1986 b) Nitrogen fixation by the unicellular cyanobacterium *Gloeothece*. Nitrogenase synthesis is only transiently repressed by oxygen. FEMS Microbiol Lett 34: 251-255

Mitsui A, Kumizawa S, Takahashi A, Ikemoto H, Cao S, Arai T (1986) Strategy by which nitrogen-fixing unicellular cyanobacteria grow photoautotrophically. Nature (Lond) 323: 720-722

Mullineaux PM, Gallon JR, Chaplin AE (1981) Acetylene reduction (nitrogen fixation) by cyanobacteria grown under alternating light-dark cycles. FEMS Microbiol Lett 10: 245-247

Ohki K, Fujita Y (1988) Aerobic nitrogenase activity measured as acetylene reduction in the marine non-heterocystous cyanobacterium *Trichodesmium* spp. grown under artificial conditions. Mar Biol 98: 111-114

Paerl HW, Carlton RG (1988) Control of nitrogen fixation by oxygen depletion in surface-associated microzones. Nature (Lond) 332: 260-262

Prantl KAE (1889) Die Assimilation freien Stickstoffs und der Parasitismus von *Nostoc*. Hedwigia 28: 135-136

Reich S, Böger P (1989) Regulation of nitrogenase activity in *Anabaena variabilis* by modification of the Fe protein. FEMS Microbiol Lett 58: 81-86

Stal LJ, Krumbein WE (1985) Nitrogenase activity in the non-heterocystous cyanobacterium *Oscillatoria* sp grown under alternating light-dark cycles. Arch Microbiol 143: 67-71

Tabita FR, Van Baalen C, Smith RL, Kumar D (1988) *In vivo* protection and recovery of nitrogenase from oxygen in heterocystous cyanobacteria. In: Bothe H, de Bruijn FJ, Newton WE (eds) Nitrogen Fixation: Hundred Years After. Fischer, Stuttgart New York, pp 189-194

Tözüm SRD, Gallon JR (1979) The effects of methyl viologen on *Gloeothece* sp LB795 and their relationship to the inhibition of acetylene reduction (nitrogen fixation) by oxygen. J Gen Microbiol 111: 313-326

Wang JC (1985) DNA topoisomerases. Annu Rev Biochem 54: 665-697

Wu H-Y, Shyy S, Wang JC, Liu LF (1988) Transcription generates positively and negatively supercoiled domains in the template. Cell 53: 433-440

Yamamoto N, Droffner ML (1985) Mechanisms determining aerobic or anaerobic growth in the facultative anaerobe *Salmonella typhimurium*. Proc Natl Acad Sci USA 82: 2077-2081

PROPERTIES OF THREE FERREDOXINS FROM THE PHOTOSYNTHETIC BACTERIUM *RHODOBACTER CAPSULATUS*: STUDIES ON THEIR POSSIBLE ROLE IN NITROGEN FIXATION

Y. Jouanneau [1]

Introduction

Rhodobacter capsulatus is a facultative phototroph endowed with a very versatile metabolism. This bacterium is able to thrive under various conditions including fermentative, heterotrophic and autotrophic type of growth, in either light or darkness. It is able to metabolize various organic and inorganic nitrogen sources including molecular dinitrogen. These remarkable properties make such a microorganism a valuable tool for studies on nitrogen fixation and assimilation. Regulation of nitrogen fixation was investigated both at the enzymatic and genetic levels in *R. capsulatus* (reviewed in Vignais et al. 1985). Synthesis of nitrogenase, the enzyme which catalyzes N_2-fixation, is repressed by NH_4^+ and O_2 as in other diazotrophic bacteria (Hallenbeck et al. 1982 b). In addition, nitrogenase synthesis was found to be closely dependent on light intensity (Jouanneau et al. 1984, 1985). Besides, nitrogenase is subjected to a feedback inhibition by NH_4^+. This short-term regulation involves the covalent modification by ADP-ribosylation of the component 2 of nitrogenase (Jouanneau et al. 1983, 1989).

Nitrogenase is a metalloenzyme composed of two different components. Component 1 is an $\alpha_2\beta_2$ tetramer containing Mo, Fe and inorganic S atoms arranged in several clusters including the FeMoCo, a cofactor present at the active site of the enzyme. Component 2 is a holodimer containing a single [4Fe-4S] cluster and functioning as a specific reductase of component 1.

The enzyme function requires ATP and low-potential reducing equivalents. *In vivo*, only ferredoxins or flavodoxins have a midpoint potential low enough to serve as electron donors to nitrogenase. The electron transport chain providing reductant to nitrogenase is known for sure in only one bacterial species, *Klebsiella pneumoniae*. In this microorganism, a specific oxidoreductase encoded by *nifJ* uses pyruvate as substrate to reduce a flavodoxin (encoded by *nifF*), which serves as physiological electron donor to nitrogenase (Nieva-Gomez et al. 1980; Shah et al. 1983). In *R. capsulatus* and in photosynthetic bacteria in general, the electron donor to nitrogenase as well as the biochemical components involved in its reduction are still unknown.

In this article, we present some biochemical properties of the three ferredoxins so far isolated from *R. capsulatus*. The possible implication of one of them, called FdI, in

[1] Biochimie Microbienne (CNRS UA 1130), Département de Recherche Fondamentale, Centre d'Etudes Nucleaires de Grenoble, 85X, 38041 Grenoble cedex, France

the electron transport to nitrogenase is discussed in the light of our recent genetic studies.

Identification and Properties of Three Ferredoxins from *R. capsulatus*

Most bacterial ferredoxins are composed of a single polypeptide ranging in size from 6 to 13 kDa and containing one or two [4Fe-4S] or [3Fe-xS] (x = 3 or 4) cluster(s). The structure and function of these proteins have been reviewed recently (Bruschi and Guerslesquin 1988). Two ferredoxins had previously been isolated from *R. capsulatus* (Hallenbeck et al. 1982; Yakunin and Gogotov 1983). We now report the identification of a third ferredoxin and compare the properties of these three ferredoxins.

Fd I

FdI was the first ferredoxin to be isolated from *R. capsulatus* (Hallenbeck et al. 1982 a). It appeared as a polypeptide of about 12 kDa containing approximately 8 atoms each of iron and sulfur, consistent with the existence of two iron sulfur clusters (Hallenbeck et al. 1982 ; Yakunin and Gogotov 1983). The primary structure was determined as deduced from the nucleotide sequence of the structural gene and confirmed by amino acid sequencing of 80% of the purified protein (Schatt et al., submitted). The FdI poly-peptide is composed of 64 amino acids and has a calculated molecular weight of 6,728.

Table 1. Comparison of the properties of three ferredoxins from *R. capsulatus*

Properties		FdI	FdII	FdIII
Molecular weight		6,728[a]	22,000	10,690[a]
Atom content mol/mol	Fe	8.0 0.2	nd[b]	7.6 0.2
	S	7.0 0.5	nd	6.5 0.5
Absorbance ratio		A385/A285=0.81	A400/A280=0.54	A385/A282=0.66
In vitro electron carrier to nitrogenase		yes	yes	no
Synthesis repressed by NH_4^+		yes	no	yes

[a] Value calculated from the amino acid sequence of the apoprotein ; [b] not determined

The discrepancy between the apparent molecular weight of 12,000 determined by gel filtration and the actual value of 6,728 may be due to the atypical behaviour of FdI towards the molecular sieve matrix. We re-estimated the atom content of iron and inorganic sulfur based on a molecular weight of 6,728 and found values close to 8 (Table 1). This result suggests the presence of two [4Fe4S] clusters. In accordance, the pri-

```
A.sp.   AYTI-TSQCISCKLCSSVCPTGAIKIAENG
        |  |    | *|  |  *||* ||
R.m.    AFKIIASQCTQCGACEFECPRGAINFKGE
        * ||    || |* ||  ||  ||  |
B.j.    PFKIIASQCTSCSACEPLCPNVAISEKGG
          ||    ||||  |||  ||  ||  | |
R.c.    AMKIDPELCTSCGDCEPVCPTNAIAPKKG
        || ||   || |||| |||||  |    |
A.c.    AMAIDGYECTVCGDCKPVCPTGSIVLQGG
        | *|  * |  |  |  |||  *|
A.v.    ALKI-VESCVNCWACVDVCPSEAISLAGP

        1   5    10   15   20   25

                    V              C
A.sp.   QHWIDSELCTNCDTVYTVPQCKAGCPTDG-CVKVPS---    36%
           |    |* |      *||   | **  |
R.m.    KYVIDPTKCNECKGGFDTQQCASVCPVSNTCV-PA        45%
         *||    | |* |  |  ||   |   | |  **
B.j.    NFVIEAAKCSELRGHFDEPQCAAACPVDNTCVVDRA---    48%
          || |  | |  |  | |||  ||  || |    |
R.c.    VYVINADTCTECEGEHDLPQCVNACMTDN-CINPAA      100%
         ||| || | ||      | |   |  |  || |
A.c.    IYVIDADSCNEC-ADLGEPRCLGVCPVDF-CIQPLDD      50%
           | |  |*|| *     *|   |      |  *
A.v.    HFEISASKCTECDGDYAEKQCASICPVEG-AILLAD---    37%

        30   35   40   45   50   55     60  64
```

Fig. 1. Comparison of the *R. capsulatus* FdI sequence with the deduced sequences of ferredoxin-like genes located next to *nif* genes. A. sp.: *Anabaena* sp. PCC 7120; R.m.: *Rhizobium meliloti* (Mulligan et al. 1988) ; B.j.: *Bradyrhizobium japonicum* (Ebeling et al. 1988) ; R.c.: *R. capsulatus* ; A.c.: *Azotobacter chroococcum* (Robson et al. 1986); A.v.: *Azotobacter vinelandii* (Joerger and Bishop 1988). Bars and stars indicate identical amino acids between FdI and other ferredoxins. Percentages on the right are relative homologies

mary structure contains two sets of 4 cysteines that are likely ligands for two such clusters. Sequence comparisons indicate that FdI has characteristics common to ferredoxins from other photosynthetic bacteria. Maximal homology (60%) was observed with a ferredoxin from *Rhodopseudomonas palustris* (Minami et al. 1984). Besides, FdI shows significant homology with polypeptides deduced from genes found next to *nif* genes in other diazotrophic bacteria (Fig. 1). FdI shares up to 50% identical amino acids with the product of a gene of *Azotobacter chroococcum*, located next to *nifH**, the structural gene for the component 2 of the vanadium-nitrogenase in this bacterium (Robson et al. 1986). Ferredoxin-like proteins encoded by genes adjacent to *nifB* in various N_2-fixing species present homologies ranging from 36 to 48% with FdI (Fig.1).

FdI was shown to function as an electron carrier to nitrogenase in *in vitro* experiments (Hallenbeck et al. 1982a ; Yakunin and Gogotov 1983). We failed to detect FdI in cells grown with excess NH_4^+ suggesting that its synthesis is repressed by fixed nitrogen (Table I). FdI synthesis was also reported to be dependent on light (Yakunin and Gogotov 1983).

Fd II

We recently isolated FdII from N deficient and N sufficient grown cultures of *R. capsulatus*. It appears as a polypeptide with a molecular weight of about 22,000 as determined by gel filtration and SDS-polyacrylamide gel electrophoresis. FdII exhibits a characteristic absorption spectrum with maxima at 280 and 400 nm and an absorbance ratio of 0.54 in the oxidized state (Table 1). It was found to be active *in vitro* as an electron donor to nitrogenase. These properties are very similar to those of a ferredoxin (FdII) previously described (Yakunin and Gogotov 1983) and therefore it is probably the same protein. The N-terminal amino acid sequence was determined up to the thirtieth residue (unpublished data). It shows 86% homology with the corresponding part of the sequence of *Azotobacter vinelandii* FdI (Howard et al. 1983). Av FdI comprises 106 amino acids and contains two unequal [4Fe-4S] and [3Fe-xS] clusters. This 7Fe-ferredoxin has a molecular weight of about 14,000 by gel filtration (Yoch and Arnon 1972), much smaller than that of FdII. However, the techniques we used might have overestimated the molecular weight of FdII. Preliminary data indicate that FdII is a very acidic protein, suggesting that it may migrate atypically on polyacrylamide gel and on gel filtration columns. Contrary to Av FdI, FdII was reported to contain a single [4Fe-4S] cluster (Yakunin and Gogotov 1983). The nature and the number of iron sulfur clusters present in FdII should be re-examined.

Fd III

FdIII was isolated together with FdI from N deficient grown cells. It is present in 4 to 5 times lesser amounts than FdI in N limited cells. The absorption spectrum of FdIII is typical of [4Fe-4S] ferredoxins with a maximum for the chromophore at 385nm and an absorbance ratio A385/A282 of 0.66. The extinction coefficient at 385nm, $\varepsilon_{385} = 29$ mM^{-1} cm^{-1}, and the atom content of iron and sulfur are consistent with the presence of two [4Fe-4S] clusters (Table 1). FdIII has no detectable activity as electron carrier to nitrogenase. Its synthesis is apparently repressed when cells are grown with an excess of fixed nitrogen. Amino acid analysis and partial sequencing showed that FdIII is composed of 100 residues including 3 Tyr, 1 Trp and 9 Cys (unpublished data). The primary structure was partially established by direct sequencing of 75% of the residues. Two sets of 4 cysteines that may accomodate two [4Fe-4S] clusters were observed. This is in accordance with the spectral and chemical data (Table 1). The product of a recently identified ORF, called ORF5, has a deduced sequence that perfectly matches that of FdIII indicating that it is probably the structural gene (Moreno-Vivián et al. 1989). This gene is located next to *nifQ* and is probably transcribed together with other *nif* genes, including *nifE*, *nifN*, *nifX* that are known to be involved in the synthesis or processing of the FeMoCo. Hence, FdIII is likely to be implicated in the processing of the component I of nitrogenase.

Function of FdI in N_2 Fixation: A Genetic Analysis

We have recently cloned and sequenced the structural gene encoding FdI (Schatt et al., submitted). It was called *fdxN*, like the ferredoxin-like genes found next to *nifB* in *Rhizobium meliloti* and *Anabaena* PCC7120 (Mulligan et al. 1988). In *R. capsulatus*, there are two copies of *nifB* and no ferredoxin coding sequence was found in the vicinity of either copy (Masepohl et al. 1988). However, FdI is almost 50% homologous to the products of *fdxN* and to the products of other *nif*-linked, ferredoxin-like genes (Fig.1). *R. capsulatus fdxN* was mapped in a region of the chromosome containing a cluster of *nif* genes. Transcriptional analysis revealed two *fdxN*-specific mRNAs of 330 and 750 bases, respectively. None of these transcripts was detected in NH_4^+-grown cells, providing direct evidence that synthesis of FdI is repressed by excess fixed nitrogen. A consensus sequence typical of *nif* promoters was detected 100 bp upstream of *fdxN*. However, a detailed transcription mapping suggested that *fdxN* might be transcribed from a promoter further upstream (Schatt et al., submitted). Indeed, an ORF was identified immediately upstream of *fdxN* (unpublished data).

These data strongly support the idea that FdI, the product of *fdxN*, is implicated in N_2-fixation. FdI and FdII can potentially serve as physiological electron donors to nitrogenase (Table 1). However, FdI appears to be more specific to N_2-fixation since its synthesis is regulated like *nif* genes and its structural gene is located within a *nif* region on the chromosome.

Acknowledgements. I am grateful for the contributions of E. Schatt, C. Meyer and C. Duport to this work. I thank R. Cesarini for typing the manuscript.

References

Bruschi M, Guerlesquin F (1988) Structure, function and evolution of bacterial ferredoxins. FEMS Microbiol Rev 54: 155-176

Ebeling S, Noti JD, Hennecke H (1988) Identification of a new *Bradyrhizobium japonicum* gene (*frxA*) encoding a ferredoxin-like protein. J Bacteriol 170: 1999-2001

Hallenbeck PC, Jouanneau Y, Vignais PM (1982 a) Purification and molecular properties of a soluble ferredoxin from *Rhodopseudomonas capsulata*. Biochim Biophys Acta 681: 168-176

Hallenbeck PC, Meyer CM, Vignais PM (1982 b) Regulation of nitrogenase in the photosynthetic bacterium *Rhodopseudomonas capsulata* as studied by two-dimensional gel electrophoresis. J Bacteriol 151: 1612-1616

Howard JB, Lorsbach TW, Ghosh D, Melis K, Stout CD (1983) Structure of *Azotobacter vinelandii* 7Fe-ferredoxin. Amino acid sequence and electron density maps of residues. J Biol Chem 258: 508-522

Joerger RD, Bishop PE (1988) Nucleotide sequence and genetic analysis of the *nif B-nifQ* region from *Azotobacter vinelandii*. J Bacteriol 170: 1475-1487

Jouanneau Y, Lebecque S, Vignais PM (1984) Ammonia and light effect on nitrogenase activity in nitrogen limited continuous cultures of *R. capsulatus*. Role of glutamine synthetase. Arch Microbiol 139: 326-331

Jouanneau Y, Meyer C, Vignais PM (1983) Regulation of nitrogenase activity through iron protein interconversion into an active and an inactive form in *Rhodopseudomonas capsulata*. Biochim Biophys Acta 749: 318-328

Jouanneau Y, Meyer C, Vignais PM (1989) ADP ribosylation of dinitrogenase reductase in *Rhodobacter capsulatus*. Biochemistry 28: 6524-6530

Jouanneau Y, Wong B, Vignais PM (1985) Stimulation by light of nitrogenase synthesis in cells of *Rhodopseudomonas capsulata* growing in N-limited continuous cultures. Biochim Biophys Acta 888: 149-155

Masepohl B, Klipp W, Pühler A (1988) Genetic characterization and sequence analysis of the duplicated *nifA/nifB* gene region of *Rhodobacter capsulatus*. Mol Gen Genet 212: 27-37

Minami Y, Wakabayashi S, Yamada F, Wada K, Zumft WB, Matsubara H (1984) Ferredoxins from the photosynthetic purple non-sulfur bacterium *Rhodopseudomonas palustris*. Isolation and amino acid sequence of ferredoxin I. J Biochem 96: 585-592

Moreno-Vivián C, Hennecke S, Pühler A, Klipp W (1989) Open reading frame 5 (ORF5) encoding a ferredoxin-like protein, and *nifQ* are cotranscribed with *nifE*, *nifN*, *nifX* and ORF4 in *Rhodobacter capsulatus*. J Bacteriol 171: 2591-2598

Mulligan ME, Buikema WJ, Haselkorn R (1988) Bacterial-type ferredoxin genes in the nitrogen fixation regions of the cyanobacterium *Anabaena* sp. strain PCC7120 and *Rhizobium meliloti*. J Bacteriol 170: 4406-4410

Nieva-Gomez D, Roberts GP, Klevickis S, Brill WJ (1980) Electron transport to nitrogenase in *Klebsiella pneumoniae*. Proc Natl Acad Sci 77: 2555-2558

Robson RL, Woodley P, Jones R (1986) Second gene (*nifH**) coding for a nitrogenase iron protein in *Azotobacter chroococcum* is adjacent to a gene coding for a ferredoxin-like protein. EMBO J 5: 1159-1163

Shah VK, Stacey G, Brill WJ (1983) Electron transport to nitrogenase. Purification and characterization of pyruvate: flavodoxin oxidoreductase, the *nifJ* gene product. J Biol Chem 258: 12064-12068

Vignais PM, Colbeau A, Willison JC, Jouanneau Y (1985) Hydrogenase, nitrogenase and hydrogen metabolism in the photosynthetic bacteria. Adv Microb Physiol 26: 155-234

Yakunin AF, Gogotov IN (1983) Properties and regulation of synthesis of two ferredoxins from *Rhodopseudomonas capsulata*. Biochim Biophys Acta 725: 298-308

Yoch DC, Arnon DI (1972) Two biologically active ferredoxins from the aerobic nitrogen fixing bacterium *Azotobacter vinelandii*. J Biol Chem 247: 4514-4520

Part 5 Enzymes: Genetics and Molecular Biology

BIOCHEMISTRY, MOLECULAR GENETICS AND REGULATION OF NITRATE REDUCTASE IN *NICOTIANA PLUMBAGINIFOLIA*, TOBACCO AND TOMATO

P. Rouzé [1], I. Chérel [1], F. Daniel-Vedele [1], M. Deng [1], J. Gabard [1], M. Gonneau [1], M. Kavanagh [1], J. Levin [2], A. Marion-Poll [1], C. Meyer [1], T. Moureaux [1], F. Pelsy [1], S. Pouteau [1], H. Vaucheret [1], M. Vincentz [1] and M. Caboche [1]

Introduction

The nitrate assimilation pathway is known to be tightly controlled in plants by many endogenous and environmental factors (Beevers and Hageman 1983), the most important being the nature and availability of the nitrogen source, and light. Nitrate compartmentalization and flux between plant tissues and inside plant cells are clearly the first critical steps in that control, but are, as yet, refractory to molecular analysis. The second step, nitrate reduction into ammonium, is catalysed by two enzymes, a likely cytosolic nitrate reductase (NR), and a plastidial nitrite reductase, both induced by nitrate. Interestingly, each factor known to control the overall pathway appears to control NR, which is thus thought to play a central role in regulation of the nitrate assimilation pathway.

In *Nicotianae* only one NR isoform (NADH:NR, EC 1.6.6.1) has ever been found. Genetic analysis strongly suggests that there is only one NR structural gene per haploid genome in these species, since, among the nitrate reductase deficient mutants (NR^-), only one complementation group attributable to the NR structural gene (*nia*) was found in a true diploid species as *N. plumbaginifolia* (Gabard et al. 1987). Tobacco, an amphidiploid species thought to be a hybrid between *N. sylvestris* and *N. tomentosiformis*, possesses two functional homeologous *nia* loci, *nia-1* and *nia-2* (Müller 1983). The recent molecular data obtained for tobacco and tomato in our laboratory agree with genetics, and suggest that a single *nia* gene per haploid genome could be the rule in *Solanaceae*. These species that may also be regenerated from protoplasts into plants, are therefore particularly well suited to study the regulation of NR activity using cellular genetics and molecular biology.

Bio- and Immunochemistry of Nitrate Reductase from *Nicotianae*

Active NADH:NR has only recently been purified from higher plants. Isolation of active NR remains a difficult task, because of its very low concentration and stability,

---

[1] Laboratoire de Biologie Cellulaire, INRA, Centre de Versailles, 78026 Versailles cedex, France
[2] Laboratoire d'Ingénierie des Protéines, INRA, Domaine de Vilvert, 78350 Jouy en Josas, France

especially in *Solanaceae*. NADH:NR was purified from *N. plumbaginifolia* leaves in our laboratory. Using 5'AMP-sepharose affinity chromatography, a 1000-fold enrichment was obtained. The enriched preparation remained heavily contaminated by various proteins, the main one being ribulose 1,5-bisphosphate carboxylase/oxygenase (Rubisco), and contained about 6 - 8% of pure NR. It has been used to raise and test monoclonal antibodies (mAb) against *N. plumbaginifolia* NR (Meyer et al., submitted) and, as a by-product, against the contaminating Rubisco.

A 20,000-fold purification was obtained through a two-step immunoaffinity protocol, using at first an anti-maize NR mAb column (ZM.96:9:25), then an anti-Rubisco mAb column (NP 7-64). The purified preparation has a NADH:NR specific activity of 80 units/mg protein. In SDS-PAGE, this preparation constantly shows two bands corresponding to apparent molecular masses of 107 and 99.5 kDa. These bands are in variable proportions, depending on the physiological state of the leaves, and are both related to an active form of NR. These two forms probably come from an *in vivo* post-translational modification of NR. Nevertheless, an artefactual *in vitro* modification of the native enzyme during the isolation process cannot be totally excluded (Moureaux et al. 1989).

From the estimated size of the native protein, and the finding of a single NR apoenzyme gene, *Nicotiana* NR is thus shown to be a homodimer, as proposed for other well-characterized higher plant NRs (Campbell 1988). Each monomer contains the three prosthetic groups, FAD, heme, and molybdopterin (MoCo). Electrons are transferred from NADH to nitrate through these three successive cofactors. In addition to this physiological NADH:NR activity, *Nicotiana* NR displays the usual *in vitro* partial activities associated with NR, and classified either as dehydrogenase when NADH is utilized to reduce ferricyanide and cytochrome *c* as electron acceptors, or as terminal activity when nitrate is reduced by electron donors such as reduced viologens (MV), flavin mononucleotide (FMN) and bromophenol blue (BPB). This recently discovered BPB:NR activity (Hoarau et al. 1986) was shown to differ in several respects (inhibition by antibodies, redox centers involved) from the two other terminal activities.

Nine mAb were first obtained against maize NR (ZMmAb). They were classified into six epitopic classes, according to reciprocal competition for maize NR (Chérel et al. 1985). Among these, only one was shown to recognize any dicot NR (Chérel et al. 1986). In this latter class, mAb ZM96:9:25 inhibited NADH:NR and all partial activities except BPB:NR which was activated (Meyer et al. 1987). This mAb was utilized thereafter as reagent for wild-type or mutant NR protein measurement as first antibody in a two-site ELISA (see below) and for immuno-affinity purification of the wild-type or mutant NR.

Eighteen mAb were later raised against *N. plumbaginifolia* NR (NPmAb). Unlike to ZMmAb, all NPmAb, except one, share their epitopes with another one (Fig. 1). These epitopes (and the ZM 96:9:25 one) thus form a continuous structure on the *N.*

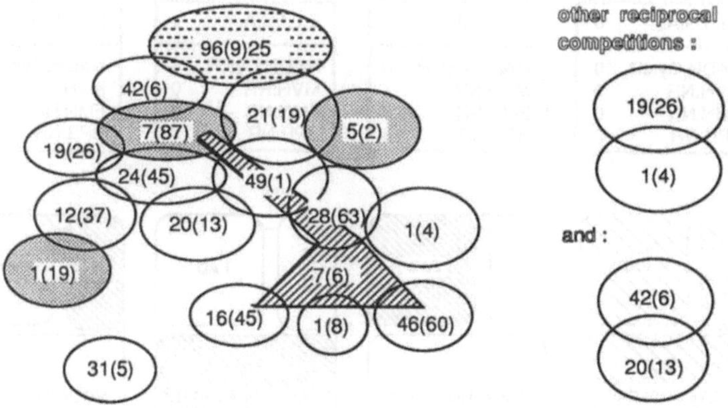

Fig. 1. Epitopic map of *N. plumbaginifolia* nitrate reductase (Shaded: epitopes for inhibitory mAb)

plumbaginifolia NR molecule. Four NPmAb were shown to partially inhibit NADH: NR, one of these (NP17:7:6) inhibits all terminal activities, including BPB:NR. None interferes with NR-associated dehydrogenase activities (Meyer et al., submitted). These antibodies are valuable probes of NR surface structure and, in the absence of crystallographic data, could be useful tools for structure-function relationships.

Nitrate Reductase Deficient Mutants: Genetical and Biochemical Analysis

More than 200 NR-deficient mutants (NR^-) have been selected from mutagenized *N. plumbaginifolia* protoplast cultures by chlorate selection and have been regenerated into plants. All NR^- mutants exhibit a characteristic chlorotic phenotype, probably coming from imbalance in carbon and nitrogen metabolism (Saux et al. 1987). Since they grew very poorly on their own roots in mineral nutrient culture, they were grafted on tobacco plants. According to their xanthine dehydrogenase activity (XDH, another MoCo-requiring enzyme) and to the phenotype of F1 crosses between them, mutant plants were classified into seven complementation groups, one *nia* (apoenzyme structural gene) (Gabard et al. 1987), and six *cnx* (*cnxA* to *F*, molybdopterin synthesis and/or insertion genes) (Gabard et al. 1988).

The amount of NR protein was measured in *cnx* mutants in the two-site ELISA. Compared to the wild-type, *cnxB* to *F* mutants show a reduced NR protein level. On the contrary, *cnxA* mutants, in which NR activity can be restored *in vivo* and *in vitro* by high molybdenum concentrations, exhibit a very high level of NR protein, up to 50 times the wild-type. The molecular weight of the NR from *cnx* mutants, estimated by

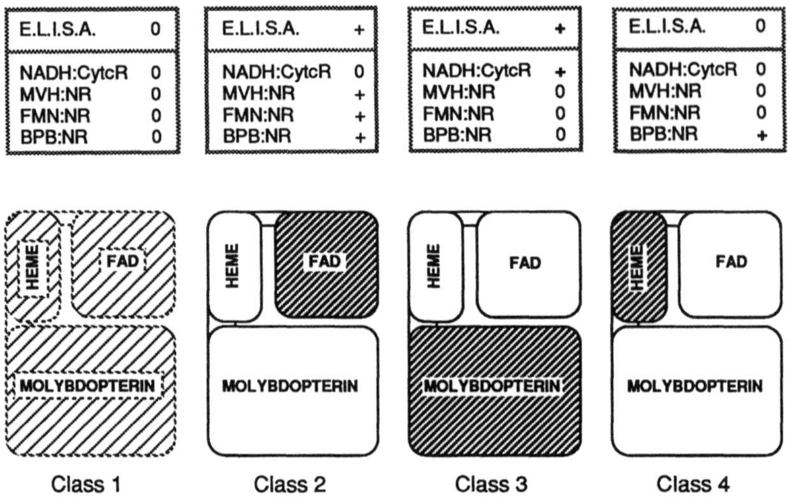

Fig. 2. Partial activities and presumed mutated NR domains (hatched) in *N. plumbaginifolia nia* plants

HPLC gel-filtration, ranks them into two groups: (i) *cnx A*, *C* and *D* with MW comparable to the wild-type NR, indicating that a native, dimeric, structure is maintained in these mutants, in spite of a modified (or absent) MoCo; (ii) *cnx B*, *E* and *F* with NR cross-reacting materials of MW about 100 kDa and lower, indicating either a monomeric structure or proteolysis (Marion-Poll et al., submitted).

The NR protein amount and partial activities in the *nia* mutant plants were measured. According to these biochemical tests, *nia* mutants were classified into four classes (Fig. 2). No NR protein can be seen by the ELISA test in most *nia* mutants (class 1). Most of the remaining *nia* mutants fall into three other classes according to which the NR redox center, FAD, heme or MoCo, appears to have a functional failure (Chérel et al. 1990). The analysis of *nia* mutants also suggests, in addition to immunological data, that BPB:NR does not need a functional heme domain, and would therefore be a true terminal activity, only requiring MoCo as cofactor.

Functional complementation was shown between some *nia* mutants, either *in vivo* (F1 crosses are able to grow on nitrate) or *in vitro* (NR activity appears after mixing extracts from each mutant). This complementation only occurs between mutants belonging to different classes.

Molecular Cloning of Tobacco and Tomato *nia* Genes

A tobacco NR apoenzyme partial cDNA was obtained by screening a λgt11 cDNA library with anti-NR polyclonal antibodies (Calza et al. 1987). The *nia* genes were isola-

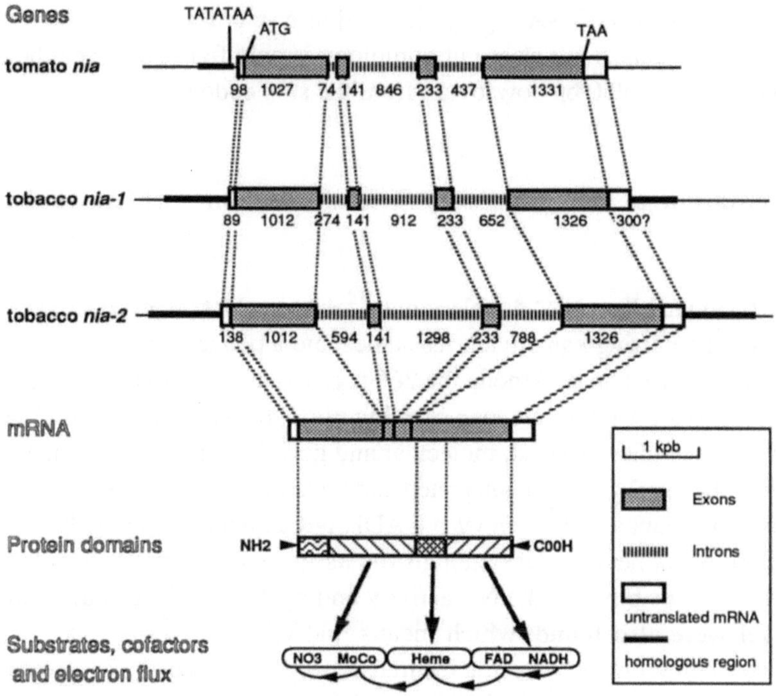

Fig. 3. Structure of tobacco and tomato *nia* genes in relation to NR domains

ted from tobacco and tomato genomic libraries, using this cDNA as probe, and sequenced. As expected from genetics, one single gene was found in tomato (Daniel-Vedele et al. 1989), and two in tobacco. Each of the tobacco genes can be attributed to either one of the tobacco ancestors, according to their restriction pattern, *nia-1* to *N. tomentosiformis*, *nia-2* to *N. sylvestris* (Vaucheret et al. 1989 a).

The structure of these three genes is essentially similar (Fig. 3): the highly homologous coding sequence comprises four exons of identical (tobacco *nia-1*/*nia-2*) or similar length (tobacco/tomato) encoding a deduced 904-aa or 911-aa polypeptide (see below). The three introns fall into the same position in the coding sequence, but vary in length between the three genes and show less homology in their common sequence, compared to the coding sequence.

An extensive homology between the two tobacco genes was found in the 5' sequence, up to 800 bp upstream the transcription start point. This region is thus thought to contain all the promoter regulatory elements (Vaucheret et al. 1989 b). The homology with the tomato *nia* gene, in this region, is restricted to a 250-bp sequence comprising the putative CAAT and TATA boxes. The untranslated mRNA leader sequence varies a lot between the three genes, and is predicted to form stable secondary structures in tobacco *nia-1* (14 kcal/mol) and *nia-2* (38 kcal/mol). In the 3' non-coding sequence, a canonical AATAAA polyadenylation box is found near the TAA stop codon for the three genes, but clear-cut homology is only found between the tobacco genes, extending 1150/850 bp dowstream from the stop codon.

Complementation of NR⁻ Plants by Transformation with the Tobacco *nia* Genes

Protoplasts from *nia*130, a NR^- double-mutant tobacco plant, were co-electroporated with a plasmid conferring kanamycin resistance and a phage genomic clone carrying either the *nia-1* or *nia-2* gene. Among the 266 regenerated kanamycin-resistant plants, three were able to grow on nitrate, one carrying *nia-1*, the two other *nia-2*. One transformant was further characterized: molecular and genetical analysis demonstrate that a single copy of the *nia-2* gene was integrated and transmitted as a mendelian character to the progeny. This plant shows an *in vitro* NADH:NR activity averaging 3% of the wild type, which appears nevertheless enough for its growth on nitrate as sole nitrogen source. Nitrate inducibility of the NR activity and nycthemeral fluctuation of the NR mRNA level were also found, which means that all the necessary *cis* regulatory sequences (see below) were present in the transferred *nia-2* gene.

Similar results were obtained when E23, a *N. plumbaginifolia nia* mutant without any detectable NR mRNA, was transformed by an *Agrobacterium* binary vector carrying the tobacco *nia-2* gene.

Nitrate Supply, Circadian Rhythm, and Light Regulation of *nia* Gene Expression

To study the molecular regulation of NR activity, the NR mRNA level (Northern blots), the NR protein amount (in a two-site ELISA using ZM96-9-25 mAb as the first step, and a polyclonal anti-maize NR as second antibody) and NADH:NR *in vitro* activity were measured in the same tobacco and tomato leaf extracts, from wild-type or mutant plants, under controlled environmental conditions.

Nitrogen starvation of 5-leaf tomato and tobacco plants resulted in a rapid fall in NR protein and activity (80 % in one day) but a very delayed decline in mRNA level (80% in 14 days). When one week N-starved plants received 10 mM NO_3^-, NR mRNA increased in less than 2 hours (x 2-3, compared to unstarved control), NR protein and activity increasing more slowly (more than 24 h are needed to reach the amount in the unstarved control). This argues for nitrate being a transcriptional inducer of the *nia* gene, as also suggested from Northern blot induction experiments in other higher plants. On the other hand, the starvation experiment tells that nitrate must also control either NR mRNA translation or NR protein turnover (Galangau et al. 1988). Tungstate-grown plants remain nitrate inducible (Deng et al. 1989), as do the *N. plumbaginifolia* or tobacco NR^- mutants (Pouteau et al., submitted), indicating that NR activity (which is prevented by W instead of Mo in the enzyme) is not involved in this transcriptional regulation, and excluding that a product of nitrate reduction would be the true inducer. Nitrate induction in all tested *cnx* and *nia* mutants also speaks against an autoregulation of the *nia* gene similar to the one hypothesized in fungi.

Over a 24 h light-dark cycle, NR mRNA shows a hundred-fold variation in tobacco and tomato, with a maximum at the end of the night and a minimum at the end of the day. NR protein and activity show a lesser variation with maxima in the day and minima before the night. During the day, noticeable discrepancies occur between the amount of NR protein and NR activity, leading to an apparent variation in NR specific activity. This indicates a post-translational activation/inactivation of NR, the mechanism of which remains to be found (Galangau et al. 1988). The day-night oscillations are maintained when plants are transferred to continuous light. This observation indicates that a circadian rhythm controls *nia* gene expression (Deng et al. 1990). Unlike nitrate induction, NR activity appears to be necessary for these circadian variations, since it is abolished in NR^- mutants and tungstate-grown plants, in which a high constant NR mRNA level is observed (Pouteau et al., submitted). This is in favour of a negative transcriptional control of the circadian rhythm, modulated by a reduced nitrogen metabolite.

If plants are transferred to continuous darkness, the circadian rhythm fades down and NR mRNA and protein progressively fall off to undetectable levels after 56 hours of dark. Light is needed to restore both mRNA level (50% of maximum control level after 4 h of light) and NR protein amount (Deng et al. 1990). This light transcriptional control was also observed in squash cotyledons and was claimed to be phytochrome dependent (Rajasekhar et al. 1988).

Nitrate Reductase Protein Structure: Amino Acid Sequence and Catalytic Domains

The 904/911aa protein sequence was deduced from the sequence of the tobacco and tomato *nia* genes, and compared to other NR aa complete or partial sequences. By homology with proteins in the sequence data banks, two domains are clearly defined: a central 85aa heme domain, homologous to cytochrome b_5, and a C-terminal 275aa NADH/FAD domain, homologous to cytochrome b_5 reductase, separated by a 25aa hinge (Calza et al. 1987). Some catalytically essential residues can be predicted from these homologies. In the heme domain, eleven residues are conserved in each protein of the cytochrome b_5 superfamily, among which the two iron liganding histidines. In the FAD/NADH domain, only one cysteine is conserved, which is thus supposed to carry the catalytically essential thiol (Barber and Solomonson 1986). A lysine, involved in cytochrome b_5 reductase NADH binding (Hackett et al. 1988), is conserved and is thus candidate to play a similar role in NR (Daniel-Vedele et al. 1989). Since the X-ray crystallographic structure has been obtained for two proteins of the cytochrome b_5 superfamily, a model of the 3D-structure of the tobacco NR heme domain was built by computer from the coordinates of the more homologous of them, bovine cytochrome b_5 (Mathews et al. 1971), by aa replacement and minimum energy calculations. The structural elements remained very much conserved, despite a one aa deletion. Interestingly, charged residues thought to be involved in interaction with cytochrome b_5 reductase, are replaced by hydrophobic aa, suggesting a different kind of interaction between the covalently linked heme and FAD/NADH domains of NR.

The molybdopterin domain, located in the N-terminal half of the sequence, does not show significant homology to any protein sequence (even to molybdoenzymes like *Drosophila* XDH or *E. coli* NR) except sulfite oxidase for which sequence data are limited to two short peptides. The 80aa N-terminal region, which contains a striking polyacidic cluster, is less conserved and is thus suggested to be a separate domain with a structural (or regulatory) function. The aa sequence alignment and NR^- mutant biochemical analysis are in fair agreement with limited proteolysis experiments performed with NR from spinach (Kubo et al. 1988), giving evidence that the predicted catalytic domains are true structural and functional units separated by labile hinge regions.

Conclusions

The aa sequence together with biochemical analysis of the wild-type and mutant NR have provided information on the functional structure of NR (domain organization and structure, essential residues) that will help in physiological studies (*in vivo* proteolysis and/or other post-translational modifications, isoforms, NR-associated activities, molecular interaction) and in protein chemistry molecular approaches (site-directed mutagenesis analysis of catalysis and quaternary structure).

On the other hand, molecular investigation in NR activity regulation reveals an intricate control. The first results have pinpointed which step(s) of gene expression are involved in these regulations, in response to essential environmental conditions, nitrogen supply, light, nycthemeral cycle. Analysis of the *nia* gene *cis* sequences and of the *trans*-acting factors involved in transcriptional control of *nia* gene expression (nitrate, light, circadian rhythm) is now the next step in molecular investigation of NR regulation.

References

Barber MJ, Solomonson LP (1986) The role of the essential sulfhydryl group in assimilatory NADH: nitrate reductase of *Chlorella*. J Biol Chem 261: 4562-4567

Beevers L, Hageman RH (1983) Uptake and reduction of nitrate: Bacteria and higher plants. In: Läuchli A, Bieleski RL (eds) Inorganic Plant Nutrition. Encyclopedia of Plant Physiology, NS vol 15A. Springer, Berlin Heidelberg New York, pp 351-375

Calza R, Huttner E, Vincentz M, Rouzé P, Galangau F, Vaucheret H, Chérel I, Meyer C, Kronenberger J, Caboche M (1987) Cloning of DNA fragments complementary to tobacco nitrate reductase mRNA and encoding epitopes common to the nitrate reductases of higher plants. Mol Gen Genet 209: 552-562

Campbell WH (1988) Nitrate reductase and its role in nitrate assimilation in plants. Physiol Plant 74: 214-219

Chérel I, Grosclaude J, Rouzé P (1985) Monoclonal antibodies identify multiple epitopes on maize leaf nitrate reductase. Biochem Biophys Res Commun 129: 686-693

Chérel I, Marion-Poll A, Meyer C, Rouzé P (1986) Immunological comparisons of nitrate reductases of different plant species using monoclonal antibodies. Plant Physiol 81: 376-378

Chérel I, Gonneau M, Meyer C, Pelsy F, Caboche M (1990) Biochemical and immunological characterization of nitrate reductase deficient mutants of *Nicotiana plumbaginifolia*. I. *nia* mutants. Plant Physiol (in press)

Daniel-Vedele F, Dorbe MF, Caboche M, Rouzé P (1989) Cloning and analysis of the tomato nitrate reductase-encoding gene: Protein domain structure and amino acid homologies in higher plants. Gene, in press

Deng M, Moureaux T, Caboche M (1989) Tungstate, a molybdate analog inactivating nitrate reductase, deregulates the expression of the nitrate reductase structural gene. Plant Physiol 91: 304-309

Deng M, Moureaux T, Leydecker MT, Caboche M (1990) Nitrate reductase expression is under the control of a circadian rhythm and is light inducible in *Nicotiana tabacum* leaves. Planta 180: 257-261

Gabard J, Marion-Poll A, Chérel I, Meyer C, Müller A, Caboche C (1987) Isolation and characterization of *Nicotiana plumbaginifolia* nitrate reductase-deficient mutants: Genetic and biochemical analysis of the NIA complementation group. Mol Gen Genet 209: 596-606

Gabard J, Pelsy F, Marion-Poll A, Caboche M, Saalbach I, Grafe R, Müller A (1988) Genetic analysis of nitrate reductase deficient mutants of *Nicotiana plumbaginifolia*: Evidence for six complementation groups among 70 classified molybdenum cofactor deficient mutants. Mol Gen Genet 213: 206-213

Galangau F, Daniel-Vedele F, Moureaux T, Dorbe MF, Leydecker MT, Caboche M (1988) Expression of leaf nitrate reductase genes from tomato and tobacco in relation to light-dark regime and nitrate supply. Plant Physiol 88 : 383-388

Hackett CS, Novoa WB, Read Kensil C, Strittmatter P (1988) NADH binding to cytochrome b_5 reductase blocks the acetylation of lysine 110. J Biol Chem 263: 7539-7543

Kubo Y, Ogura N, Nakagawa H (1988) Limited proteolysis of the nitrate reductase from spinach leaves. J Biol Chem 263: 19684-19688

Mathews FS, Levine M, Argos P (1971) The structure of calf liver cytochrome b_5 at 2.8Å resolution. Nature New Biol 233:15-16

Meyer C, Chérel I, Moureaux T, Hoarau J, Gabard J , Rouzé P (1987) Bromophenol blue:nitrate reductase activity in *Nicotiana plumbaginifolia*: An immunochemical and genetic approach. Biochimie 69: 735-742

Moureaux T, Leydecker MT, Meyer C (1989) Purification of nitrate reductase from *Nicotiana plumbaginifolia* by affinity chromatography using 5'AMP-Sepharose and monoclonal antibodies. Eur J Biochem 179: 617-620

Müller AJ (1983) Genetic analysis of nitrate reductase-deficient tobacco plants regenerated from mutant cells. Evidence for duplicate structural genes. Mol Gen Genet 161: 67-76

Rajasekhar VK, Gowri G, Campbell WH (1988) Phytochrome-mediated light regulation of nitrate reductase expression in squash cotyledons. Plant Physiol 88: 242-244

Saux C, Lemoine Y, Marion-Poll A, Valadier MH, Deng M, Morot-Gaudry JF (1987) Consequence of absence of nitrate reductase activity on photosynthesis in *Nicotiana plumbaginifolia* plants. Plant Physiol 84: 67-72

Vaucheret H, Vincentz M, Kronenberger J, Caboche M, Rouzé P (1989 a) Molecular cloning and characterisation of the two homeologous genes coding for nitrate reductase in tobacco. Mol Gen Genet 216: 10-15

Vaucheret H, Kronenberger J, Rouzé P, Caboche M (1989 b) Complete nucleotide sequence of the two homeologous tobacco nitrate reductase genes. Plant Mol Biol 12: 597-600

MOLECULAR CHARACTERIZATION OF MAIZE NADH:NITRATE REDUCTASE

W.H. Campbell [1]

Introduction

The long-term goals of research on higher plant nitrate reductase (NR) are: (1) to understand the structure of the enzyme in relation to its mechanism of action; and (2) to explain how NR is regulated in relation to plant metabolism. Both goals are highly complex, but when they are pursued simultaneously, the most rapid progress is achieved. We have pursued the biochemistry of NR in squash (*Cucurbita maxima*), since it has been easy to obtain large quantities of homogeneous enzyme from its cotyledons (Campbell 1989). At the same time, we have studied the physiology of NR in maize (*Zea mays*), since it is a plant with agronomic importance and has been studied for a long time (Campbell 1988). At first, we bridged the gap between NR in these plant systems using immunochemical methods and now, a much stronger bridge has been established with molecular biology methods. Cloning and sequencing of NR has provided new insight into the structure of the enzyme (Crawford and Davis 1988; Galangau et al. 1988). At the same time, the availability of molecular probes for NR has made it possible to study the regulation of NR at the level of its mRNA (Kleinhofs et al. 1988; Gowri and Campbell 1989). Here I describe our recent progress toward gaining a unified understanding of NR biochemistry and physiology in maize.

Purification of Maize NR by Immunoaffinity Chromatography

The preparation of homogeneous NR has remained a problem for many years. Of course, the use of blue Sepharose as a purification tool for NR was a major breakthrough (Campbell and Smarrelli 1986). However, in order to prepare homogeneous NR, additional purification steps were needed (Redinbaugh and Campbell 1985). To overcome these problems, we have developed an immunoaffinity chromatographic method, based on monoclonal antibodies (McAb), for the purification of higher plant NR (Hyde et al. 1989). Our approach to the selection of McAb for NR was different from other groups and as a consequence, we obtained a different assortment of McAb. Basically, we used ELISA, where NR was purified on the microtiter plate with rabbit monospecific polyclonal anti-NR, to identify mouse hybridomas secreting McAb for

[1] Department of Biological Sciences, Michigan Technological University, Houghton, Michigan 49931 USA

NR (Miller and Campbell 1986). This eliminated the need for a second screening procedure. We also selected for IgG McAb during this screen, which made it easy to purify the McAb on protein-A Sepharose (Remmler and Campbell 1986). Consequently, all our studies have been done with purified McAb of known protein concentration. when our collection of McAb was tested for their effect on NADH:NR activity, we found three types: activators, inhibitors and activity-neutrals (Hyde et al. 1989). A comparison between an activator and inhibitor McAb is presented in Figure 1.

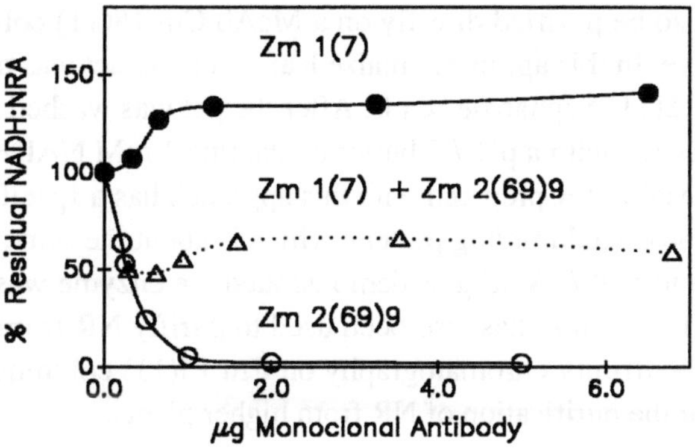

Fig. 1. Titration of 0.4 μg of purified maize leaf NADH:NR with two McAb. The specified amounts of McAb were mixed with NR in the presence of 0.5 mg of BSA, allowed to stand for 3 min at 30°C and then the NR activity was assayed. For the middle curve, the NR was first mixed with 0.3 μg of McAb Zm 2(69)9 and then titrated with McAb Zm 1(7)

McAb Zm 2(69)9 produces 50% inhibition at 0.5 antibody combining-site/NR subunit, which indicates only one antibody molecule need bind to an NR dimer to totally inhibit that enzyme molecule. McAb Zm 1(7) yields 50% activation at 1 antibody combining-site/NR subunit. The activator titration curve is sigmoidal, which is characteristic of cooperative binding; therefore, both subunits of an NR dimer probably bind McAb before activation is achieved. Activation of NR by Zm 1(7) is probably due to a conformation change in enzyme structure, which is consistent with this McAb activating all reactions catalyzed by NR (Campbell 1989). Competitive ELISA assays have shown that these two McAb cannot bind to NR at the same time, which is illustrated here by the middle curve of Figure 1. After maize NR was 50% inhibited by McAb Zm 2(69)9, only the residual NADH:NR activity was increased by addition of McAb Zm 1(7). Another NR activator McAb, Cm 15(11), has also been identified (Hyde et al. 1989). We have found that activator McAbs, which apparently induce conformational change in NR and depend on this change for binding, make the best ligands for immunoaffinity chromatography.

I used McAb 2(69)9 to make an immunoaffinity column for the purification of maize NR, but I found the NR was held very tightly and difficult to elute. If a pH 11 buffer was used to elute NR, 40% of the activity was recovered (Campbell). When crude extract of maize leaves was applied directly to a Zm 2(69)9 Sepharose column and the NR eluted at pH 11, two bands were found in an SDS/PAGE gel. To further purify the maize NR, a second immunoaffinity column, prepared with squash McAb Cm 15(11), which has an affinity for both corn and squash NR, was used. Homogeneous maize NR was obtained after the second column as judged by a single band of 100 kD on an SDS/PAGE gel. Two hundred μg of maize NR was prepared by this two column method and used for determination of a partial amino acid sequence for the enzyme.

Maize NR can also be purified directly on a McAb Cm 15(11) column, which has proved more effective. In this approach, maize leaf crude extract was mixed in a batch mode with the Cm 15(11) Sepharose beads. After the gel was washed and poured into a column, NR was eluted with a pH 7.5 buffer containing 1 mM NADH and 1 M NaCl (Hyde et al. 1989). Maize NR prepared with this approach has a specific activity of 80 to 100 μmol nitrite produced/min/mg protein, which is about the same as for homogeneous squash NR, and SDS/PAGE gels demonstrated the enzyme was homogeneous. The McAb Cm 15(11) column has also been used to purifiy NR from squash, barley and tobacco. Immunoaffinity chromatography on Cm 15(11) columns may provide a universal method for the purification of NR from higher plants.

Characterization of a cDNA Clone for Maize Leaf NADH:NR

A lambda gt11 cDNA library was prepared with poly(A)$^+$RNA from nitrate-treated maize leaves. This expression library was screened with rabbit monospecific polyclonal antibodies for maize NADH:NR and positive clones selected (Gowri and Campbell 1989). A clone containing a 2.1 kb insert, which is designated Zmnr1, hybridized to a 3.2 kb mRNA from nitrate-treated leaves and did not hybridize to mRNA from untreated leaves. Maize leaf NADH:NR with a polypeptide of 100 kDa would be expected to have an mRNA of about 3 kb (Campbell 1988). The nucleotide sequence of Zmnr1 was determined and it had an open reading frame of 617 amino acids. Amino acid sequences had been determined for 11 peptides of maize NR and the sequences of 10 of these peptides were virtually identical to portions of the deduced amino acid sequence of Zmnr1 (Gowri and Campbell 1989). Thus, Zmnr1 is an authentic cDNA clone for maize leaf NADH:NR, but it encodes only about two thirds of the NR polypeptide. The partial sequence for maize leaf NADH:NR was homologous to the deduced amino acid sequences of NR clones from *Arabidopsis* and tobacco (Crawford and Davis 1988; Galangau et al. 1988). The nucleotide sequence of Zmnr1 contains the code for the heme-Fe, FAD and NADH binding sites of the NR polypeptide as well as the 3' untranslated region. Thus Zmnr1 is a very specific probe for maize NADH:NR and under stringent hybridization conditions (i.e. high salt concentration at 60°C), should not cross-hybridize to any other NR form expressed in maize tissues.

Identification of Two Domains of Maize NADH:NR

The partial amino acid sequence of maize leaf NADH:NR was compared to a protein sequence database. It was found that regions of maize NR were homologous to mammalian cytochrome b_5 and NADH:cytochrome b_5 reductase. The regions of homology are shown diagrammatically in Figure 2. *Arabidopsis* and tobacco NR are also homologous to these same two mammalian proteins (Crawford and Davis 1988; Galangau et al. 1988). Thus it is now possible to identify the amino acid sequence regions of NR which represent the domains of the enzyme for binding its heme-Fe and FAD (Campbell 1988). Furthermore, since the NADH binding site of NADH:cytochrome b_5 reductase has been located in the carboxy-terminal region of its amino acid sequence, the NADH binding site of NR can also be localized to its carboxy-terminal region.

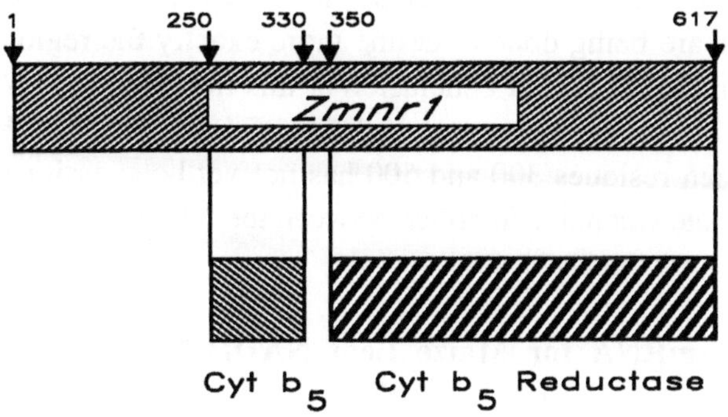

Fig. 2. Representation of the regions of homolgy between maize NADH:NR and mammalian cytochrome b_5 and NADH:cytochrome b_5 reductase. The numbers above the Zmnr1 block show the amino acid residues of the maize NADH:NR polypeptide, which is only partially complete at this time. Residue 1 of maize NR corresponds to residue 294 of the amino acid sequence of *Arabidopsis* NR, which contains a total of 917 residues. Residue 617 of maize NR is the carboxy-terminal amino acid of the enzyme's polypeptide, which corresponds to residue 917 of *Arabidopsis* NR. The cytochrome *b* domain of maize NR is defined by residues 250 to 330, which have homology to mammalian cytochrome b_5. A short hinge region of 20 amino acids joins the cytochrome *b* domain to the reductase domain of NR, which begins at residue 350 and ends at the carboxy-terminal residue as defined by homology to mammalian NADH:cytochrome b_5 reductase

Definition of the Binding Site of McAb Cm 15(11) on NR

We have demonstrated that McAb Cm 15(11) has an affinity for both maize and squash NR and can be used as a ligand for purification of these enzymes (Hyde et al. 1989). We have also tested if this McAb can bind to the lacZ fusion proteins made by the lambda gt11 clones isolated for these two enzymes (Crawford et al. 1986; Gowri and Campbell 1989). Our results show that McAb Cm 15(11) binds to the Zmnr1 fusion protein, but does not bind to the Cmc1 fusion protein. Although the lack of affinity for the Cmc1 fusion protein could be for several reasons, we believe it is due to the absence of the McAb binding site in the section of the NR encoded by Cmc1. We have recently sequenced the 5' end of the Cmc1 insert DNA and determined that the first amino acid encoded corresponds to residue 498 of *Arabidopsis* NR. Thus, since the first amino acid residue encoded by Zmnr1 corresponds to residue 294 of *Arabidopsis* NR, we have localized the binding site of McAb Cm 15(11) to the region of NR between residues 294 and 498. This locale for the McAb Cm 15(11) binding site has been supported by experiments where NR was partially degraded by proteolysis and the NR fragments passed over the McAb Cm 15(11) immunoaffinity column. The small fragments containing the cytochrome *b* and dehydrogenase function of NR did not bind to this column, while a larger fragment of approximately 55 kDa did bind. These results are consistent with our assignment of the McAb Cm 15(11) binding site to a region of NR not containing the cytochrome *b* and the dehydrogenase function. Further experiments are being done to define more exactly the region of NR which binds to Cm 15(11). However, our conclusion at this time is that McAb Cm 15(11) binds to a region of NR on the amino-terminal side of the cytochrome *b* domain. This region of NR between residues 300 and 500 has not yet been assigned a functional role, but may be the site where the interface between the NR subunits is located.

Expression of the mRNA for Maize Leaf NADH:NR in Maize Tissues

To study NR expression, maize plants were grown in the dark without nutrients prior to being transferred to light and given 50 mM NH_4NO_3. Leaf tissue harvested over 12 hours time course was assayed for NR mRNA and activity (Gowri and Campbell 1989). In Figure 3, data obtained in this experiment are compared to previous data on the expression of NR activity and NR protein in maize leaves treated in the same manner (Remmler and Campbell 1986). NR mRNA is low or absent from the etiolated, N-starved leaves, but is expressed at a high level 2 hours after nitrate and light are given. NR protein could be detected at 1 hour, but its initial rate of appearance was slower than NR mRNA. Finally NR activity could be detected at 2 hours, but its initial rate of appearance was the slowest. The NR protein and activity do not appear simultaneously because NR needs to have cofactors integrated into the polypeptide prior to the appearance of activity. The slow conversion of the NR apo-protein into active enzyme is due either to limited availability of cofactors in greening leaves or to a rate-limiting assembly of the cofactors into the apo-protein.

One of the key questions yet to be answered is whether nitrate is the only regulator of NR or if light has a role in the regulation process. To address this issue, we treated etiolated and green maize plants with nitrate in the dark and evaluated NR mRNA and activity levels in leaves over a 12 hour time course (Gowri and Campbell 1989). In

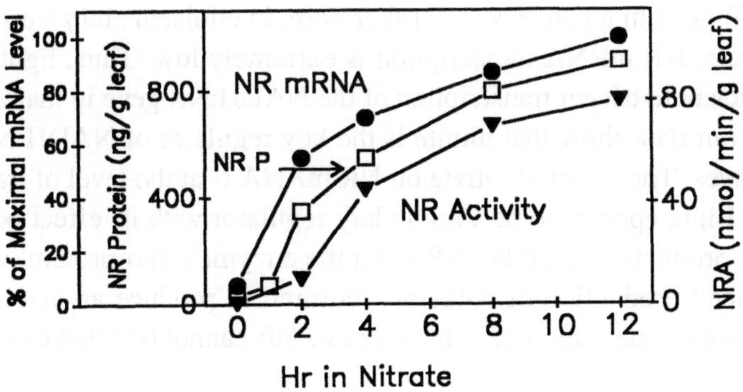

Fig. 3. Expression of NR in maize leaves. Nitrate and light were provided to maize plants, which had been grown in the dark without nutrients for 5 days. NR mRNA was assayed as described (Gowri and Campbell 1989). NR protein (NRP) was measured by ELISA and NR activity was assayed with an *in vitro* method (Remmler and Campbell 1986)

both etiolated and green leaves, nitrate induced NR mRNA and activity in the dark. NR mRNA and activity levels were lower in these dark-treated leaves than in similar light-treated leaves, but the time courses for their appearance were nearly identical (Gowri and Campbell 1989). Thus, nitrate appears to be the key regulator of NR mRNA and activity expression in maize leaves.

Since NR is expressed in non-green tissues of maize, we evaluated the levels of NADH:NR mRNA and activity in maize scutellum and demonstrated that NADH:NR mRNA is expressed in maize roots. Scutella were excised from 3-day-old maize plants grown in the dark without nutrients. When the excised scutella were treated with nitrate *in vitro*, NADH:NR mRNA was detected within 15 min and reached its highest level in 2 hours. NADH:NR activity appeared after 2 hours of nitrate treatment and continued to increase for 10 hours more. Clearly, only nitrate is required for expression of NR in this storage organ of maize. For roots of 6-day-old maize treated with nitrate for 24 hours, a Northern blot of total root RNA demonstrated the presence of only the 3.2 kb NADH:NR mRNA, when the blot was hybridized to Zmnr1. If not treated with nitrate, the roots did not contain this mRNA. Thus, nitrate induces NADH:NR mRNA in both scutella and roots of maize, which have previously been shown to express this enzyme activity (Campbell and Smarrelli 1986).

These expression experiments have evaluated the steady-state level of NR mRNA in maize tissues. To establish if nitrate treatment results in increased transcription of NR mRNA or a decreased degradation of NR mRNA at a constant level of transcription, we isolated maize leaf nuclei and evaluated the level of NR mRNA transcription in nitrate-treated and untreated green leaves (Lillo and Campbell, unpublished data). In untreated leaves, virtually no NR mRNA is transcribed, while in nitrate-treated leaves the level of NR mRNA transcription is high. Thus, nitrate appears to regulate transcription of the NADH:NR gene in green leaves. Furthermore, in etiolated and green leaves not treated with nitrate, NR mRNA transcription is extremely low. Thus, light treatment alone is not sufficient to trigger transcription of the NADH:NR gene in maize leaves.

In summary, our data show that nitrate is the key regulator of NADH:NR expression in maize leaves. The effect of nitrate on NR mRNA is at the level of transcription of the NR gene. Light appears to be a secondary regulator with its effect being manifested mainly in production of active NR from the enzyme's apo-protein, which may be due to enhanced production of cofactors required to produce an active enzyme. However, other stabilizing effects of light on active NR cannot be ruled out.

Acknowledgments. This research was supported by grants from the U.S. Department of Agriculture and the National Science Foundation.

References

Campbell WH (1988) Higher plant nitrate reductase: Arriving at a molecular view. Curr Top Plant Biochem Physiol 7: 1-15

Campbell WH (1989) Structure and synthesis of higher plant nitrate reductase. In: Wray JL, Kinghorn JR (eds) Molecular and Genetic Aspects of Nitrate Assimilation. Oxford Univ Press, Oxford, UK, pp 125-154

Campbell WH, Smarrelli J (1986) Nitrate reductase: Biochemistry and regulation. In: Neyra CA (ed) Biochemical Basis of Plant Breeding, vol 2. CRC Press, Boca Raton, FL, pp 1-39

Crawford NM, Campbell WH, Davis RW (1986) Nitrate reductase from squash: cDNA cloning and nitrate regulation. Proc Natl Acad Sci USA 83: 8073-8076

Crawford NM, Davis RW (1988) Plant nitrate reductase is a tripartite electron transfer protein that belongs to the cytochrome b_5 superfamily of enzymes. Curr Top Plant Biochem Physiol 7: 16-25

Galangau F, Chérel I, Deng M, Meyer C, Moureaux T, Rouzé P, Vaucheret H, Vedele F, Vincentz M, Caboche M (1988) Nitrate reductase expression in tobacco and tomato. Curr Top Plant Biochem Physiol 7: 26-34

Gowri G, Campbell WH (1989) cDNA clones for corn leaf NADH:nitrate reductase and chloroplast NAD(P)$^+$:glyceraldehyde-3-phosphate dehydrogenase. Plant Physiol 90: 792-798

Hyde GE, Wilberding JA, Meyer AL, Campbell ER, Campbell WH (1989) Monclonal antibody-based immunoaffinity chromatography for purifying corn and squash NADH:nitrate reductases. Plant Mol Biol 13: 233-246

Kleinhofs A, Warner RL, Hamat HB, Juricek M, Huang C, Schnorr K (1988) Molecular genetics of barley and rice nitrate reductases. Curr Top Plant Biochem Physiol 7: 35-42

Miller AM, Campbell WH (1986) Monoclonal antibodies for higher plant nitrate reductase. Curr Top Plant Biochem Physiol 5: 192

Redinbaugh MG, Campbell WH (1985) Quaternary structure and composition of squash NADH:nitrate reductase. J Biol Chem 260: 3380-3385

Remmler JL, Campbell WH (1986) Regulation of corn leaf nitrate reductase. II Synthesis and turnover of the enzyme's activity and protein. Plant Physiol 80: 442-447

GLUTAMINE SYNTHETASE OF FRENCH BEAN: FROM GENES TO ISOENZYMES

J.V. Cullimore [1], J.M. Cock [1], M.P. Robbins [2] and M.J. Bennett [1]

Introduction

Glutamine synthetase (GS) occupies a key position in the nitrogen metabolism of higher plants. The enzyme is responsible for the assimilation of ammonium which arises from a number of reactions in different parts of the plant (Miflin and Lea 1980). In roots, ammonium occurs from direct uptake from the growth medium or reduction of exogenous nitrate, whereas in legume root nodules it arises primarily from dinitrogen fixation carried out by symbiotic *Rhizobium.* In leaves the major flux of ammonium is produced by photorespiration but other sources include reduction of nitrate and utilization of organic nitrogen transport compounds such as ureides and amides. In addition ammonium is released by amino acid catabolism and phenylpropanoid metabolism in many parts of the plant. GS activity must therefore be sufficient to serve these many physiological functions.

In this article we aim to review the nature of GS in nodules, roots and leaves of the legume *Phaseolus vulgaris* and show how the structure of the enzyme in these different organs of the plant is determined by differential expression of a small GS gene family.

Isoenzymes and Subunits

In higher plants GS is an octameric enzyme of M_r 300-360,000 with subunits of M_r 38-45,000. The enzyme is located in both the chloroplasts and cytosol and these two activities can be separated from crude cell-free extracts by ion-exchange chromatography (Stewart et al. 1980). We have used ion-exchange high performance liquid chromatography (IEX-HPLC) (Bennett and Cullimore 1989) to separate GS isoenzyme activities from crude extracts of nodules, roots and leaves of *Phaseolus vulgaris* (Fig. 1a). Two activity peaks are evident in the leaf extracts (GS_1 and GS_2) which correspond to the cytosolic and chloroplast GS activities, respectively. Roots contain only a single activity peak (GS_r) similar in elution position to the cytosolic GS in leaves. Nodules however contain a complicated pattern of isoenzymes including two activity peaks, not evident in the root or leaf extracts. These two peaks probably represent the isoenzyme termed GS_{n1}, previously identified by Lara et al. (1983) using

[1] Department of Biological Sciences, University of Warwick, Coventry, CV4 7AL, England
[2] Present address: Plant and Cell Biology Department, Welsh Plant Breeding Station, Plas Gogerddan, Aberysthwyth, Dyfed, SY23 3SB, Wales

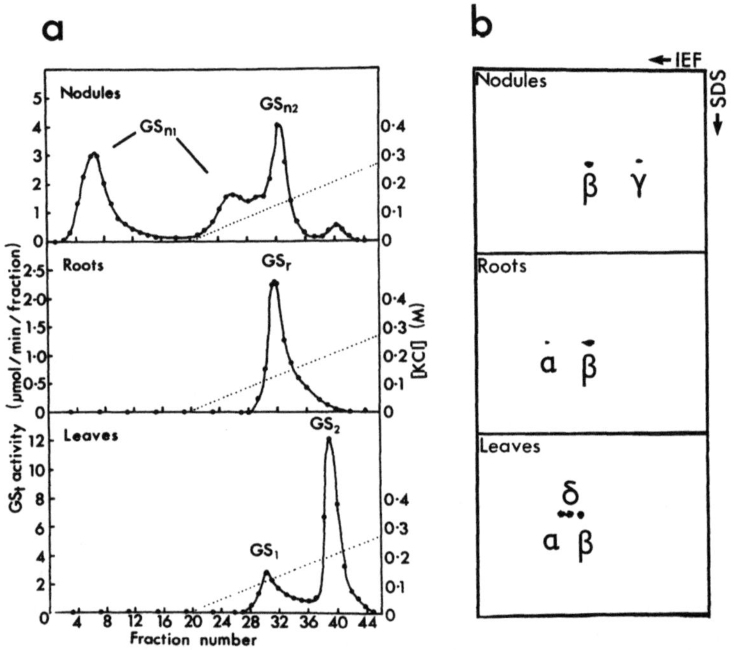

Fig. 1. GS isoenzymes and polypeptides in nodules, roots and leaves of *P. vulgaris*
a. Isoenzyme activity profiles following separation by IEX-HPLC
b. Polypeptides following 2D-gel electrophoresis (adapted from Lara et al. 1984)

conventional IEX, whereas the third peak is referred to as GS_{n2}.

Lara et al. (1984) showed that *P. vulgaris* GS could be rapidly purified from a number of organs by chromatography on an anthranilate affinity column. They then separated GS polypeptides using 2D-gel electrophoresis with isoelectric focusing in the first dimension and SDS-polyacrylamide gel electrophoresis in the second. GS from nodules, roots and leaves, purified and separated in this way, and then stained with Coomassie blue, is shown in Figure 1b. The root GS consists of two cytosolic polypeptides termed α and β. Ortega et al. (1985) have found that the proportion of these two polypeptides in roots varies during root development, with the α polypeptide occurring predominantly in radicles and very young roots and the β polypeptide constituting the major GS activity in developed root systems. The α and β polypeptides also occur in leaves (almost undetectable in this figure) but the major GS in this organ was found to consist of four polypeptides of a slightly higher M_r corresponding to the chloroplast GS. It is now known that all these polypeptides relate to a single gene product and so have collectively been renamed δ (Lightfoot et al. 1988). It is not clear whether the multiple polypeptides arise by post-translational modifications or are artifacts occurring during the tissue extraction procedures. In nodules one of the two major GS polypeptides present in this organ is not detectable in leaves and roots and has been termed γ. Nodules also contain high levels of the β polypeptide and recently low levels of the α and δ GS polypeptides have also been detected (Padilla et al. 1986; Bennett and Cullimore 1989). Interestingly, in nodules, the δ polypeptide occurs as only two of the four leaf polypeptides (data not shown).

Recently the polypeptide composition of the four nodule GS activity peaks shown in Figure 1a have been determined by 2D-gel analysis of the polypeptides and detection by Western blotting (Bennett and Cullimore 1989). The first three peaks consist of differing proportions of the γ and β polypeptides and represent γ-rich, γ/β, and β-rich isoenzymes respectively whereas the fourth peak represents the plastid-located δ polypeptides. As shown by Robert and Wong (1986) using native polyacrylamide gel electrophoresis, all nine possible octamers with differing proportions of the γ and β polypeptides probably occur in nodule tissue, but we are able to separate only some of these by IEX-HPLC. Similarly the α and β polypeptides have been shown to form mixed active octamers in plumules (Bennett and Cullimore 1989) and therefore the GS_r and GS_1 isoenzymes shown in Figure 1a probably represent isoenzymes containing both of these two polypeptides. It is therefore suggested that the three cytosolic GS polypeptides can assemble randomly into a variety of hetero-octamers (Bennett and Cullimore 1989). However this can only occur if the polypeptides are in the same location at the same time and therefore differences in the spatial and temporal production of the polypeptides may determine the range of isoenzymes observed. It is notable in this context that the chloroplast GS always forms a separate homo-octameric isoenzyme because of its subcellular compartmentation away from the cytosolic GS polypeptides.

cDNA Clones and Genes

In order to study the genetic origin of the different GS polypeptides a programme was initiated to isolate cDNA and genomic clones related to the enzyme. Antibodies were raised against purified nodule GS and used to immunopurify polysomes. Purified GS mRNA was then isolated, radiolabelled, and used to screen a nodule cDNA library. In this way a partial GS cDNA clone was isolated (Cullimore et al. 1984) and this was then used to isolate a longer homologous version of this clone (Bennett et al. 1989). Similarly, two related but not identical clones were isolated from a root library (Gebhardt et al. 1986). These three clones were found to encode GS polypeptides of 356 amino acids corresponding to M_rs of about 39,000. Bennett et al. (1989) then synthesized the polypeptides encoded by these cDNAs by *in vitro* transcription and translation. When these products were run on 2D-gels it was found that each of them

Table 1. *Phaseolus vulgaris* GS cDNA clones

Clone	cDNA library	Encoded polypeptide	% amino acid identity / % nucleotide identity				Reference
			pcGS-α1	pcGSβ1	pcGSγ1	pcGSδ1	
pcGS-α1	root	356 aa (α)		88	85	78	Gebhardt et al. (1986)
pcGS-β1	root	356 aa (β)	81		89	79	Gebhardt et al. (1986
pcGS-γ1	nodule	356 aa (γ)	79	86		78	Bennett et al. (1989)
pcGS-δ1	leaf	429 aa (δ)	71	70	70		Lightfoot et al. (1988)

co-ran with one or other of the three authentic cytosolic GS polypeptides isolated from the plant tissue. In this way the plant polypeptide that the cDNA encoded was identified and the clones were then renamed accordingly (see Table 1). A fourth GS cDNA was isolated from a leaf library and found to encode a polypeptide of 429 amino acids (M_r 47,000) with extensions to the N- and C-termini of 57 and 16 amino acids respectively. This polypeptide was then synthesized *in vitro* and was capable of being imported into isolated pea chloroplasts where it was cleaved to two smaller polypeptides of about 42 kDa (Lightfoot et al. 1988). Moreover, when the N-terminal 68 amino acids of the chloroplast GS sequence was fused to the N-terminus of the β polypeptide by genetic engineering techniques the fusion product was also imported and processed by isolated chloroplasts. These experiments suggest that the leaf GS cDNA encodes a higher M_r precursor (pre-δ) to the chloroplast GS and this is then post-translationally imported into the chloroplast by the cleavable N-terminal extension. Note that this procedure is followed by the majority of chloroplast proteins (Ellis and Robinson 1987).

Sequencing of the four GS cDNAs (Gebhardt et al. 1986; Lightfoot et al. 1988; Bennett et al. 1989) has revealed that the cytosolic GS polypeptides are about 90% identical, whereas they show only about 80% identity to the chloroplast GS polypeptide. Lower % identities are seen at the nucleotide level (Table 1). The isolation of the four different cDNA clones have shown that the α, β, γ and δ GS polypeptides are encoded by four different nuclear genes which have now been termed *gln*-α, *gln*-β, *gln*-γ and *gln*-δ. In addition a fifth GS gene has been identified (*gln*-ε) closely linked to *gln*-γ (Forde et al. 1989) but the lack of detection of mRNA related to this gene in a variety of organs (Bennett, Swarup and Cullimore, unpublished) may suggest that this gene is not expressed. Southern blot analysis of *P. vulgaris* genomic DNA using the GS cDNA clones as probes has suggested that these are the only GS genes in the *P. vulgaris* genome.

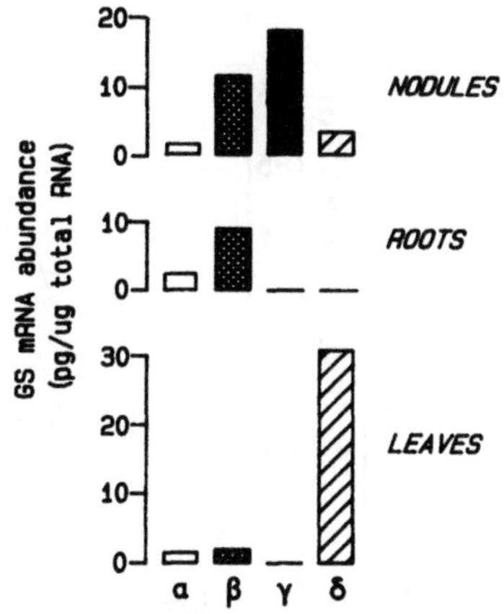

Fig. 2. Abundances of GS mRNAs in nodules, roots and leaves of *Phaseolus vulgaris* (reproduced with permission from Forde and Cullimore 1989)

Differential Expression

The abundance of mRNA of the four expressed GS genes has been determined in nodules, roots and leaves (Fig. 2) by a specific and quantitative RNase protection technique (Bennett et al. 1989). The chloroplast GS (δ) mRNA is most highly abundant in leaves but is also detected in nodules. The α and β mRNAs are present in all three organs but at different abundances. The γ mRNA appears to be nodule-specific as it is not detectable in roots and leaves but more recently this mRNA has been detected at lower levels in stems, petioles and germinating cotyledons (Bennett et al. 1989). The relative abundances of these mRNAs in the three organs is very similar to the distribution of the GS polypeptides and related isoenzymes shown in Figure 1. These results suggest that the production of the GS polypeptides and isoenzymes is largely determined by the abundances of the mRNAs. This was investigated more closely during the development of root nodules (Bennett et al. 1989; Bennett and Cullimore 1989). The abundances of the four GS mRNAs and the relative activities of the GS isoenzymes shown in Figure 1a, were determined in nodules at different stages of development (Fig. 3). The most abundant GS mRNA and isoenzyme in the early nodulating roots were from the *gln*-β gene and this remained approximately constant throughout nodule development. The *gln*-α mRNA and isoenzyme (data not shown) were most abundant only in the early samples. The increase in GS activity during nodulation was accounted for by the strong, but transient, induction of the *gln*-γ mRNA and associated isoenzymes. The γ polypeptide formed hetero-octamers with β

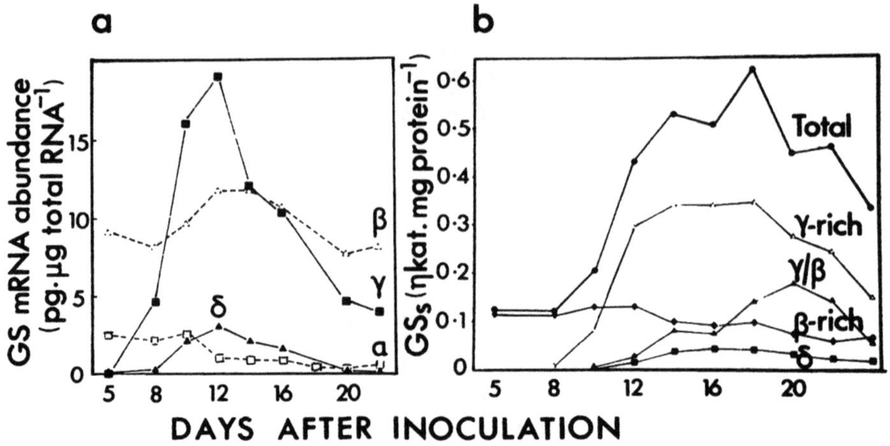

Fig. 3. Changes in GS mRNAs and isoenzymes during nodule development.
a) Abundances of GS mRNAs (reproduced with permission from Forde and Cullimore 1984).
b) Activities of the four nodule GS activity peaks shown in Fig. 1a

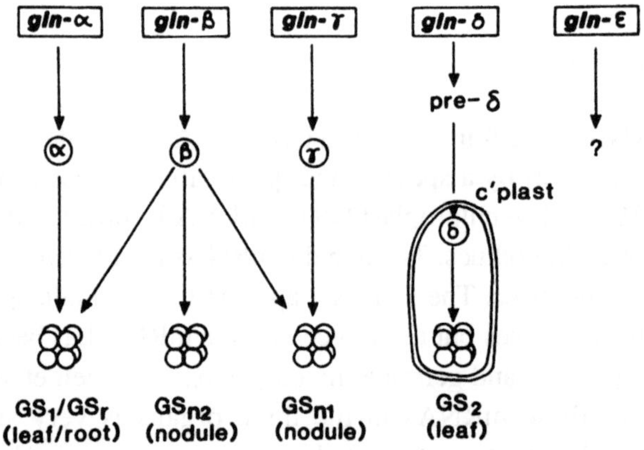

Fig. 4. Model of GS expression in *Phaseolus vulgaris* (reproduced with permission from Forde and Cullimore 1989)

showing that these two polypeptides were to some extent being expressed in the same cell type. Recently work with gene fusions has shown that the *gln*-γ gene is expressed only in the rhizobially infected cells whereas *gln*-β expression is more widespread infected cells whereas *gln*-β expression is more widespread throughout mature nodules (Forde et al. 1989) and the β polypeptide has been detected in the nodule cortex (Chen and Cullimore 1989). The *gln*-δ gene also showed a transient but low expression during nodulation, leading to detection of the plastid located isoenzyme.

These experiments have thus suggested that changes in the abundances of the different GS mRNAs are the major factor in determining the production of the GS polypeptides and isoenzymes. These changes in the mRNAs probably reflect changes in gene expression although at this stage it is not possible to eliminate changes occurring in mRNA stability. The production of different GS isoenzymes in *P. vulgaris* therefore appears to be controlled largely by the differential expression of the small GS gene family.

A Model for GS Expression in *Phaseolus vulgaris*

A model for the coding and production of GS isoenzymes in *P. vulgaris* is shown in Figure 4. Five GS genes occur in the *P. vulgaris* genome (Fig. 4) and four of these are expressed in nodules, roots or leaves, but to differing extents (Fig. 2). The mRNAs from these four genes are translated to produce three cytosolic GS polypeptides (α, β and γ) and a precursor to the chloroplast GS (pre-δ) (Fig. 1). The precursor is post-translationally imported into the chloroplasts and cleaved to produce the mature, δ polypeptide. The cytosolic polypeptides appear to be able to assemble randomly into a variety of hetero-octamers if they are expressed in the same cell at the same time, whereas the chloroplast GS polypeptide always assembles into a discrete homo-octamer (Fig. 3). The regulation of the expression of these genes and the functions of the different isoenzymes are presently under investigation.

References

Bennett MJ, Cullimore JV (1989) Glutamine synthetase isoenzymes of *Phaseolus vulgaris* L.: Subunit composition in developing root nodules and plumules. Planta 179: 433-440

Bennett MJ, Lightfoot DA, Cullimore JV (1989) cDNA sequence and differential expression of the gene encoding the glutamine synthetase γ polypeptide of *Phaseolus vulgaris* L. Plant Mol Biol 12: 553-565

Chen F-L, Cullimore JV (1989) Location of two isoenzymes of NADH-dependent glutamate synthase in root nodules of *Phaseolus vulgaris* L. Planta 179: 441-447

Cullimore JV, Gebhardt C, Saarelainen R, Miflin BJ, Idler BK, Barker RF (1984) Glutamine synthetase of *Phaseolus vulgaris* L.: Organ-specific expression of a multigene family. J Mol Appl Genet 2: 589-599

Ellis RJ, Robinson C (1987) Protein targeting. Adv Bot Res 14: 1-24

Forde BG, Cullimore JV (1989) The molecular biology of glutamine synthetase in higher plants. In: Miflin BJ (ed) Oxford Surveys of Plant Molecular and Cell Biology, Vol 6. Oxford University Press, Oxford, UK (in press)

Forde BG, Day HM, Turton JF, Shen W-J, Cullimore JV, Oliver JE (1989) Two glutamine synthetase genes from *Phaseolus vulgaris* L. display contrasting developmental and spatial patterns of expression in transgenic *Lotus corniculatus* plants. The Plant Cell 1: 391-401

Gebhardt C, Oliver JE, Forde BG, Saarelainen R, Miflin BJ (1986). Primary structure and differential expression of glutamine synthetase genes in nodules, roots and leaves of *Phaseolus vulgaris*. EMBO J 5: 1429-1435

Lara M, Cullimore JV, Lea PJ, Miflin BJ, Johnston AWB, Lamb JW (1983) Appearance of a novel form of plant glutamine synthetase during nodule development in *Phaseolus vulgaris* L. Planta 157: 254-258

Lara M, Porta H, Padilla J, Folch J, Sánchez F (1984) Heterogeneity of glutamine synthetase polypeptides in *Phaseolus vulgaris* L. Plant Physiol 76: 1019-1023

Lightfoot DA, Green NK, Cullimore JV (1988) The chloroplast-located glutamine synthetase of *Phaseolus vulgaris* L.: nucleotide sequence, expression in different organs and uptake into isolated chloroplasts. Plant Mol Biol 11: 191-202

Miflin BJ, Lea PJ (1980) Ammonia assimilation. In: Miflin BJ (ed) The Biochemistry of Plants, Vol 5. Academic Press, New York, pp 169-202

Ortega JL, Campos F, Sánchez F, Lara M (1986) Expression of two different glutamine synthetase polypeptides during root development in *Phaseolus vulgaris* L. Plant Physiol 80: 1051-1054

Padilla JE, Campos F, Conde V, Lara M, Sánchez F (1987) Nodule-specific glutamine synthetase is expressed before the onset of nitrogen fixation in *Phaseolus vulgaris* L. Plant Mol Biol 9: 65-74

Robert FM, Wong PP (1986) Isoenzymes of glutamine synthetase in *Phaseolus vulgaris* L. and *Phaseolus lunatus* L. root nodules. Plant Physiol 81: 142-148

Stewart GR, Mann AF, Fentem PA (1980) Enzymes of glutamate formation: Glutamate dehydrogenase, glutamine synthetase and glutamate synthase. In: Miflin BJ (ed) The Biochemistry of Plants, Vol 5. Academic Press, New York, pp 271-327

MOLECULAR BIOLOGY OF NITRATE ASSIMILATION IN *CHLAMYDOMONAS REINHARDTII*

E. Fernández [1], R. Schnell, L.P.W. Ranum, P.B. Savereide, C.D. Silflow and P.A. Lefebvre [2]

Introduction

As in other photosynthetic eukaryotes, nitrate assimilation in the green alga *Chlamydomonas* is a highly regulated process which requires transport of the anion into the cell and its reduction to ammonium catalyzed by nitrate reductase (NR) and nitrite reductase (NiR). Several genes have already been identified and related with NR structure, regulation of NR synthesis and molybdopterin cofactor (MoCo) biosynthesis. The regulation of the amount of NR enzyme seems to occur at the levels of transcription, translation and degradation. The structural gene for NR, *nit*-1, has been cloned and its identity demonstrated by restriction fragment length polymorphism analysis, sequencing data and complementation of a *nit*-1 mutation by transformation. It has been shown that the *nit*-1 transcript levels are repressed in ammonium medium. NR enzyme activity and the NR transcript are derepressed in nitrate medium with similar kinetics. In nitrogen-free medium, however, derepression of the NR transcript occurred but NR activity levels were very low. The *nit*-2 gene is involved in the control of NR mRNA levels. By using molecular approaches, several of the previously reported suggestions for NR regulation have been confirmed.

Chlamydomonas reinhardtii is a unicellular haploid green alga with many experimental advantages for molecular and genetic studies (Harris 1989). The analysis and characterization of wild-type strains and NR-deficient mutants have greatly contributed to a better understanding of the structure and regulation of the NR enzyme complex (Fernández and Cárdenas 1989).

At least six genes are required for the synthesis of active NR (Fernández and Matagne 1984). Among them *nit*-1 is the structural locus for NR. It is composed of two complementation groups designated *nit*-1*a* and *nit*-1*b*, separately responsible for the two partial activities of the complex: diaphorase and terminal-NR (Fernández and Matagne 1986). On the basis of *in vitro* and *in vivo* complementation results between mutant strains and biochemical data, a heteromultimeric structure for the enzyme complex was proposed (Franco et al. 1984, Fernández and Matagne 1986). *nit*-2 is a putative regulatory locus for NR (Fernández and Matagne 1986). Four genes: *nit*-3,

[1] Departamento de Bioquímica, Biología Molecular y Fisiología, Facultad de Ciencias, Universidad de Córdoba, Córdoba 14071 (Spain)

[2] Department of Genetics and Cell Biology and Plant Molecular Genetics Institute, University of Minnesota, St. Paul, MN-55108 (USA)

nit-4, *nit*-5 and *nit*-6 are needed for molybdopterin cofactor production. *nit*-3 is involved in an undetermined step in the biosynthesis of the pterin, and the other three loci are involved in the processing of molybdate before its insertion in the NR apoprotein (Fernández and Aguilar 1987). Two additional genes (*ma*-1 and *ma*-2) are separately responsible for the two transport systems of ammonium in *Chlamydomonas*. These systems control intracellular ammonium concentrations which, in turn, regulate nitrate reductase expression (Franco et al. 1988).

By comparing wild-type and the above mutant strains of *Chlamydomonas* with respect to expression and half-lives of NR under different physiological conditions, the following regulatory mechanisms were proposed (Franco et al. 1987, Fernández and Cárdenas 1989): (i) the expression of the structural gene for NR is controlled by ammonium, both at the transcriptional and post-transcriptional levels; (ii) once NR has been synthesized, the enzyme becomes reversibly inactivated, if nitrate is not present, the inactive form of the enzyme being the target of a specific degradation system; and (iii) the extent of NR interconversion between active and inactive forms appears to be involved in the autoregulation of NR enzyme levels. A molecular study and verification of all these proposals required first the cloning of the NR structural gene.

Cloning of the NR Structural Gene

The structural gene *nit*-1 of *Chlamydomonas* has been isolated from a genomic DNA library (constructed in lambda EMBL4 by J. Youngblom and C. D. Silflow) using the barley bNRp10 cDNA probe kindly provided by Howard Goodman (Cheng et al. 1986). Under medium stringency conditions (6 x SSC, 60°C for hybridization and 0.8 x SSC, 55°C for washes), four positive clones were isolated among 6 - 7 genome equivalents. Three of the clones (designated B6a) were identical, and overlapped with a fourth clone; they span a genomic region of about 24 kb (Fernández et al. 1989). Both the frequency of isolated positives and the hybridization analysis of genomic DNA digested with a number of restriction enzymes showed that a single copy genomic region had been cloned.

That the isolated DNA contains the structural gene for NR was demonstrated by several lines of evidence (Fernández et al. 1989). First, restriction fragment length polymorphism (RFLP) analysis was performed to map the isolated DNA fragments relative to the *nit*-1 locus using tetrad progeny from crosses between two interfertile strains of *Chlamydomonas* (*C. reinhardtii* carrying a *nit*-1 mutation and *C. smithii* carrying a wild-type *nit*-1 gene). This powerful technique has been used successfully to map different molecular markers on the *Chlamydomonas* genome (Ranum et al. 1988). It was shown that the *nit*-1$^-$ allele and a *C. reinhardtii* polymorphic DNA fragment hybridizing to a B6a probe co-segregated in 67 progeny corresponding to 22 tetrads, indicating a close linkage between *nit*-1 and the B6a sequence on linkage group IX (Fernández et al. 1989).

Of seven fragments of B6a produced by digestion with Sma I and Eco RI, only one (B6a-2) of about 1.8 kb cross-hybridized to the barley NR probe. DNA sequence was obtained for two Pst I fragments of B6a-2, which hybridized strongly with the barley NR probe. The predicted amino acid sequences for both of them were compared with amino acid sequences reported for NR in tobacco (Calza et al. 1987) and *Arabidopsis* (Crawford et al. 1988) and with sequences of b_5-type heme containing proteins like yeast flavocytochrome b_2 (Lederer et al. 1985) and NADH /FAD binding proteins like human cytochrome b_5 reductase (Yubisui et al. 1986). It was revealed that the two sequenced fragments showed a striking degree of identity at the amino acid level with each of the above sequences; the two sequenced fragments corresponded to the heme and NADH/FAD binding domains (Fernández et al. 1989). The relative position of the two domains (from the 5' to the 3' ends of the transcript) is the same as in higher plants. However, the similarity between the *Chlamydomonas* and the higher plant NR sequences is not as great as between the higher plant sequences, possibly reflecting a divergence between the NR genes in *Chlamydomonas* and higher plants (Fernández et al. 1989).

A plasmid (pMN24) containing the whole NR gene complemented a *nit*-1 mutation when introduced into cells by transformation, confirming the identity of the cloned gene of NR. Transformation of the *nit*-1 mutant strain (which carried a cell-wall deficiency, CW 15) was performed by a simple method based on a procedure described previously for fusing *Chlamydomonas* protoplasts (Matagne et al. 1979) involving polyethylene glycol and calcium chloride. It was shown that DNA sequences from the pMN24 plasmid integrated non-homologously into the host genome, restoring a nit^{+} phenotype (Fernández et al. 1989). This transformation procedure worked with low efficiency, only one transformant being obtained with 8 μg of plasmid DNA. A reproducible and efficient protocol for transformation involving the gun particle procedure has been recently achieved by using pMN24 and the *nit*-1 305 mutant (Kindle KL, Schnell R, Fernández E, Lefebvre PA, submitted).

Expression of the *nit*-1 Gene

Repression by ammonium and derepression in nitrate medium of NR was analyzed at the transcriptional level by using the B6a clone to probe RNA gel blots of poly $(A)^{+}$ RNA isolated from cells under repression and induction conditions. It was observed that the NR transcript is not detected in ammonium-grown cells, and that in nitrate medium a single RNA species of about 3.4 kb accumulates. This transcript size is large enough to encode a protein of about 110 kDa (Fernández et al. 1989). The result suggests that the complementation previously observed between *nit*-1 mutants (Fer-

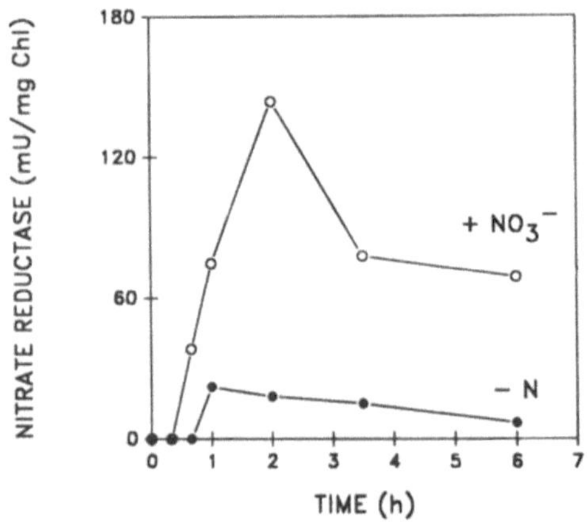

Fig. 1. Time course for derepression of NR activity in *Chlamydomonas reinhardtii*. Ammonium-grown wild-type cells of *Chlamydomonas* were transferred to medium either containing 4 mM KNO_3 or lacking a nitrogen source. The reduced benzylviologen NR activity was assayed *in situ* at the indicated times as previously reported (Fernández and Matagne 1986)

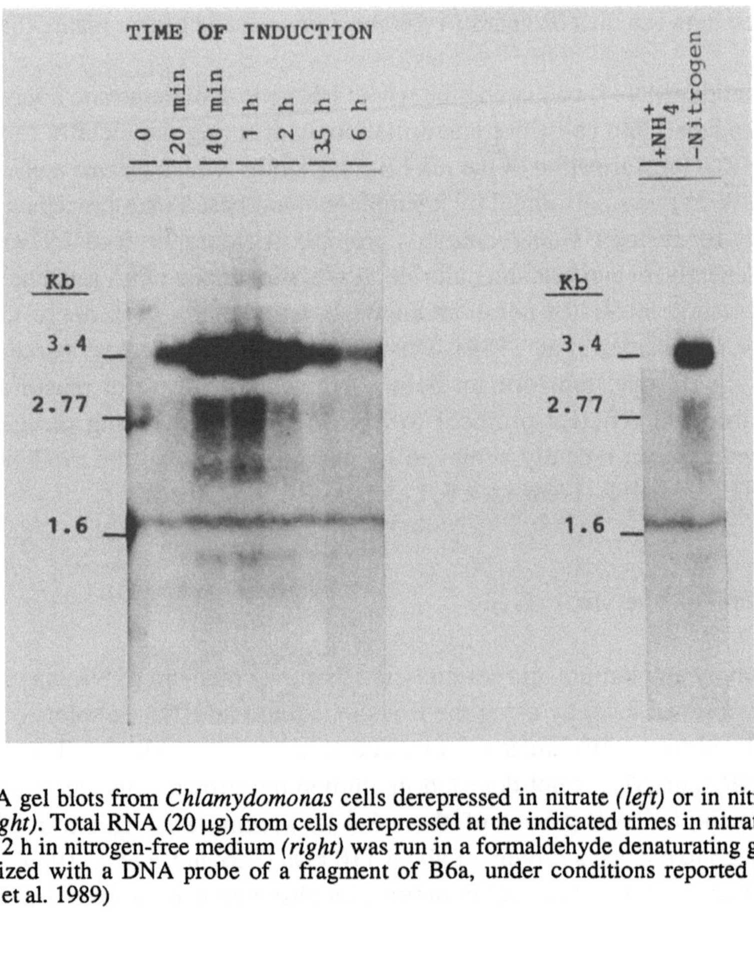

Fig. 2. RNA gel blots from *Chlamydomonas* cells derepressed in nitrate *(left)* or in nitrogen-free medium *(right)*. Total RNA (20 μg) from cells derepressed at the indicated times in nitrate medium *(left)* or for 2 h in nitrogen-free medium *(right)* was run in a formaldehyde denaturating gel, blotted and hybridized with a DNA probe of a fragment of B6a, under conditions reported elsewhere (Fernández et al. 1989)

nández and Matagne 1986) is intragenic. The levels of NR activity achieved by this complementation are relatively high in *Chlamydomonas* in contrast to other systems and has not yet been explained in molecular terms (Fernández and Matagne 1986). The smaller subunit size of 45 and 67 kDa reported for the NR complex in *Chlamydomonas* (Franco et al. 1984) could be accounted for either by post-transductional cleavage of precursor subunits or by proteolytic cleavage in protein extracts.

When ammonium-grown cells of *Chlamydomonas* are transferred to nitrate medium, NR activity is derepressed (Fernández and Cárdenas 1989). In a representative experiment shown in Figure 1, NR activity reached a maximum level at about 2 hours and then decreased. In nitrogen-free medium, a very small amount of NR activity was derepressed. RNA gel blots using total RNA isolated from each time point showed that, under the indicated conditions, a very similar derepression pattern for the NR transcript (Fig. 2, left) and NR activity (Fig. 1) occurred. The non-regulated 1.6 kb transcript in such a lane represents nonspecific hybridization to ribosomal RNA. Derepression of the NR transcript also took place in nitrogen-free medium (Fig. 2, right) without a concomitant appearance of the enzyme activity. These results indicate that regulatory mechanisms operate at both the transcriptional and post-transcriptional levels for the synthesis of NR, as previously proposed (Fernández and Cárdenas 1989).

In RNA gel blots of poly(A)$^+$ RNA from a mutant *nit*-2 of *Chlamydomonas*, it has been found that the 3.4 kb message is only present in nitrate-induced cells, at very low levels. Thus the *nit*-2 locus is involved in the regulation of *nit*-1 transcript levels. *Chlamydomonas* is the first photosynthetic eukaryote in which a regulatory gene for the nitrate assimilatory pathway has been identified genetically (Fernández et al. 1989). The isolation of the NR structural gene from *Chlamydomonas reinhardtii* will allow a molecular analysis of the interesting regulatory circuit for NR in photosynthetic eukaryotes. In addition, the development of the transformation system using NR as a selectable marker will provide an additional tool for characterizing NR structure and regulation.

Acknowledgments. The authors thank the following institutions for financial support: U.S.D.A. Competitive Grants Program (88-37234-4007, P.A.L.), National Institute of General Medical Sciences (GM-34437, P.A.L.; and GM-31159, C.D.S.), Junta de Andalucía (Spain) (E.F.), Fulbright-MEC (Spain) (E.F.), American Cancer Society (R.S.), and N.I. H. Training Grant (GM-07323) (L.P.W.R.).

References

Calza R, Huttner E, Vincentz M, Rouzé P, Galangau F, Vaucheret H, Cherl I, Meyer C, Kronenberger J, Caboche M (1987) Cloning of DNA fragments complementary to tobacco nitrate reductase mRNA and encoding epitopes common to the nitrate reductases from higher plants. Mol Gen Genet 209: 552-562

Cheng CL, Dewdney J, Kleinhofs A, Goodman HM (1986) Cloning and nitrate induction of nitrate reductase mRNA. Proc Natl Acad Sci USA 83: 6825-6828

Crawford NM, Smith M, Bellissimo D, Davis RW (1988) Sequence and nitrate regulation of the *Arabidopsis thaliana* mRNA encoding nitrate reductase, a metalloflavoprotein with three functional domains. Proc Natl Acad Sci USA 85: 5006-5010

Fernández E, Cárdenas J (1989) Genetics and regulatory aspects of nitrate assimilation in algae. In: Wray J, Kinghorn JR (eds) Nitrate Assimilation in Plants: Genetic and Molecular Aspects. Oxford University Press, Oxford, pp 101-124

Fernández E, Aguilar MA (1987) Molybdate repair of molybdopterin-deficient mutants from *Chlamydomonas reinhardtii*. Curr Genet 12: 349-355

Fernández E, Matagne RF (1984) Genetic analysis of nitrate reductase-deficient mutants in *Chlamydomonas reinhardtii*. Curr Genet 8: 635-640

Fernández E, Matagne RF (1986) *In vivo* complementation analysis of nitrate reductase-deficient mutants in *Chlamydomonas reinhardtii*. Curr Genet 10: 397-403

Fernández E, Schnell R, Ranum LPW, Hussey SC, Silflow CD, Lefebvre PA (1989) Isolation and characterization of the nitrate reductase structural gene of *Chlamydomonas reinhardtii*. Proc Natl Acad Sci USA 86: 6449-6453

Franco AR, Cárdenas J, Fernández E (1984) Heteromultimeric structure of the nitrate reductase complex of *Chlamydomonas reinhardii*. EMBO J 3: 1403-1407

Franco AR, Cárdenas J, Fernández E (1987) Involvement of reversible inactivation in the regulation of nitrate reductase enzyme levels in *Chlamydomonas reinhardtii*. Plant Physiol 84: 665-669

Franco AR, Cárdenas J, Fernández E (1988) Two different carriers transport both ammonium and methylammonium in *Chlamydomonas reinhardtii*. J Biol Chem 263: 14039-14043

Harris EH (1989) A *Chlamydomonas* Sourcebook. Academic Press, New York

Lederer F, Cortial S, Becam AM, Haumont PY, Perez L (1985) Complete amino acid sequence of flavocytochrome b_2 from baker's yeast. Eur J Biochem 152: 419-428

Matagne RF, Deltour R, Ledoux L (1979) Somatic fusion between cell wall mutants of *Chlamydomonas reinhardi*. Nature 278: 344-346

Ranum LPW, Thompson MD, Schloss JA, Lefebvre PA, Silflow CD (1988) Mapping flagellar genes in *Chlamydomonas* using restriction fragment length polymorphism. Genetics 120: 109-122

Yubisu T, Miyata T, Iwanaga S, Tamura M, Takeshita M (1986) Complete amino acid sequence of NADH-cytochrome b_5 reductase purified from human erythrocytes. J Biochem 99: 407-422

REGULATORY, STRUCTURAL AND BIOTECHNOLOGICAL ASPECTS OF THE *ASPERGILLUS NIDULANS* GENE CLUSTER FOR NITRATE ASSIMILATION

J.R. Kinghorn, P. Montague and S. E. Unkles [1]

Introduction

The genes for a nitrate permease (*crnA*), nitrite reductase (*niiA*) and nitrate reductase (*niaD*) in the filamentous fungus *Aspergillus nidulans* have been found, by classical genetic analysis, to map together as a gene cluster on linkage group VIII (Cove 1979; Tomsett and Cove 1979; Brownlee and Arst 1979). The order of the genes is *crnA-niiA-niaD* with the biochemical and genetical relationship a follows:

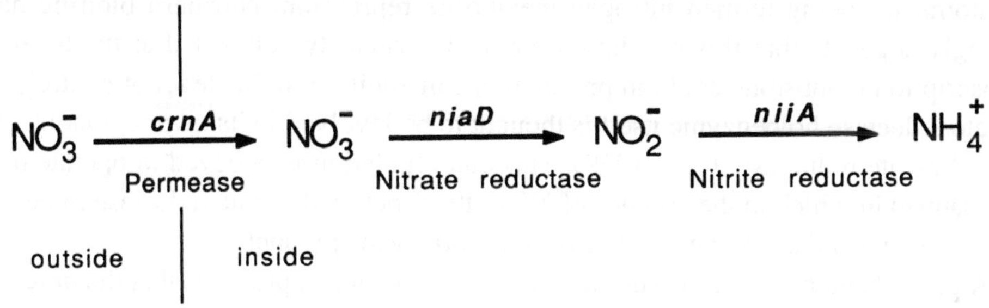

Nitrate reductase shares a common molybdenum cofactor with purine hydroxylase I (hypoxanthine dehydrogenase) an enzyme required for purine catabolism (Scazzocchio 1980; Scazzocchio and Arst 1989). Mutatiuons within the number of genes (designated *cnxA-J*) result in the impairment of the biosynthesis of this cofactor. Consequently *cnx* mutants lack both nitrate reductase and hypoxanthine dehydrogenase activities. *cnx* mutants therefore can be easily distinguished from *niaD* mutants since the latter are able to utilize hypoxanthine as a sole source of nitrogen.

The *crnA-niiA-niaD* gene cluster has recently been isolated by molecular cloning techniques and the *niiA* and *niaD* genes characterized at the nucleotide level (Greaves 1989; Johnstone et al., in preparation; and reviewed in Kinghorn 1989). DNA sequence determination of the *crnA* gene is in progress.

Such molecular genetic experimentation (A) provides preliminary insights into the control of *niiA* and *niaD* genes, (B) permits us to infer the primary amino acid sequence and to compare *A. nidulans* nitrate reductase and nitrite reductase proteins with counterparts from distant organisms as well as with physiologically unrelated proteins,

[1] Plant Molecular Genetics Unit, Sir Harold Mitchell Building, University of St. Andrews, St. Andrews, Fife, KY16 9TH, UK

(C) generates a marker and selection system for the development of gene mediated transformation in filamentous fungi based on nitrate reductase.

Regulatory Aspects

The expression of the *niiA* and *niaD* genes is under tight genetic control. This regulation is affected by the action of two unlinked positively acting regulatory genes, *nirA* and *areA* (Scazzocchio and Arst 1989, and references therein).The former gene is pathway-specific and in the absence of a good nitrogen source such as glutamine or NH_4^+ and presence of its co-inducer NO_3^-, causes a dramatic increase in the levels of nitrite reductase and nitrate reductase enzyme activities. In contrast, the *areA* product in conjunction with its co-repressor (thought to be glutamine) switches off this pathway in addition to a number of systems which utilize alternative nitrogen sources, this phenomenon being termed nitrogen metabolite repression. Northern blotting data strongly suggests that this regulatory circuit is primarily controlled at the level of transcription (Johnstone et al., in preparation). In addition to this level of control, the nitrate reductase holoenzyme itself is thought to be involved in its own regulation and that of nitrite reductase (Cove 1969). This autoregulation is believed to operate by a mechanism in which in the absence of NO_3^-, the functional nitrate reductase molecule is able to bind and subsequently inactivate the *nirA* gene product.

A possible mechanism explaining transcriptional control proposes that the products of regulatory genes (in this case *nirA* and *areA*) which can be considered 'transcriptional factors' have two functionally different and physically separate domains on the protein molecule. These are responsible for DNA binding and subsequent transcriptional activation. The DNA binding domain directly or as a complex with another molecule binds non-covalently to *cis*-acting motifs such as promoter elements or UASs, usually, but not exclusively, located in the 5' non-coding region of the structural gene. Such protein:DNA interaction leads to a local conformational change in the chromatin. The 'activator' domain which is characterized by a region rich in acidic residues (Giniger and Ptashne 1987) binds to the promoter transcriptional complex. This protein:protein interaction results in an increase in density of RNA polymerase II molecules around the gene (Treisman and Maniatis 1985). The ubiquity of this as a unifying mechanism is illustrated by a series of experiments showing that yeast regulatory genes can function efficiently in mammalian cells (Kakidani and Ptashne 1988).

Leaving aside the phenomenon of autoregulation and accepting that such a sequence of events may serve as a model to explain the regulated expression of the *niiA* and *niaD* genes, a detailed examination of the 5' non-coding region should reveal promoter elements and putative receptor or binding sites for the products of the *nirA* and *areA* genes in the 1267 bp intergenic region of the divergently transcribed *niiA* and *niaD* genes.

Figure 1 shows the nucleotide sequence of the intergenic region between the respective ATG's. As has been detailed for a number of cases in surveys of fungal Pol II gene structure (Gurr et al. 1987) both genes have the putative core promoter elements, a CT block and a TATA box. The CT blocks adjacent to the transcriptional start sites are 7 and 6 nucleotides in length for the *niiA* and *niaD* genes respectively. Possible TATA box candidates map at -72 and -58 from the site of RNA initiation for the *niiA* and *niaD* genes. Other elements associated with the core promoter of eukaryotic genes include potential CCAAT motifs and Sp1 binding sites, i.e. GGCGGG which although present are not found at corresponding positions in *niaD* and *niiA* and so their functional significance is unclear. Comparative sequence analysis has identified seven blocks of DNA, within 300 bp of the RNA initiation site, ranging from 9 to 21 bp in length and displaying varying degrees of sequence similarity viz. 64, 62, 67, 67, 80, 64 and 70 % respectively. This leaves a gap of around 600 bp with no significant stretches of sequence similarity. Each pair as depicted in Figure 1 occupies a similar position relative to the RNA start site. However, and rather surprisingly, none of these regions is characterized by having direct/indirect repeats or hyphenated-dyad symmetry which are a hallmark of the binding sites for some yeast and fungal regulatory genes (Arndt and Fink 1986; Baum et al. 1987). The significance or functional requirement for each of these seven regions is difficult to reconcile with the classical genetic evidence that only two loci, i.e. *nirA* and *areA*, are responsible for the regulation of the *niiA* and *niaD* genes. Further information regarding *areA* gene product binding may be gained by comparing the intergenic region of *niiA* and *niaD* with other *A..nidulans* genes which are under nitrogen metabolite repression. A survey of the *prnB* gene encoding a proline transport protein (Sophianopoulou and Scazziocchio 1989) and the *amdS* gene encoding acetamidase (Hynes et al. 1988) reveals a short motif, CACCC, common to all four genes. Rather surprisingly this conserved region was initially identified as an indispensible component of the promoter of β-type globin genes situated at around -100 (Myers et al. 1986). Extended studies have shown that this *cis*-acting element is recognized by a family of transcriptional factors, termed CON, from different tissues (Montovani et al. 1988). This motif is located adjacent to the TATA box in the *prnB* gene at -24 bp, -85 bp in *amdS*, at -103 bp in *niaD* and -104 bp in *niiA*. It may be relevant that the CACCC element is contained within one of the seven conserved regions, box number 3 a 9-mer with 67% sequence similarity. However correlation of this signal with the binding of the *areA* protein should take into account the finding reported by Hynes et al. (1988) who demonstrated that a strain carrying a deletion mutation including the CACCC region was still subject to nitrogen metabolite repression.

Presumably one of the remaining six sequence blocks is associated with *nirA* gene product binding. However, assignment armed only with this information is wildly speculative. The positioning of Box 1 between the ATG and the transcription start site may favour involvement in ribosome binding. Box 2 is adjacent to the putative TATA

```
niiA
AGTGATTTCA CCGGCGCTTT TTGGCAATTC CGATGACTCT CGATCGTGTC TCATAGATAG
        10         20         30         40         50         60
AGATAGGGCT TGATTATCGT TGATTCACTG TCCAATCAGA GCCTCGGAAT CTCTTATCTC
        70         80         90        100        110        120
TGGTGGGAAG AAGGCGGTAG ATCATGGCAC CATTATGCCC AATCAGAACG CTGCCCTGAG
       130        140        150        160        170        180
CCGTGGCCAG ACTTCCCAAA TATCATCATT CGATCTCCAG CCCAACTCCG CGGAAATTCA
       190        200        210        220        230        240
GGCAGTGCAT CGAAGCCATC CACGATGATC CACATCCACC AAACAATGAC AGCGCCAACA
       250        260        270        280        290        300
TCATGCTTAC TGCCGGCTCA TTATCGTCAA GCGGCTCTAG CCACACTCCA GATCCTAAAC
       310        320        330        340        350        360
CCCCGTTCTT TGTGGCGTTC AATATCGATG ACGGTCGTCC ATTGCGAGAT CGTGATTCGC
       370        380        390        400        410        420
GTGTTACCAC ATAGTCATTG GCCCATTTAA TTATCTCATG CTCCGATGCC ACGGAGATAC
       430        440        450        460        470        480
AATCACCAAG GCGTTTGGTG ACTCGTTGTG CCACACTATG GATTCCCGCT ATATGAAGCT
       490        500        510        520        530        540
GGCAGCGTCG CCCGCGGCCG GCTCTTGTGG GACTTGTGTT TTCTCAGACC CCCTGCTCAA
       550        560        570        580        590        600
TCACCGCGCC CGCCATCATG
       610        620

niaD
TTAAATATTA CATCGACTGT GAGTAATAGC CATTGATCAT ATTCTGGACA CTTCCCAGTT
        10         20         30         40         50         60
CCCGGAACAT TCAGGATCGC GTAGGTGGCA GCAGTGACTG TACACTATGA GCGGACTCCT
        70         80         90        100        110        120
ACCCAGATAC GCTCACGCCA TAACCCTATT GCCACTAGAT GACAGTACGA GCCTCTGACC
       130        140        150        160        170        180
GTCTGGCTTG CCGTTGACCC GGGCTGCAGT ATTCTTGCTG TGTAATTACT TGATATAGTA
       190        200        210        220        230        240
CTGGTCAGGG CTACACTTGC AGCCGGTTCG CGGCCCGCGG AGATGGAGCC CTGAAGTGGA
       250        260        270        280        290        300
TCGTAGGATC CAGCATATTG CCTGCTGTCC TTAGAATTAG AAAACGAAAA TACTCTGTTT
       310        320        330        340        350        360
CTCCGAGGCT GAAAGGCAGC GGGCGGCACG CAACTGTATT TCGTCTTATT TTGTCTCGTT
       370        380        390        400        410        420
TTGTCTTGAT TTGGTCTTAT TCCATCTGTT TTCGTCTCAT TTCGTCAGAT CCCGCCGGCC
       430        440        450        460        470        480
TTTGTATTTG ACGTTCTACA TATCCCCCAC GTCAGCCTGG AGCCTGAACC CGTGCCCTAT
       490        500        510        520        530        540
ACTATCTAAT CGACCTTGAT CTGGCATATC TACCAGTTCA TGCCGTGGTC GCTGAGAATT
       550        560        570        580        590        600
CTTCAGTAAT CTGTACCTCT ATATTATTTT CCTATCCCAT ACTCTCACAA TG
       610        620        630        640        650
```

Fig. 1. Comparative sequence analysis of the 5' non-coding regions of the divergently transcribed *niiA* (top) and *niaD* (bottom) genes. Sequences shown are those of the sense strand for both genes. At an arbitrary point midway between the genes, therefore, the sequence has been switched from one strand to the other. ATG's are thick underlined. Major transcriptional start sites are depicted by directional arrows. CT blocks are shown by broken lines. TATA boxes are underlined. Regions of sequence similarity are boxed and boxes numbered 1 to 7

motif. Sequence similarity between *niiA* and *niaD* in Box 4 and Box 7 is confined to one end of the DNA stretch while in Box 6 the sequence correspondence has an interrupted pattern. This leaves Box 5 with an 80% sequence similarity as a likely potential candidate for *nirA* product binding.

Such sequence comparison exercises must be accompanied by functional studies, i.e. gel shift assays, DNAase I footprinting and site directed mutagenesis before any of these regions can attain classification status. However further light may be shed on the functional makeup of the intergenic region by conducting a sequence comparison with the recently cloned equivalent regions from *A. niger* (Unkles et al. 1989 a) and *A. oryzae* (Unkles et al. 1989 b).

Structural Aspects

Nitrate permease. Little biochemical information is available for the uptake system (Brownlee and Arst 1979). The *crnA* transcript size of 1.6 kb would suggest a protein size of around 60 kDa (Johnstone et al., in preparation). DNA sequencing of the *crnA* is in progress and should yield valuable information on the general nature of anion transport mechanisms.

Nitrite reductase (EC 1.6.6.4). As for the nitrate permease, little or no formal biochemical data is available for nitrite reductase. The DNA sequence data predicts that the *niiA* gene encodes a polypeptide of 1104 amino acids (approximately 123 kDa) (Johnstone et al., in preparation). Surprisingly *A. nidulans* nitrite reductase shares little amino acid homology with higher plant nitrite reductase such as spinach (Back et al. 1989) suggesting perhaps very different origins. Other major distinctions are that (1) the plant polypeptide is much smaller at 566 amino acids, (2) plant nitrite reductase utilizes ferredoxin whilst *A. nidulans* utilizes NADPH as a cofactor and (3) *A. nidulans* contains a FAD domain. It is speculated that the discrepancy in size may be due to the presence of this nucleotide domain in the *A. nidulans* protein. In contrast *A. nidulans* nitrite reductase shows considerable amino acid sequence similarity to the *Escherichia coli* protein (Kinghorn and Campbell 1989) which is also NADP dependent (Cole 1989). All nitrite reductases studied so far however contain sirohaem-Fe and Fe_4S_4 centres (Siegel et al. 1987; Back et al. 1989; Kinghorn and Campbell 1989).

Nitrate reductase (EC 1.6.6.3). More is known about the biochemistry of *A. nidulans* nitrate reductase than the other two steps although this does not match contributions in certain other organisms including *Neurospora crassa*. The molecular weight of *A. nidulans* nitrate reductase has been shown to be 182 kDa (Cooley and Tomsett 1985, and references therein) consisting of a homodimer. The DNA sequence for the *niaD* gene predicts a polypeptide of 873 amino acids (98 kDa) (Johnstone et al., in preparation). Extensive sequence similarity has been observed between the *niaD* protein of fungi (*A. nidulans*) and higher plants (*Arabidopsis thaliana*, *Nicotiana tabacum*) reflecting a common evolutionary origin. In contrast no obvious homology exists between the fungal protein and *E. coli* nitrate reductase as exemplified by the *narG* gene. It should be noted that the *E. coli* enzyme, which is respiratory in function is encoded by three genes or cistrons (Boxer 1989; Wootton, unpublished data). No similarity is observed between *A. nidulans* nitrate reductase and nitrite reductase proteins indicating that their flavoreductases have independent origins (Kinghorn and Campbell 1989).

Comparison of fungal or plant nitrate reductase protein sequences with those of functionally unrelated proteins which have molybdenum, haem or reductase domains, has revealed the likely positions of equivalent domains in nitrate reductase (Crawford et al. 1989; Kinghorn and Campbell 1989).

Biotechnological Aspects

A prerequisite for the biotechnological manipulation and exploitation of filamentous fungi using recombinant DNA techniques, is a gene-mediated transformation system. Many such systems are now available for a variety of fungi (Hynes 1986) and rely mainly on either acquisition of antibiotic resistance or reversal of auxotrophy. However, the use of nitrate reductase as the selectable marker in transformation has several distinct advantages. Most filamentous fungi can grow on nitrate as sole nitrogen source and structural gene mutants can be isolated easily and economically by selection for chlorate resistance. Isolation of mutants by positive selection is of particular relevance to those fungal species in which genetic characterization is poor. This includes many of the current production strains. Additionally as mutagenesis is not involved, introduction of gratuitous and possibly deleterious mutations into production strains is not a problem. Mutants in *niaD* can usually be distinguished easily from other chlorate resistant mutants by growth tests (Cove 1979). An exception to this has been *Cephalosporium acremonium*, the major producer of cephalosporin antibiotics. Unlike *A. nidulans* (see introduction), this organism is unable to grow on hypoxanthine as a sole nitrogen source, and so it is not possible to distinguish *niaD* from *cnx* mutants in this way. However, inducible purine hydroxylase I activity could be detected in the wild type and putative *niaD* mutants but not *cnx* mutants (Whitehead et al., in preparation). Mutant designation was confirmed when it was shown that *niaD* mutants could grow on inosine, a catalytic product of hypoxanthine, in the presence of quinic acid as the carbon source, while *cnx* mutants did not grow. Isolation and characterization of chlorate resistant mutants has now been described in a number of fungal species (reviewed in Unkles 1989).

A further advantage of the nitrate system particularly for industrial organisms is that the nitrate assimilation pathway is entirely dispensable under most growth conditions and so mutations within it should not cause disturbances to general metabolism resulting in altered growth or fermentation characteristics.

Finally, the isolation of the *A. nidulans niaD* gene (Malardier et al. 1989; Johnstone et al., in preparation) has facilitated the cloning by cross-hybridization of the *niaD* genes from the commercially important fungal species *A. niger* (Unkles et al. 1989 a) and *A. oryzae* (Unkles et al. 1989 b), major producers of primary and secondary metabolites as well as extracellular enzymes, *C. acremonium* (Whitehead et al., in preparation) and the plant pathogen *Ustilago maydis* (Smith et al. 1988). Using these genes, homologous transformation has been developed for *A. nidulans* (Johnstone et al., in preparation), *A. niger* (Campbell et al. 1989), *A. oryzae* (Unkles et al. 1989) and *C. acremonium* (Whitehead et al., in preparation) with transformation frequencies as high as 1100 per µg DNA being reported (Table 1). Heterologous transformation of *niaD* genes has also been reported for the penicillin producer *Penicillium chrysogenum* (Whitehead et al. 1989) and the plant pathogen *Fusarium oxysporum* (Malardier et al. 1989) although the frequencies of heterologous transformation are generally much lower (Table 1). In addition, heterologous expression of *niaD* genes has been observed in *A. nidulans*, *A. niger* and *A. oryzae* (Table 1) but not in *C. acremonium*.

Table 1. Transformation frequencies using *niaD* genes

Recipient	Maximum transformation frequency [a] with *niaD* from			
	A. nidulans	*A. niger*	*A. oryzae*	*C. acremonium*
A. nidulans	40	13	10	NT[c]
A. niger	4	1176	43	NT
A. oryzae	1	43	1009	NT
C. acremonium	> 0.1[b]	> 0.1	> 0.1	50
P. chrysogenum	20	60	2	NT
F. oxysporum	10	NT	NT	NT

[a] number of transformants per µg DNA; [b] no transformants detected; [c] NT: not tested

Notwithstanding the apparent fastidiousness of *C. acremonium*, it would appear that most genomic *niaD* clones are functionally interchangeable, increasing the attractiveness of the system as a general selective marker for heterologous transformation in fungi from which *niaD* has not been cloned. As in most other fungal systems, transformation occurs by integration which should result in greater stability of incoming genes (Unkles et al. 1989 a; Malardier et al. 1989). Other notable features of *niaD* transformation include the very low background growth allowing clean selection of transformant colonies and, with the exception of *F. oxysporum*, the absence of abortive transformants. Additionally, in homologous transformations, the incoming *niaD* is targeted to the resident *niaD* locus in the majority of integrations, resulting in either gene conversion or low copy tandem duplications. This greatly aids physical analysis of transformants and allows study of gene regulation following *in vitro* mutagensis.

Co-transformation of several bacterial and fungal genes has also been demonstrated following selection on nitrate. Frequencies of co-transformation vary from 5 to 16% and illustrate the potential of the nitrate system for the introduction of non-selectable genes. Co-transformation can thus be used to introduce recombinant plasmids carrying heterologous mammalian genes such as those encoding interferons, interleukins, bovine chymosin etc. into filamentous fungi which may prove to be efficient expression and secretion vehicles for at least some of these proteins.

The nitrate reductase promoter itself may be of interest for expression of foreign genes, especially those which may be toxic for the heterologous host. As the promoter is inducible, it should be possible to grow an organism to a sufficient biomass in medium containing ammonium, switch to nitrate and induce expression of the foreign gene.

Acknowledgment. JRK wishes to thank the SERC and EEC for support for these research studies. He also acknowledges contributions by past research workers.

References

Arndt K, Fink GR (1986) GCN4 protein, a positive transcription factor in yeast, binds general control promoters at all 5'-TGACTC-3' sequences. Proc Natl Acad Sci USA 83: 8516-8520

Back E, Burkhart W, Moyer M, Privalle L, Rothstein S (1989) Molecular cloning of spinach nitrite reductase. In: Wray JL, Kinghorn JR (eds) Molecular and Genetic Aspects of Nitrate Assimilation. Oxford Univ Press, Oxford, pp 284-296

Baum JA, Geever R, Giles NH (1987) Expression of a qa-1F activator protein: Identification of upstream binding sites in the *qa* gene cluster and localisation of the DNA binding domain. Mol Cell Biol 7: 1256-1266

Boxer D (1989) Involvement of the chlorate resistance loci and the molybdenum cofactor in the biosynthesis of the *Escherichia coli* respiratory nitrate reductase. In: Wray JL, Kinghorn JR (eds) Molecular and Genetic Aspects of Nitrate Assimilation. Oxford Univ Press, Oxford, pp 27-39

Brownlee AG, Arst HN jr (1983) Nitrate uptake in *Aspergillus nidulans* and involvement of the third gene of the nitrate assimilation gene cluster. J Bacteriol 115: 1138-1146

Caboche M, Chérel I, Galangau F, Grandbastian M-A, Meyer C, Moureaux T, Pelsy F, Rouzé P, Vaucheret H, Vedele F, Vincentz M (1989) Molecular genetics of nitrite reduction in *Nicotiana*. In: Wray JL, Kinghorn JR (eds) Molecular and Genetic Aspects of Nitrate Assimilation. Oxford Univ Press, Oxford, pp 186-190

Campbell EI, Unkles SE, Macro JA, van den Hondel CAMJJ, Contreras R, Kinghorn JR (1989) Improved transformation in *Aspergillus niger* using *niaD*. Curr Genet 16: 53-56

Cole JA (1989) Physiology, biochemistry and genetics of nitrite reduction by *Escherichia coli*. In: Wray JL, Kinghorn JR (eds) Molecular and Genetic Aspects of Nitrate Assimilation. Oxford Univ Press, Oxford, pp 229-243

Cooley RN, Tomsett AB (1985) Determination of the subunit size of NADPH-nitrate reductase from *Aspergillus nidulans*. Biochim Biophys Acta 831: 89-93

Cove DJ (1969) Evidence for a near limiting intracellular concentration of a regulator substance. Nature 224: 272-273

Cove DJ (1979) Genetic studies of nitrate assimilation in *Aspergillus nidulans*. Biol Rev 54: 291-327

Crawford NM, Smith M, Bellissimo D, Davis RW (1988) Sequence and nitrate regulation of the *Arabidopsis thaliana* mRNA encoding nitrate reductase, a metalloflavoprotein with three functional domains. Proc Natl Acad Sci USA 85: 5006-5010

Giniger E, Ptashne M (1987) Transcription in yeast activated by a putative amphipathic helix linked to a DNA binding unit. Nature 330: 670-672

Greaves P (1989) Molecular genetical analysis of the nitrate utilisation gene cluster of *Aspergillus nidulans*. PhD Thesis, Univ of St. Andrews, UK

Gurr SJ, Unkles SE, Kinghorn JR (1987) The structure and organisation of nuclear genes of filamentous fungi. In: Kinghorn JR (ed) Gene Structure in Eukaryotic Microbes. SGM Special Publications. IRL Press, Oxford, UK, pp 93-139

Hynes MJ (1986) Transformation in filamentous fungi. Exp Mycology 10: 1-8

Hynes MJ, Corrick CM, Kelly JM, Littlejohn TG (1988) Identification of the sites of action for regulatory genes controlling the *amdS* gene of *Aspergillus nidulans*. Mol Cell Biol 8: 2589-2596

Kakidani H, Ptashne M (1988) GAL4 activates gene expression in mammalian cells. Cell 52: 161-167

Kinghorn JR (1989) Genetical, biochemical and structural organisation of the *Aspergillus nidulans crnA-niiA-niaD* gene cluster. In: Wray JL, Kinghorn JR (eds) Molecular and Genetic Aspects of Nitrate Assimilation. Oxford Univ Press, Oxford, pp 69-87

Kinghorn JR, Campbell EI (1989) Amino acid comparisons between bacterial, fungal and plant nitrate and nitrite reductase. In: Wray JL, Kinghorn JR (eds) Molecular and Genetic Aspects of Nitrate Assimilation. Oxford Univ Press, Oxford, pp 385-404

Malardier L, Daboussi MJ, Julian J, Roussel F, Scazzocchio C, Brygoo Y (1989) Cloning of the nitrate reductase gene (*niaD*) of *Aspergillus nidulans* and its use for transformation of *Fusarium oxysporum*. Gene 78: 147-156

Mantovani R, Malgaretti N, Niscolis S, Giglioni B, Comi P, Cappellini N, Bertero MT, Caligaris-Cappio F, Ottolenghi S (1988) An erythroid specific nuclear factor binding to the proximal CACCC box of the β-globulin gene promoter. Nucl Acids Res 16: 4299-4313

Myers RM, Tilly K, Maniatis T (1986) Fine structure genetic analysis of a β-globin promoter. Science 232: 613-618

Scazzocchio C (1980) The genetics of the molybdenum containing enzymes. In: Coughlan MP (ed) Molybdenum and Molybdenum-Containing Enzymes. Pergamon Press, Oxford, pp 487-515

Scazzocchio C, Arst HN jr (1989) Regulation of nitrate assimilation in *Aspergillus nidulans*. In: Wray JL, Kinghorn JR (eds) Molecular and Genetic Aspects of Nitrate Assimilation. Oxford Univ Press, Oxford, pp 298-313

Siegel LM, Wilkerson JO, Janick PA (1987) Structural studies on siroheme (4Fe-4S) cluster active centers of spinach ferredoxin-nitrate reductase and *Escherichia coli* sulfite reductase. In: Ullrich WR, Aparicio PJ, Syrett PJ, Castillo F (eds) Inorganic Nitrogen Metabolism. Springer, Berlin Heidelberg New York, pp 118-122

Smith TM, Spanos A, Banks GR (1988) Differential expression of *Ustilago maydis* DNA sequences during induction of nitrate enzyme activity. Curr Genet 14: 457-460

Sophianopoulou V, Scazzocchio C (1989) The proline transport protein of *Aspergillus nidulans* is very similar to amino acid transporters of *Saccharomyces cerevisiae*. Mol Microbiol 3: 705-714

Tomsett BA, Cove DJ (1979) Deletion mapping of the *niiA-niaD* gene region of *Aspergillus nidulans*. Genet Res 34: 19-32

Treisman R, Maniatis T (1985) Simian virus 40 enhancer increases number of RNA polymerase II molecules on linked DNA. Nature 315: 75-77

Unkles SE (1989) Fungal biotechnology and the nitrate assimilation pathway. In: Wray JL, Kinghorn JR (eds) Molecular and Genetic Aspects of Nitrate Assimilation. Oxford Univ Press, Oxford, pp 341-363

Unkles SE, Campbell EI, Carrez D, Grieve C, Contreras R, Fiers W, van den Hondel CAMJJ, Kinghorn JR (1989 a) Transformation of *Aspergillus niger* using *niaD* for nitrate reductase. Gene 78: 157-166

Unkles SE, Campbell EI, de Ruiter-Jacobs YMJT, Broekhuijzen M, Macro JA, Carrez D, Contreras R, van den Hondel CAMJJ, Kinghorn JR (1989 b) The development of a homologous transformation system for *Aspergillus oryzae* based on the nitrate assimilation pathway: A convenient and general selection system for filamentous fungi. Mol Gen Genet 218: 99-104

Whitehead MP, Unkles SE, Ramsden M, Campbell EI, Gurr SJ, Spence D, van den Hondel CAMJJ, Contreras R, Kinghorn JR (1989) Transformation of a nitrate reductase deficient mutant of *Penicillium chrysogenum* with the corresponding *Aspergillus niger* and *A. nidulans niaD* genes. Mol Gen Genet 216: 408-411

GENETIC REGULATION OF NITROGEN METABOLISM IN *NEUROSPORA CRASSA*

G.A. Marzluf and Y.-H. Fu [1]

Nitrate Assimilation

The filamentous fungus *Neurospora crassa* preferentially uses certain favored nitrogen sources, ammonia, glutamine, or glutamate. However, in their absence, a substantial number of secondary nitrogen sources can be utilized, including nitrate, purines, amino acids, and proteins. The use of the secondary nitrogen sources is dependent upon *de novo* synthesis of nitrogen catabolic enzymes whose formation is governed by the nitrogen regulatory circuit (Marzluf 1981). The nitrogen regulatory circuit consists of an entire set of unlinked structural genes whose expression is highly regulated by major and minor regulatory genes and by nitrogen catabolite repression and specific patterns of induction. Here we will focus primarily on the control of nitrate reductase as a representative, well studied enzyme of the nitrogen control circuit.

Nitrate is converted to nitrite in a two electron transfer reaction catalyzed by nitrate reductase. Nitrate reductase of *Neurospora* is a dimeric protein, containing two identical subunits, each of molecular weight 114 kDa (Bahns and Garrett 1980). Nitrate reductase contains a molydenum containing cofactor, which also occurs in other molybdenum enzymes, such as xanthine dehydrogenase (Scazzocchio 1979). In the subsequent step, the enzyme nitrite reductase converts nitrite into ammonia in a single complex 6-electron transfer reaction. The synthesis of nitrate reductase and of nitrite reductase is dependent upon relief from nitrogen catabolite repression and also requires induction by nitrate, as well as specific genetic signals. A loss of the ability of *Neurospora* to utilize nitrate results from mutation in at least eight different unlinked genes. The subunit of nitrate reductase is specified by the *nit*-3 gene, which is unlinked to *nit*-6, which encodes nitrite reductase. Four genes, *nit*-1, *nit*-7, *nit*-8, and *nit*-9, are required for the synthesis and assembly of the molybdenum cofactor (Tomsett and Garrett 1980). The cofactor is believed to consist of a molybdenum ligand and a polypeptide component, both of which are necessary for the catalytic function of the molybdoenzymes, nitrate reductase and xanthine dehydrogenase. Activity of both of these enzymes is missing in mutants of any of these cofactor genes. Two separate regulatory genes, designated as *nit*-2 and *nit*-4, regulate nitrate metabolism. Mutants of either of these two regulatory genes lack both nitrate and nitrite reductase and cannot utilize nitrate or nitrite for growth (Marzluf 1981). These two control genes act in a positive manner and each is presumed to encode a specific regulatory protein which is

[1] Department of Biochemistry, The Ohio State University, Columbus, OH USA

required for turning on the expression of the structural genes which encode nitrate reductase and nitrite reductase.

Nitrogen Repression and Nitrate Induction

Although nitrate, purines, proteins, and amino acids comprise excellent nitrogen sources for *Neurospora*, they are not utilized if certain favored nitrogen sources, such as ammonium ion, glutamate, or glutamine are present. The presence of any one of these preferred nitrogen sources leads to catabolite repression which prevents the synthesis of nitrate reductase, xanthine dehydrogenase, L-amino acid oxidase, and many other nitrogen related enzymes. The major nitrogen regulatory gene of *Neurospora*, designated *nit*-2, plays a central role in nitrogen repression control. The *nit*-2 gene product is required for expression of nitrate reductase and nitrite reductase and the other enzymes within the nitrogen control circuit. A distinct negative-acting regulatory gene, designated *nmr* (for nitrogen metabolic regulation) also appears to have a major role in nitrogen repression (Dunn-Coleman et al. 1981; Premakumar et al. 1980 b; Debusk and Ogilvie 1984). Thus, nitrogen catabolite repression in *Neurospora* appears to be mediated by two separate regulatory genes, the positive-acting *nit*-2 gene and the negative-acting *nmr* gene. In contrast, in *Aspergillus nidulans*, a single regulatory gene known as *areA* mediates nitrogen repression (Caddick et al. 1986). Glutamine is the effector molecule responsible for nitrogen catabolite repression (Premakumar et al. 1979, 1980 a). The mechanism of nitrogen repression is unknown, but it may occur by the binding of glutamine to a regulatory protein specified by *nmr* or by *nit*-2 or even a multimeric control protein comprised of *nmr* and *nit*-2 subunits.

Synthesis of the enzymes involved in use of secondary nitrogen sources requires relief from nitrogen catabolite repression but is also dependent upon induction by specific substrates. This feature allows expression of only particular subsets of the nitrogen circuit enzymes, when their specific substrates are available. Thus, synthesis of nitrate reductase and nitrite reductase requires nitrogen derepression and specific induction by nitrate. Nitrate induction is mediated by a minor positive-acting regulatory gene, *nit*-4. The action of *nit*-4 is pathway-specific and mutants of it lack nitrate reductase and nitrite reductase but are unaffected for other nitrogen-related enzymes. The *nit*-4 gene apparently specifies a regulatory protein, which activates the *nit*-3 and *nit*-6 structural genes (Marzluf 1981; Tomsett and Garrett 1980). The *nit*-4 protein might directly bind nitrate and thus attain an active conformation for turning on the expression of *nit*-3 and *nit*-6. Alternatively, the *nit*-4 activator might be involved in a protein-protein interaction with nitrate reductase, in an autogenous control mechanism as described below.

Synthesis of Nitrate Reductase

It is of considerable interest to determine the exact step at which each of the several controlling factors, induction, derepression, and genetic regulation of nitrate reductase occurs; it is conceivable that these multiple signals could function at several different levels, e.g., in transcription, translation, and post-translational modifications. Sorger and his colleagues carried out studies which suggested that control of nitrate reductase synthesis may occur at transcription (Premakumar et al. 1979, 1980 a). These experiments depended upon the apparent accumulation *in vivo* of nitrate reductase mRNA under conditions that prevented active enzyme synthesis, followed by the subsequent translation of the accumulated mRNA to yield active nitrate reductase. The results suggested that the regulation of nitrate reductase by nitrogen repression and by nitrate induction both occurred at the transcriptional level. Recent studies have provided direct evidence that nitrate reductase expression is subject to transcriptional control. The structural gene which specifies nitrate reductase, *nit*-3, was isolated by screening a *Neurospora* genomic library for a recombinant clone that complemented a *nit*-3 mutant (Fu and Marzluf 1987 b). The cloned *nit*-3 gene provides a molecular probe to directly examine the content of nitrate reductase mRNA in various wild-type and mutant cells grown under contrasting metabolic conditions. Northern blot analyses revealed that the *nit*-3 gene is expressed to give a single Poly $(A)^+$ RNA transcript of approximately 3.4 kb, the size required to encode the 114 kDa nitrate reductase subunit. Results were obtained which clearly demonstrated that the regulation of the *nit*-3 gene occurs at the level of transcription. Thus, wild-type cells did not contain any detectable *nit*-3 mRNA when subject to nitrogen repression or when uninduced with nitrate. Upon simultaneous nitrate induction and N-derepression, wild-type cells possessed a high level of *nit*-3 mRNA. Furthermore, neither *nit*-2 mutant cells nor *nit*-4 mutant cells contained any detectable *nit*-3 mRNA, even when derepressed and induced. Thus, control of *nit*-3 expression by nitrogen derepression and also by nitrate induction occurs at the level of accumulation of messenger RNA, which is also the step at which the two positive-acting control genes, *nit*-2 and *nit*-4 act. Although these results suggest that *nit*-3 gene expression is controlled at transcription, it is possible that a closely related step such as RNA processing or messenger RNA stability could be involved. The *nmr* mutant displays a partial constitutive synthesis of nitrate reductase, suggesting that *nmr* negatively regulates *nit*-3 expression. Northern blot analyses revealed that *nmr* cells possessed a significant level of *nit*-3 mRNA even in the presence of sufficient glutamine to fully repress expression in wild-type, i.e, the expression of *nit*-3 mRNA was partially insensitive to nitrogen catabolite repression in the *nmr* mutant strain.

Autogenous Control of Nitrate Reductase

It has long been speculated that nitrate reductase expression is subject to autogenous control, in which the enzyme itself plays a negative role in controlling its own synthesis and that of nitrite reductase (Cove 1986; Tomsett and Garrett 1981). Most *nit*-3 mutants are fully constitutive and synthesize nitrite reductase and nitrate reductase partial activities or cross-reacting material in the absence of any added inducer (Tomsett and Garrett 1981), suggesting that the mutant nitrate reductase has lost its autoregulatory function as well as its catalytic activity. A possible autoregulatory role suggested for the nitrate reductase protein is that it is the element that senses the presence of the inducer nitrate. It was of interest to determine whether any evidence for autogenous control could be obtained at the level of nitrate reductase mRNA synthesis. In wild type, expression of nitrate reductase mRNA requires release from nitrogen catabolite repression and specific induction by nitrate; nitrogen derepression alone does not lead to any detectable synthesis of *nit*-3 messenger RNA. In marked contrast, three different *nit*-3 mutant strains constitutively expressed *nit*-3 mRNA; i.e., these mutant strains fully expressed the *nit*-3 gene in the complete absence of any nitrate added as inducer (they do require N-derepression). The *nit*-1 mutant strain produces wild-type nitrate reductase subunits, which apparently fail to dimerize because of the absence of the molybdenum cofactor. The *nit*-1 mutant strain also was found to contain a fully constitutive level of *nit*-3 mRNA without any need for induction. These results imply that the nitrate reductase protein is indeed involved in autoregulation and that this control function occurs at the transcriptional level. The nitrate reductase protein might carry out its postulated autogenous control by interacting with the *nit*-4 regulatory gene product in direct protein-protein binding. Such a protein interaction could prevent the *nit*-4 protein from entering the nucleus to turn on the expression of *nit*-3 and *nit*-6. In the presence of inducer, nitrate reductase could bind nitrate, and release the *nit*-4 protein, which can then turn on *nit*-3 and *nit*-6. Because the *nit*-4 gene has now been cloned (Fu, Kneesi and Marzluf, in press), direct studies of possible interactions of a *nit*-4 regulatory protein with nitrate reductase should soon be possible, and may provide new insight into the postulated autoregulatory mechanism.

Expression and Function of the *nit*-2 Gene

The *nit*-2 gene is the major positive-acting nitrogen regulatory gene, whose product is required to turn on the expression of the various structural genes of the nitrogen control circuit (Stewart and Vollmer 1986; Fu and Marzluf 1987). An intriguing question is whether expression of the *nit*-2 regulatory gene is itself regulated. Indeed, it has been suggested that *nit*-2 is controlled at the transcriptional level by the negative-acting *nmr* gene. Northern blot analyses demonstrated that the *nit*-2 gene is transcribed to give a single large messenger RNA of approximately 3.5 kb. Moreover, the expression of the *nit*-2 gene is regulated and its transcript was present at a 3- to 4-fold higher level when wild-type cells were subjected to nitrogen limitation. However, *nit*-2 expression was not constitutive in *nmr*; rather *nit*-2 mRNA was present at only a low level in *nmr* mutant cells, whether N-repressed or N-derepressed, similar to the level present in N-repressed *nmr* cells (Fu and Marzluf 1988). These results imply that the *nmr* gene does not control *nit*-2 gene transcription in a simple negative fashion as suggested by DeBusk and Ogilvie (1984), but must exert its negative effect in some other fashion. It remains possible that *nmr* might have both a negative and positive effect upon *nit*-2 expression. Finally, the *nit*-2 gene does not appear to be controlled by autogenous regulation since *nit*-2 mutants display the same transcript patterns during repressed and derepressed conditions as found with the wild-type strain. Another way in which the nitrogen regulatory genes might work in a sequential fashion, rather than independently, is that the *nit*-2 gene might control the expression of *nit*-4 which in turn directly activates the nitrate and nitrite reductase structural genes. In this case, transcription of *nit*-4 should be subject to nitrogen repression and should be expressed in *nit*-2^+ but not in *nit*-2 mutant strains. However, Northern blot analyses have shown that *nit*-4 expression is not controlled by *nit*-2 nor by nitrogen limitation; *nit*-4 is expressed constitutively to give a rare transcript of approximately 2.2 kb. These results imply that *nit*-2 and *nit*-4 act independently to turn on expression of the structural genes which encode nitrate and nitrite reductase. Thus, we anticipate that separate binding sites for *nit*-2 and *nit*-4 regulatory proteins will lie upstream of both *nit*-3 and *nit*-6. It is interesting that detectable expression of *nit*-3 requires the simultaneous presence of functional *nit*-2 and *nit*-4 regulatory proteins, neither alone being sufficient to elicit even a low level of expression.

The complete oligonucleotide sequence of genomic and cDNA copies of the *nit*-2 gene has been ascertained. Two short introns of 98 and 78 kbp with good splice site consensus sequences are located near the 5' end of the gene within the protein coding sequence. The *nit*-2 gene contains a single open reading frame, which has 5 in-frame closely spaced AUG codons near the amino terminal portion of the translated protein. It is presently unknown which of these is the natural initiation codon. If translation initiates at the first AUG, *nit*-2 encodes a protein comprised of 1,036 amino acids, with a molecular weight of approximately 110 kDa. Deletion mutants of *nit*-2 which lack the carboxy terminal 205 amino acids are still functional, whereas deletions which remove 280 or more amino acids lead to complete loss of function. The *nit*-2 protein contains exactly four cysteine residues which appear to specify a single zinc finger element which begins at amino acid 743 and has the arrangement: CYS - 2 AAs - 17 AAs - CYS - 2 AAs - CYS. A 3' deletion whose 5' end occurs within the 17 amino acid loop of the putative zinc finger causes complete loss of *nit*-2 function. Furthermore, amino acid substitutions for two of the cysteine residues were accomplished by site-directed mutagenesis and also destroyed *nit*-2 function, which adds support to the possibility that this region of the protein represents a zinc finger element that functions in DNA-binding. In a region immediately downsteam of the zinc finger, the *nit*-2 protein has a cluster of basic amino acids (net charge of +10), which may contribute to specificity of DNA binding. The translated protein also contains two other basic regions and two acidic regions, but their function, if any, is unknown.

The *nmr* Regulatory Gene

Is the negative-acting *nmr* gene itself subject to nitrogen regulation or expressed constitutively? The *nmr* gene was found to be expressed in a constitutive manner to give a single transcript of approximately 1.8 kb (Fu et al. 1988). Comparison of the oligonucleotide sequence of a genomic clone and a full-length cDNA clone has shown that *nmr* lacks any introns and can be translated to yield a protein which contains 479 amino acids, with a molecular weight of 54 kDa. UAG nonsense codons were placed at various locations within the *nmr* coding region by site-directed mutagenesis and the mutated genes were assayed for function with transformation assays. An *nmr* mutant protein which lacked 70 amino acids at its carboxy terminus was active but a protein which was missing the carboxy terminal 121 amino acids was nonfunctional. Thus, a sizeable amount of both the *nit*-2 protein (approximately 22%) and of the *nmr* protein (approximately 15%) appears to be dispensable, i.e., in each case, the protein lacking the above-mentioned region still functions in regulation, comparable to the wild-type protein. A general feature which appears to be emerging is that relatively large segments of regulatory proteins can be deleted without functional consequences; thus, it is not at all obvious why these regions have been retained unless they play subtle roles in the activities of these control proteins.

References

Bahns M, Garrett RH (1980) Demonstration of *de novo* synthesis of *Neurospora crassa* nitrate reductase during induction. J Biol Chem 255: 690-693

Caddick MX, Arst HN Jr, Taylor LH, Johnson RI, Brownless AG (1986) Cloning of the regulatory gene *areA* mediating nitrogen metabolite repression in *Aspergillus nidulans*. EMBO J 5: 1087-1090

Cove DJ (1976) Chlorate toxicity in *Asperigillus nidulans*. Study of mutants altered in nitrate assimilation. Mol Gen Genet 145: 147-159

DeBusk RM, Ogilvie S (1984) Nitrogen regulation of amino acid utilization by *Neurospora crassa*, J Bacteriol 160: 656-661

Dunn-Coleman NS, Tomsett AB, Garrett RH (1981) The regulation of nitrate assimilation in *Neurospora crassa*: Biochemical analysis of the *nmr*-1 mutants. Molec Gen Genet: 182: 234-239

Fu YH, Marzluf GA (1987 a) Characterization of *nit*-2, the major nitrogen regulatory gene of *Neurospora crassa*. Molec Cell Biol 7: 1691-1696

Fu YH, Marzluf GA (1987 b) Molecular cloning and analysis of the regulation of *Nit*-3, the structural gene for nitrate reductase in *Neurospora crassa*. Proc Natl Acad Sci USA 84: 8243-8247

Fu YH, Marzluf GA (1988) Metabolic control and autogenous regulation of *Nit*-3, the nitrate reductase structural gene of *Neurospora crassa*. J Bacteriol 170: 657-661
Fu YH, Young JL, Marzluf GA (1988) Molecular cloning and characterization of a negative-acting nitrogen regulatory gene of *Neurospora crassa*. Mol Gen Genet 214: 74-79
Marzluf GA (1981) Regulation of nitrogen metabolism and gene expression in fungi. Microbiol Rev 45: 437-61
Premakumar R, Sorger GJ, Gooden D (1979) Nitrogen metabolite repression of nitrate reductase in *Neurospora crassa*. J Bacteriol 137: 1119-1126
Premakumar R, Sorger GJ, Gooden D (1980 a) Repression of nitrate reductase in *Neurospora* studied by using L-methionine-DL-sulfoximine and glutamine auxotroph *gln*-1b. J Bacteriol 143: 411-415
Premakumar R, Sorger GJ, Gooden D (1980 b) Physiological characterization of a *Neurospora crassa* mutant with impaired regulation of nitrate reductase. J Bacteriol 144: 542-551
Scazzocchio C (1979) The genetics of molybdenum-containing enzymes. In: Coughlan W (ed) Molybdenum Containing Enzymes. Pergamon Press, Oxford, pp 487-515
Stewart V, Vollmer SJ (1986) Molecular cloning of *nit*-2, a regulatory gene required for nitrogen metabolite repression in *Neurospora crassa*. Gene 46: 291-295
Tomsett AB, Garrett RH (1980) The isolation and characterization of mutants defective in nitrate assimilation in *Neurospora crassa*. Genetics 95: 649-660
Tomsett AB, Garrett RH (1981) Biochemical analysis of mutants defective in nitrate assimilation in *Neurospora crassa*: Evidence for autogenous control by nitrate reductase. Molec Gen Genet 184: 183-190
Tomsett AB, Dunn-Coleman NS, Garrett RH (1981) The regulation of nitrate assimilation in *Neurospora crassa*: The isolation and genetic analysis of *nmr*-1 mutants. Molec Gen Genet 182: 229-235

ANALYSIS OF NITRATE REDUCTION GENES IN CYANOBACTERIA

X. Andriesse [1], H. Bakker [1] and P. Weisbeek [1,2]

Introduction

Cyanobacteria form a very divergent group of unicellular or filamentous microorganisms, which constitute one of the largest subgroups of Gram-negative prokaryotes. They have very few nutrient requirements due to their capacity to carry out a plant-like, oxygen-evolving photosynthesis, so mainly carbon dioxide and nitrogen are needed in more than small amounts. Some strains can fix atmospheric nitrogen, but the majority relies on nitrate, nitrite or ammonia. Nitrate reduction in cyanobacteria is tightly coupled to photosynthesis by ferredoxin, which provides both nitrate reductase (NR) and nitrite reductase (NiR) with electrons. In contrast to higher plant NR, cyanobacterial NR consists of only a single polypeptide of about 80 kDa with one $[Fe_4S_4]$ or two $[Fe_2S_2]$ clusters and a molybdenum cofactor (Guerrero et al. 1981; Mikami and Ida 1984). Kuhlemeier (1984 a,b) desribed the construction of nitrate reduction mutants of *Synechococcus* PCC 7942 (formerly called *Anacystis nidulans* R2) by chemical mutagenesis with N-methyl-N'-nitro-N-nitrosoguanidine (MNNG) or transposon Tn901 mutagenesis (Tandeau de Marsac et al. 1982). Phenotypical characteristics of these mutants include: (a) slow growth and yellow color on nitrate medium, (b) complete restoration of growth and color after addition of nitrite to the medium, (c) no detectable nitrate reductase activity, even in the presence of 1 mM sodium molybdate in the medium.

Two genes involved in nitrate reduction were cloned by complementation of the corresponding mutants, a third gene was obtained via the transposon, which caused the mutation. These are the only genes known to be directly involved in the reduction step from nitrate to nitrite. Madueño et al. (1988) found two different nitrate assimilation mutants of *Synechococcus* PCC 7942 but both mutants, although unable to grow on nitrate medium, exhibit nitrate reductase activity. The exact function of the nitrate reduction genes isolated is not known, but they can code for the NR itself, for regulatory proteins, for receptor proteins or for proteins involved in molybdopterin synthesis. In this paper, we describe the analysis of one of the cloned genes, the *narB* gene.

[1] Department of Molecular Biology, University of Utrecht, Padualaan 8, 3584 CH Utrecht, The Netherlands

[2] Institute for Molecular Biology and Medical Biotechnology, University of Utrecht, Padualaan 8, 3584 CH Utrecht, The Netherlands

Sequence Analysis

To gain insight into the protein coded for by the *narB* gene we determined the nucleotide sequence. The position of the gene on the cloned fragment was mapped by localizing the sites of the mutations. Seven nitrate reduction mutants of the *narB* group were transformed with subclones of the complementing DNA fragment. By comparing the results of these complementations the mutation site could be determined for each mutant; because the mutations were assumed to map on the smallest overlapping region of the sub-clones responsible for transformation to phenotypical wild type cells (fig. 1).

A.

	R2	P11	P7	K18	K15	Y7	Y6	Y5
1	+	+	+	+	+	+	+	-
2	+	+	+	+	+	+	+	-
3	+	-	+	+	-	+	-	-
4	+	-	-	-	-	-	-	-
5	+	-	-	-	-	-	-	-
6	+	-	+	-	-	+	-	-
7	+	-	-	-	-	-	-	-
8	+	+	+	-	+	+	+	-
9	+	-	-	+	-	-	-	-
10	+	+	-	-	+	-	+	-
11	+	-	-	-	-	-	-	-
12	+	-	-	-	-	-	-	-

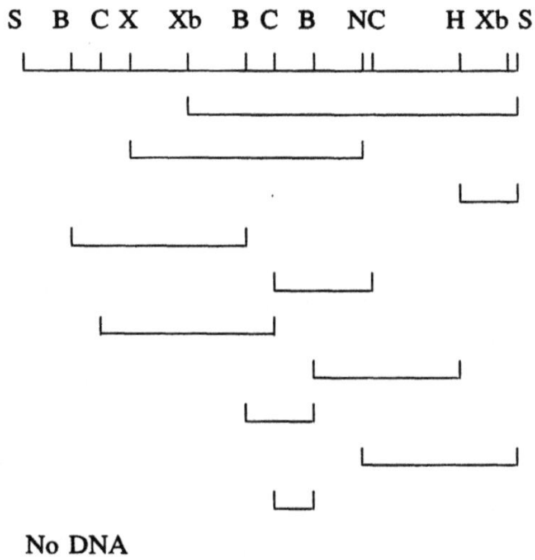

B.

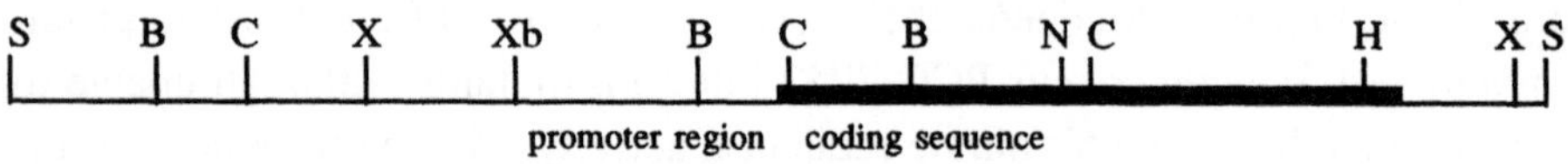

Fig. 1. Complementation of *narB* mutants. **A.** The right panel shows the subclones used for the transformation experiments. Normal growth for wild type R2 and complementation for the seven different mutants are scored by + in the left panel, no complementation by –. S = SalI, B = BglII, C = ClaI, X = XhoI, Xb = XbaI, N = NcoI and H = HindIII.
B. Position of the *narB* gene on the cloned fragment. Restriction enzymes as in A

To check these results chromosomal DNA of transposon mutants and of the wild type was digested with several restriction enzymes, which had no recognition sites in the transposon. *NarB* subclones were used as probes on Southern blots, and hybridization patterns were compared. The transposon mutations mapped on the restriction fragments with shifted molecular weights due to the presence of the transposon. The results obtained with this method correspond with those obtained with the transformations (data not shown).

The nucleotide sequence of the *narB* gene was determined in both directions with subclones in pEMBL vectors (Dente et al. 1983) using the dideoxy-method (Sanger et al. 1977). For stretches lacking suitable restriction sites, oligo-nucleotides were used as primers. As follows from the sequence analysis, the *narB* gene codes for a protein with a molecular weight of about 80 kDa. Because there are two possible start codons, it is not clear whether the protein consists of 715 or 729 amino acids. The first possible start codon is preceded by a truncated ribosome binding site, the 5' region of the second one lacks any comparable sequence.

Upstream of the coding region, a possible promoter was found with an *Escherichia coli* consensus sequence. This kind of promoters is found in more cyanobacteria; the *narB* promoter sequence differs from the predicted *Synechocystis* 6803 *psbB* promoter (Vermaas et al. 1986) only by two AT conversions (table 1).

Table 1. Promoter sequences

(Possible) promoter	-35 region	space	-10 region
E. coli consensus	TTGACA	16 - 18	TATAAT
Synechocystis 6803 *psbB*	TTGCAG	17	TACAAT
Synechococcus 7942 *narB*	TTGCAG	17	ATCAAT

Comparison of the nucleotide sequence of *narB* with the EMBL sequence library revealed two other genes that show homology with the *narB* gene. Both genes code for formate dehydrogenase polypeptides, the first for the seleno-polypeptide of FDH_F from *E. coli* (Zinoni et al. 1987), the second for one of the polypeptides of FDH_A of *Methanobacterium formicicum* (Shuber et al. 1986). Homology is only found in the coding sequences; it was as high as 80% in three clear boxes. The regions upstream of the start codon and downstream of the stop codon are completely different.

At the level of the proteins there is an overall similarity of 30% between NAR_B and FDH_F and 34% between NAR_B and FDH_A, whereas the similarity between the two formate dehydrogenases is 40%. Dot matrix comparisons (Fig. 2) show that the patterns of similarity correspond with the homology at the DNA level. The similarity is spread over the full length of the protein starting with the second possible start codon and not with the first one.

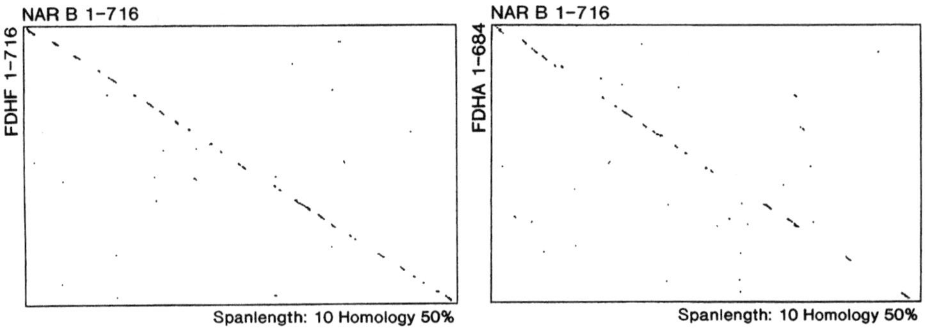

Fig. 2. Dot matrix amino acid comparisons between NAR_B and FDH_F (*left* panel) and between NAR_B and FDH_A (*right* panel)

Discussion and Conclusions

One of the nitrate reduction genes, *narB*, of the cyanobacterium *Synechococcus* PCC 7942 shows a remarkable similarity with two formate dehydrogenase genes from two bacteria, the FDH_F from *E. coli* and the FDH_A from *Methanobacterium formicicum*. The meaning of this phenomenon is unclear, although there are some common features in nitrate reduction and formate dehydrogenation in *E. coli*. The enzymes involved in these reactions are multimers with molybdenum cofactors and are induced under anaerobic conditions. In cyanobacteria, which evolve oxygen during photosynthesis, nitrate reductase is repressed by ammonium and, in the absence of ammonium, induced by nitrate and nitrite (Herrero et al. 1981, 1985). Cyanobacterial nitrate reductase consists of a single polypeptide, but also with a molybdenum cofactor. We are presently investigating the evolutionary relationship by computer comparisons of nucleotide and amino acid sequences.

The *narB* gene has two possible start codons and the similarity with the two formate dehydrogenases starts with the second one, although there is no ribosome binding site in front of it. The Shine-Delgarno sequence of the first start codon consists of only four nucleotides at a site two bases upstream. From these results it is not clear, which one is the correct start codon. The most direct way to solve this problem is to purify the protein and to determine its amino acid sequence. Because NR is present in very low amounts in the cyanobacterial cell, we did not yet succeed in obtaining enough material. Chimeric genes are now constructed which consist of the *narB* promoter and a reporter gene (*gus*) by fusing these elements at the sites of both possible start codons. Together with the results of the chemical experiments this will enlarge our knowledge of cyanobacterial nitrate reduction.

Acknowledgments. This work was supported in part by the Netherlands Organization for Chemical Research (SON), with financial aid from the Netherlands Organization for Scientific Research (NWO) and by the Consiglio Nazionale delle Ricerche (CNR) as proposed by NWO.

References

Dente L, Cesareni G, Cortese R (1983) pEMBL: A new family of single stranded plasmids. Nucleic Acids Res 11: 1644-1655

Guerrero MG, Vega JM, Losada M (1981) The assimilatory nitrate reducing system and its regulation. Annu Rev Plant Physiol 32: 169-204

Herrero A, Flores E, Guerrero MG (1981) Regulation of nitrate reductase levels in the cyanobacteria *Anacystis nidulans, Anabaena sp.* strain 7119, and *Nostoc sp.* strain 6719. J Bacteriol 145: 175-180

Herrero A, Flores E, Guerrero MG (1985) Regulation of nitrate reductase cellular levels in the cyanobacteria *Anabaena variabilis* and *Synechocystis sp.* FEMS Microbiol Lett 26: 21-25

Kuhlemeier CJ, Logtenberg T, Stoorvogel W, Van Heugten HHA, Borrias WE, Van Arkel GA (1984 a) Cloning of two nitrate reductase genes from the cyanobacterium *Anacystis nidulans* R2. J Bacteriol 159: 36-41

Kuhlemeier CJ, Teeuwsen VJP, Janssen MJT, Van Arkel GA (1984 b) Cloning of a third nitrate reductase gene from the cyanobacterium *Anacystis nidulans* R2 using a shuttle gene library. Gene 31: 109-116

Madueño F, Borrias WE, Van Arkel GA, Guerrero MG (1988) Isolation and characterization of *Anacystis nidulans* R2 mutants affected in nitrate assimilation: Establishment of two new mutant types. Mol Gen Genet 213: 223-228

Mikami B, Ida S (1984) Purification and properties of ferredoxin-nitrate reductase from the cyanobacterium *Plectonema boryanum*. Biochim Biophys Acta 791: 294-304

Sanger F, Nicklen S, Coulsen AR (1977) DNA sequencing with chain-terminating inhibitors. Proc Natl Acad Sci USA 74: 5463-5467

Shuber AP, Orr EC, Recny MA, Schendel PF (1986) Cloning, expression and nucleotide sequence of the formate dehydrogenase genes from *Methanobacterium formicicum*. J Biol Chem 261: 12942-12947

Tandeau de Marsac N, Borrias WE, Kuhlemeier CJ, Castets AM, Van Arkel GA, Van den Hondel CAMJJ (1982) A new approach for molecular cloning in cyanobacteria: cloning of an *Anacystis nidulans met* gene using a Tn901-induced mutant. Gene 20: 111-119

Vermaas WFJ, Williams JGK, Arntzen CJ (1987) Sequencing and modification of *psbB*, the gene encoding the CP-47 protein of Photosystem II in the cyanobacterium *Synechocystis* 6803. Plant Mol Biol 8: 317-326

Zinoni F, Birkmann A, Stadtman TC, Böck A (1987) Nucleotide sequence and expression of the selenocysteine-containing polypeptide of formate dehydrogenase (formate-hydrogen-lyase-linked) from *Escherichia coli*. Proc Natl Acad Sci USA 83: 4650-4654

MOLECULAR BIOLOGY OF THE ASSIMILATION OF NITROGENOUS COMPOUNDS BY CYANOBACTERIA

A. Herrero, M.A. Vega-Palas, J. Martín-Nieto, A.M. Muro-Pastor, F. Madueño and E. Flores [1]

Introduction

Cyanobacteria are prokaryotic organisms that carry out oxygenic photosynthesis and are able to utilize nitrate, nitrite or ammonium as a nitrogen source for growth. Some of their representatives are also able to fix dinitrogen or to grow on certain amino acids. Although significant information is currently available concerning the physiology of nitrate assimilation by cyanobacteria (Guerrero and Lara 1987), very little is known about the structure or the regulation of the system at the genetic level. Our aim is the study of the genetics of nitrogen assimilation by cyanobacteria, with special emphasis on nitrate assimilation and on global nitrogen regulation.

Nitrate Assimilation Mutants of *Synechococcus* R2

Nitrate assimilation in the unicellular cyanobacterium *Synechococcus* spp. is regulated mainly through ammonium repression. Mutants unable to use nitrate were isolated from *Synechococcus* sp. strain R2 after nitrosoguanidine or UV-light treatment and penicillin enrichment. In addition to mutants lacking nitrate reductase or nitrite reductase, mutants simultaneously impaired in a number of ammonium-repressible activities were isolated. These activities included nitrate reductase, nitrite reductase, glutamine synthetase (activity or protein), and methylammonium transport, the levels in derepressed cultures of the mutants being similar to those of ammonium-grown cells of the parental strain.

A 48-kDa protein present in the cytoplasmic membrane of *Synechococcus* was found by us (Madueño et al. 1988) and others (Sivak et al. 1989; Omata et al. 1989) to be subject to ammonium repression. Derepressed cultures of the pleiotropic mutants described above had low levels of this protein, similar to those of ammonium cultures of the wild-type strain. From this and the fact that the 48-kDa protein is constitutively expressed in a mutant of *Synechococcus* R2 that constitutively expresses nitrate up-

[1] Instituto de Bioquímica Vegetal y Fotosíntesis, CSIC-Universidad de Sevilla, Apartado 1113, 41080-Sevilla, Spain.

take, we suggested (Madueño et al. 1988) that this cytoplasmic-membrane protein was an element of the nitrate transport system. This is supported by the evidence provided by Omata et al. (1989).

Reversion studies showed that the phenotype exhibited by the pleiotropic mutants was the result of single mutations, which should affect a regulatory gene. From the fact that all revertants analyzed had gained wild-type levels and patterns of regulation of the mutated functions, we propose that the product of that gene is a positively acting element required for the expression of genes involved in nitrogen assimilation. The wild-type version of the regulatory gene has been cloned by complementation. Strain CS31, one of the pleiotropic mutants, was used as a recipient in transformation experiments with a cosmid gene bank of strain R2 previously described (Kuhlemeier et al. 1984). Several restriction fragments from a complementing cosmid were subcloned in pUC19 and tested for the ability to complement strain CS31. In this way the regulatory gene was located to a 3.1 kbp *Hind* III-*Sal* I fragment which is being further analyzed. We have called the cloned gene *ntc* A, "*ntc*" standing for nitrogen control.

Nitrate Assimilation Mutants of *Anabaena*

Mutants impaired in nitrogen assimilation have also been isolated from filamentous, dinitrogen-fixing cyanobacteria. In these organisms nitrate assimilation is under the control of ammonium repression and of obligate induction that can be met by either nitrate or nitrite (Martín-Nieto et al. 1989). Mutants unable to grow on nitrate which were impaired in nitrate reductase or in both nitrate reductase and nitrite reductase were isolated. *Nif*$^-$ mutants were also obtained.

A number of mutants derived from *Anabaena variabilis* ATCC 29413 (strains CS52 to CS55) were able to utilize nitrate only if the medium was supplemented with molybdate concentrations of about 50 - 70 mM. These strains lacked nitrate reductase under normal, low-molybdate culture conditions and seem to be impaired in the synthesis of the nitrate reductase molybdenum cofactor. On the other hand, strain CS60 (also derived from *A. variabilis*) exhibited low nitrate reductase and nitrogenase activities unless the medium was supplemented with moderate levels of molybdate (about 0.1 mM). This mutant appears then to be impaired in the synthesis of both the molybdenum cofactor of nitrate reductase and the iron-molybdenum cofactor of nitrogenase. Our results support the operation in *Anabaena* of a pathway for molybdate assimilation of the type proposed by Shah et al. (1984), comprising a common trunk and pathway-specific reactions for the synthesis of Mo-co and FeMoco. Strains CS52 to CS55 would in fact constitute the first evidence for the existence of genes specific for the synthesis of the molybdenum cofactor of nitrate reductase in a N_2-fixing organism.

Finally, mutants of *Anabaena* sp. PCC 7120 and of *A. variabilis* exhibiting, as the result of single mutations, low levels of nitrate reductase and nitrite reductase, but normal levels of nitrogenase and glutamine synthetase have been isolated. The gene(s) mutated in these strains might be involved in the induction of the nitrate assimilatory system. Mutant CS403 (derived from PCC 7120) has been used as a recipient for conjugation with a cosmid library from wild-type *Anabaena* (kindly supplied by W.J. Buikema and R. Haselkorn). Several complemented cyanobacterial clones were obtained from which the wild-type version of the gene altered in strain CS403 is being cloned.

Amino Acid Transport

Arginine was able to support the growth of *Synechocystis* sp. PCC 6803 (a unicellular cyanobacterium) and of *Anabaena* sp. PCC 7120. Strain PCC 7120 exhibited two transport systems for basic amino acids. The high affinity system (K_m for L-arginine or L-lysine = 2 μM) mediated an energy-dependent, concentrative transport of arginine into the cell. Canavanine-resistant mutants were isolated which lacked this transport system. On the other hand, the low-affinity transport system (K_m for L-arginine = 0.75 mM; K_m for L-lysine = 0.11 mM) mediated a non-concentrative, energy-independent transport.

In strain PCC 6803, we have detected two different kinetic components in transport studies with both L-arginine and L-lysine. Mutants simultaneously lacking one component of L-arginine transport (K_m = 0.8 μM) and one component of L-lysine transport (K_m = 58 μM), but still able to express other transport activities for these amino acids (L-arginine: K_m = 50 nM; L-lysine: K_m = 300 μM), were obtained. This shows that, at least, two transport systems are involved in basic amino acid transport in *Synechocystis*. However, these two systems should share some gene(s), since mutants lacking any activity of transport of basic amino acids have been obtained with a high frequency. The transport of arginine in strain PCC 6803 was concentrative and energy-dependent.

Concluding Remarks

Cyanobacteria are prokaryotic organisms which carry out a number of relevant processes for nitrogen assimilation, including nitrate reduction and dinitrogen fixation. Since these organisms are amenable to molecular genetic analysis, they represent an interesting model system for the study of nitrogen assimilation by photosynthetic cells. Our approach involves the isolation of mutants, their biochemical characterization, and the cloning of genes by means of mutant complementation. This has allowed us to identify and clone a gene (*ntcA*)) involved in global nitrogen control in *Synechococcus*. Mutants of this gene have helped to define a 48-kDa cytoplasmic-membrane protein as a component of the nitrate transport system.

References

Guerrero MG, Lara C (1987) Assimilation of inorganic nitrogen. In: Fay P, van Baalen C (eds) The Cyanobacteria. Elsevier, Amsterdam New York, pp 163-186

Kuhlemeier CJ, Logtenberg T, Stoorvogel W, van Heugten HAA, Borrias WE, van Arkel GA (1984) Cloning of nitrate reductase genes from the cyanobacterium *Anacystis nidulans*. J Bacteriol 159: 36-41

Madueño F, Vega-Palas MA, Flores E, Herrero A (1988) A cytoplasmic-membrane protein repressible by ammonium in *Synechococcus* R2: altered expression in nitrate-assimilation mutants. FEBS Lett 239: 289-291

Martín-Nieto J, Herrero A, Flores E (1989) Regulation of nitrate and nitrite reductases in dinitrogen-fixing cyanobacteria and *Nif* mutants. Arch Microbiol 151: 475-478

Omata T, Ohmori M, Arai N, Ogawa T (1989) Genetically engineered mutant of the cyanobacterium *Synechococcus* PCC 7942 defective in nitrate transport. Proc Natl Acad Sci USA 86: 6612-6616

Shah VK, Ugalde RA, Imperial J, Brill WJ (1984) Molybdenum in nitrogenase. Annu Rev Biochem 53: 231-257

Sivak MN, Lara C, Romero JM, Rodríguez R, Guerrero MG (1989) Relationship between a 47-kDa cytoplasmic membrane polypeptide and nitrate transport in *Anacystis nidulans*. Biochem Biophys Res Commun 159: 257-262

GENETIC CONTROL OF NITROGEN METABOLISM IN THE PHOTOSYNTHETIC BACTERIUM *RHODOBACTER CAPSULATUS*

J.C. Willison, G. Ahombo.and P.M. Vignais [1]

Introduction

The photosynthetic bacterium *Rhodobacter capsulatus* (formerly *Rhodopseudomonas capsulata*; Imhoff et al. 1984) is a facultatively anaerobic phototroph, capable of growing either aerobically or anaerobically in the dark, as well as under anaerobic or semi-aerobic conditions in the light. It is able to use a wide range of organic and inorganic N sources for growth, including NH_4^+, N_2, and, in some strains, NO_3^-.

NH_4^+ is the preferred N source, and is assimilated via glutamine synthetase (GS) and glutamine:2-oxoglutarate aminotransferase (GOGAT). GS activity seems to be regulated mainly by adenylylation/deadenylylation, since the total enzyme activity of the adenylylated plus deadenylylated forms shows little variation with the N source used for growth (Johansson and Gest 1976, 1977). In contrast, the synthesis of GOGAT appears to be repressed by glutamate.

Studies on the genetic regulation of nitrogen metabolism in *R. capsulatus* have mostly concerned the regulation of N_2 fixation. Synthesis of the key enzyme of N_2 fixation, nitrogenase, is repressed by O_2 and by sources of fixed N, such as NH_4^+, and is stimulated by light (see Vignais et al. 1985, for review). At present, very little is known about the molecular mechanisms by which light intensity influences nitrogenase synthesis. However, recent studies using molecular genetic techniques have shed light on the mechanism of regulation by fixed N and by O_2, in particular as regards the identification of the regulatory genes involved. Many of the genes found to be involved in regulation by fixed N are structurally homologous to genes found in the more extensively studied bacterium, *Klebsiella pneumoniae*. However, in some cases, the function of these genes appears to be different in the two organisms.

In the present article, we describe the genetic organization of the genes involved in N_2 fixation in *R. capsulatus*, and then discuss in more detail the role of individual genes involved in regulation by fixed N and by O_2.

Genetic Organization of N_2 Fixation Genes

The genes for N_2 fixation have been localized on the genetic map of *R. capsulatus* (Fig. 1). Mutations affecting N_2 fixation were found to map in five different regions of

[1] Biochimie Microbienne (CNRS UA 1130), Département de Recherche Fondamentale, Centre d'Etudes Nuclaires de Grenoble, 85X, F-38041 Grenoble cedex, France

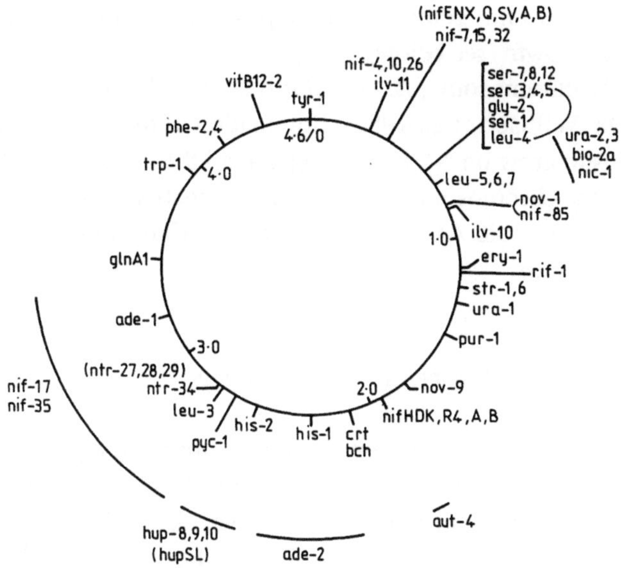

Fig. 1. Genetic linkage map of the *Rhodobacter capsulatus* chromosome. The different groups of *nif* mutations are as follows: Group I: *nifHDK*, *nifR4*; Group IIA, *nif-85*; Group IIB, *nif-7,15,32*; Group III: *nif-4,10,26*; Group IV: *nif-17,35; ntr-27,28,29,34* (*adgA27,28,29,34*) (cf. Willison et al. 1985). The map distance in arbitrary units is shown inside the circle

the chromosome (I, IIA, IIB, III and IV) (Willison et al. 1985). Analysis of cloned DNA fragments by hybridization and DNA sequencing has shown that most of the N_2 fixation (*nif*) genes are located in three regions, termed regions A, B and C (Klipp et al. 1988). Region A contains a large number of *nif* genes, including *nifE, N, X, Q, S, V* and *W*, as well as one copy of the genes *nifA* and *nifB*. Region B contains the nitrogenase structural genes (*nifH, D, K*), the regulatory gene *nifR4*, and a second copy of *nifA* and *nifB*. Region C contains the regulatory genes *nifR1* and *nifR2* (the existence of a third regulatory gene in this region, *nifR3*, was not confirmed by DNA sequence analysis; see below). On the genetic map, region A corresponds to the linkage group IIA; region B to group I; and region C probably corresponds to group IIB, on the basis that the phenotype of the sole mutation in this group (*nif-85*) corresponds to that of a regulatory mutation.

Group IV contains mutations which affect growth not only on N_2, but on all N sources other than NH_4^+. The gene involved has been cloned and is located on a 5.8 kb *Eco*RI fragment (unpublished data). These mutations were initially referred to as "*ntr*-like", due to the phenotypic resemblance of mutants to nitrogen regulatory (*Ntr*$^-$) mutants of *Klebsiella pneumoniae* (Willison et al. 1985). However, neither an *E. coli* gene which complements these mutations (Allibert et al. 1987) nor the *R. capsulatus* gene, which we have now sequenced (unpublished data), show any sequence homology to the *K. pneumoniae ntr* genes. We now propose to call this gene *adgA*, for *a*mmonia-*d*ependent *g*rowth (see below).

In addition to the chromosome, *R. capsulatus* strain B10 contains an endogenous plasmid of 86 MDa (Willison et al. 1987). Elimination of this plasmid does not affect the ability of strains to grow on N_2, suggesting that all the genes essential for N_2 fixation are located on the chromosome. In contrast, elimination of the endogenous plasmid from *R. capsulatus* strain AD2 results in the loss of the ability to grow on NO_3^- (J.C. Willison, submitted).

Regulation of *nif* Gene Expression by Fixed N

In *K. pneumoniae*, expression of *nif* operons other than the *nifLA* control operon requires the presence of the *nifA* gene product (NIFA) and the *ntrA* gene product (NTRA) (for reviews, see Gussin et al 1986; Dixon et al. 1987). The *ntrA* gene codes for a sigma factor which interacts with the RNA polymerase. NIFA is an activator protein which binds to specific sequences upstream from each *nif* operon, thus enabling the modified RNA polymerase (containing NTRA) to recognize the unusual *nif* promoters.

The *nifLA* operon is activated by the products of the *ntrC* and *ntrA* genes. NTRC is an activator protein which behaves in a manner similar to NIFA. In *K. pneumoniae*, it is required for the activation, not only of *nifLA*, but also of some other genes involved in N metabolism, including the glutamine synthetase gene (*glnA*), and the genes for some amino acid permeases (Magasanik 1982). Information about the N status of the cell is transmitted to NTRC via the product of the *ntrB* gene, NTRB. The latter protein is a protein kinase, which catalyses the reversible phosphorylation/dephosphorylation of NTRC. The activity of NTRB is, in turn, modulated by components of the cascade system for control of glutamine synthetase activity, which depends on the intracellular ratio of glutamine and α-ketoglutarate.

In *R. capsulatus*, genes homologous to the regulatory genes mentioned above have been identified by hybridization and DNA sequence analysis. However, the physiological role of some of these genes appears to be different in the two organisms, as discussed below.

nifA

R. capsulatus differs from most other N_2 fixing organisms in containing two copies of a gene homologous to *nifA* of *K. pneumoniae*. Both copies are functional, since deletion of one or the other copy does not result in a Nif^- phenotype. However, deletion of both copies results in the loss of the ability to grow on N_2 and to synthesize nitrogenase (Masepohl et al. 1988). Two copies of *nifB*, which is contiguous with *nifA*, are also found in *R. capsulatus*. *NifA* and *nifB* are also contiguous in *K. pneumoniae*.

The reason for this contiguity is not clear, since *nifB* is involved in the synthesis of the iron-molybdenum cofactor (FeMoCo), and is apparently not a regulatory gene.

It has not yet been demonstrated that NIFA functions as an activator protein in *R. capsulatus*, as it does in *K. pneumoniae*. However, sequences homologous to those of the putative NIFA-binding sites in *K. pneumoniae* have been found upstream of *nifH*, *nifB* and *nifE* in *R. capsulatus* (Pollock et al. 1988; Masepohl et al. 1988; Moreno-Vivián et al. 1989).

nifR4

The *nifR4* gene is located adjacent to, and downstream from, the *nifHDK* operon (Ahombo et al. 1986). The gene has been sequenced by Jones and Haselkorn (1989) and in our laboratory (Alias et al. 1989), and shows significant homology with the *ntrA* gene of *K. pneumoniae*. However, unlike mutations in the *ntrA* gene of *K. pneumoniae*, mutations in the *nifR4* gene of *R. capsulatus* appear to affect only N_2 fixation, with growth on other N sources being unaffected.

In other organisms, *ntrA* has been shown to be required for the expression of genes involved in processes unrelated to N metabolism, e.g. dicarboxylic acid transport (*dctA*) in *Rhizobium meliloti* (Ronson et al. 1987), formate oxidation (*fdhF*) in *E. coli* (Birkmann et al. 1987), hydrogen oxidation in *Alcaligenes eutrophus* (Römermann et al. 1989). In *R. capsulatus*, sequences homologous to the -24/-12 consensus promoter sequences recognized by NTRA-containing RNA polymerase have been found upstream of the photosynthetic *puf* operon (Bauer et al. 1988) and the hydrogenase structural genes (P. Richaud, unpublished). However, in *nifR4* mutants, both photosynthetic growth and hydrogenase synthesis appear to be normal. If these genes are controlled by an NTRA-like sigma factor, it must be encoded by a gene other than *nifR4*.

Direct evidence that NIFR4 is an RNA polymerase sigma factor in *R. capsulatus* has not yet been provided. However, two lines of evidence suggest that it plays a role in the activation of *nif* gene expression. Firstly, in *nifR4* mutants, synthesis of the nitrogenase polypeptides (the *nifHDK* gene products) is undetectable (Willison et al. 1985) and transcription of the *nifHDK* operon is also undetectable (M. Leclerc 1988, Ph.D. Thesis). Secondly, when a plasmid containing the cloned *nifR4* gene is introduced into an *R. capsulatus nifR4* mutant, the rate of derepression of nitrogenase synthesis after N starvation is significantly faster than in the wild-type strain (Fig. 2).

nifR1 and nifR2

Hybridization experiments and DNA sequence analysis have shown that the *nif R1* gene is homologous to *ntrC* and *nifR2* is homologous to *ntrB* (Kranz and Haselkorn 1985; Jones and Haselkorn 1989). The existence of a third regulatory gene, *nifR3*, upstream of *nifR2*, was suggested by results with *lacZ* gene fusions, but was not confirmed by DNA sequencing.

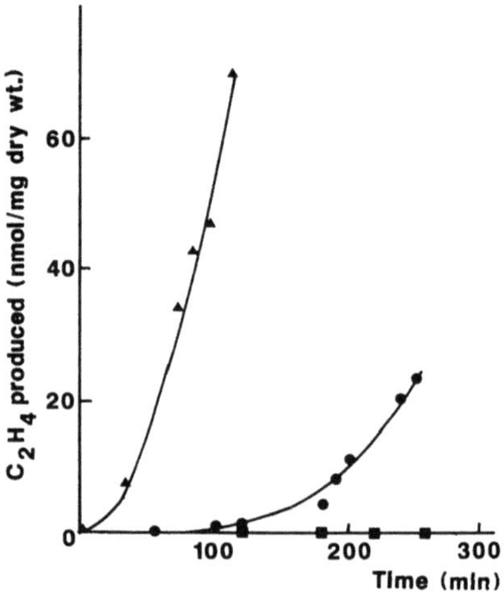

Fig. 2. Effect of a plasmid containing the *nifR4* gene on the derepression of nitrogenase synthesis in *R. capsulatus*. Precultures of the wild-type strain B10 (●), a *nifR4* mutant, GA111 (■) and GA111 containing pGA14 (▲) (Ahombo et al. 1986), were grown photosynthetically to stationary phase in minimal medium containing 15 mM NH_4^+ as N-source. Samples of culture were then centrifuged, washed three times in NH_4^+-free medium saturated with argon, and resuspended in the same medium to a density of about 1.0 mg dry wt ml^{-1}. The suspension (2 ml) was incubated in the light in an argon-filled vial containing 10% (v/v) C_2H_2. Samples of the gas phase (50 µl) were withdrawn at different times, and the amount of C_2H_4 produced by nitrogenase was assayed by gas chromatography (Willison and Vignais 1982)

As with *nifR4*, mutations in *nifR1* and *nifR2* resulted in a Nif⁻ phenotype, but did not affect growth on substrates other than N_2. Interestingly, hybridization experiments have suggested the presence of additional sequences homologous to these genes, suggesting that alternative regulatory circuits may be present (Kranz and Haselkorn 1988).

adgA

Mutations in this gene pleiotropically affect N metabolism, as might be expected for an *ntr* mutation. However, neither the *E. coli* gene complementing these mutations (Allibert et al. 1987) nor the *R. capsulatus* gene (unpublished data) show any homology to the *K. pneumoniae ntr* genes. The predicted translation products of the *E. coli* and *R. capsulatus* genes show limited homology (about 25%) with three short, highly conserved regions.

The pleiotrophic effect of *adgA* mutations suggests that *adgA* may be a regulatory gene. However, although *adgA* mutants are devoid of nitrogenase activity *in vitro*, nitrogenase synthesis is not completely abolished, being around 20% of the wild-type level. This inhibition might therefore be due to an indirect, metabolic effect.

The protein coded for by the *E. coli* gene has been purified and shown to be a dimer of M_r 60,000 (Allibert et al. 1987). Preliminary results have shown that the protein can bind nucleoside triphosphates and diphosphates, with ATP being bound more strongly than the other nucleotides. The nucleotide-binding capacity of the protein appeared to be labile, however (or may even have been due to a contaminating protein) since at saturation only about 0.1 mol of ^3H-ATP was bound per mol of protein subunits.

Regulation of *nif* Gene Expression by O_2

In *K. pneumoniae*, the regulation of *nif* gene expression by O_2 appears to be mediated by the *nifL* gene product. In the presence of O_2, NIFL appears to antagonize the action of NIFA by a still unknown mechanism (Dixon et al. 1987). Oxygen regulation at the level of the *nifLA* promoter may occur by an effect on the degree of DNA supercoiling (Dimri and Das 1988; Dixon et al. 1988). DNA supercoiling has been involved in the regulation of a large number of environmentally regulated genes, although the mechanism by which environmental stimuli affect the degree of DNA supercoiling is unknown. In *R. capsulatus*, no evidence has yet been obtained for the presence of a *nifL* gene. However, a role for DNA supercoiling in O_2 regulation has been suggested.

nifL ?

Attempts to isolate O_2-constitutive mutants using *nifH::lacZ* gene fusions were unsuccessful, suggesting that *R. capsulatus* may lack the *nifL* gene (Kranz and Haselkorn 1986) (Alternatively, two functional copies might be present, but these copies would not be linked to the *nifA* gene). In *Bradyrhizobium japonicum*, *nifL* is also thought to be absent, and it has been proposed that NIFA itself is the site of O_2 sensitivity (Fischer and Hennecke 1987). The *B. japonicum* NIFA protein contains a possible metal-bind-ing site, which is absent in the *K. pneumoniae* protein (Fischer et al.1988), and this sequence is also present in *R. capsulatus* NIFA (Masepohl et al. 1988).

DNA supercoiling

The use of DNA gyrase inhibitors such as coumermycin and novobiocin has suggested a requirement for DNA supercoiling for the derepression of nitrogenase synthesis (Kranz and Haselkorn 1986). These inhibitors also affect the expression of several photosynthesis genes (Zhu and Hearst 1988). The results shown in Figure 3 show the effect of novobiocin on the anaerobic derepression of nitrogenase synthesis in response to N starvation. In *Bradyrhizobium japonicum*, DNA gyrase inhibitors have been shown to affect the induction of uptake hydrogenase synthesis (Novak and Maier 1987). However, in a parallel experiment to that shown in Figure 3, the same concentration of novobiocin was found to have no effect on the induction of hydrogenase synthesis in *R. capsulatus* (A. Colbeau, unpublished data).

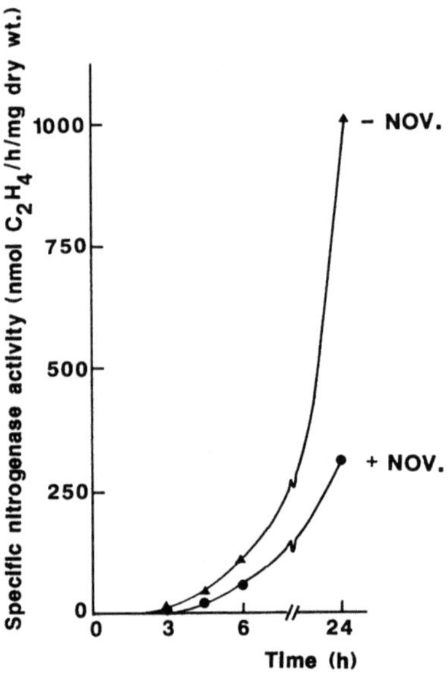

Fig. 3. Effect of novobiocin on the anaerobic derepression of nitrogenase synthesis in *R. capsulatus*. A culture of *R. capsulatus* B10 was harvested and washed as described in the legend to Fig. 2, and derepressed for nitrogenase synthesis in the presence (●) or absence (▲) of novobiocin (100 µg/ml). At the times shown, samples of the suspension were withdrawn under argon, and assayed for nitrogenase activity by measurement of the rate of C_2H_2 reduction, as described by Willison and Vignais (1982)

References

Ahombo G, Willison JC, Vignais PM (1986) The *nifHDK* genes are contiguous with a *nifA*-like regulatory gene in *Rhodobacter capsulatus*. Mol Gen Genet 205: 442-445

Alias A, Cejudo FJ, Willison JC, Vignais PM (1989) Nucleotide sequence of wild-type and mutant *nifR4* (*ntrA*) genes of *Rhodobacter capsulatus*: Identification of an essential glycine residue. Nucleic Acids Res 17: 5377

Allibert P, Willison JC, Vignais PM (1987) Complementation of nitrogen-regulatory (*ntr*-like) mutations in *Rhodobacter capsulatus* by an *Escherichia coli* gene: Cloning and sequencing of the gene and characterization of the gene product. J Bacteriol 169: 260-271

Bauer CE, Young DA, Marrs BL (1988) Analysis of the *Rhodobacter capsulatus puf* operon: Location of the oxygen-regulated promoter region and the identification of an additional *puf*-encoded gene. J Biol Chem 263: 4820-4827

Birkmann A, Sawers RG, Böck A (1987) Involvement of the *ntrA* gene product in the anaerobic metabolism of *Escherichia coli*. Mol Gen Genet 210: 535-542

Dimri GP, Das HK (1988) Transcriptional regulation of nitrogen fixation genes by DNA supercoiling. Mol Gen Genet 212: 360-363

Dixon RA, Austin S, Buck M, Drummond M, Hill S, Holtel A, MacFarlane S, Merrick M, Minchin S (1987) Genetics and regulation of *nif* and related genes in *Klebsiella pneumoniae*. Phil Trans R Soc Lond B 317: 147-158

Dixon RA, Henderson NC, Austin S (1988) DNA supercoiling and aerobic regulation of transcription from the *Klebsiella pneumoniae nifLA* promoter. Nucleic Acids Res 16: 9933-9946

Fischer HM, Bruderer T, Hennecke H (1988) Essential and non-essential domains in the *Bradyrhizobium japonicum nifA* protein: Identification of indispensable cysteine residues potentially involved in redox reactivity and/or metal binding. Nucleic Acids Res 16: 2207-2224

Fischer HM, Hennecke H (1987) Direct response of *Bradyrhizobium japonicum nifA*-mediated *nif* gene regulation to cellular oxygen status. Mol Gen Genet 209: 621-626

Gussin GN, Ronson CW, Ausubel FM (1986) Regulation of nitrogen fixation genes. Annu Rev Genet 20: 567-591

Imhoff JF, Trüper HG, Pfennig N (1984) Rearrangement of the species and genera of the phototrophic "purple nonsulfur bacteria". Internat J Syst Bacteriol 34: 340-343

Johansson BC, Gest H (1976) Inorganic nitrogen assimilation by the photosynthetic bacterium *Rhodopseudomonas capsulata*. J Bacteriol 128: 683-688

Johansson BC, Gest H (1977) Adenylylation/deadenylylation control of glutamine synthetase of *Rhodopseudomonas capsulata*. Eur J Biochem 81: 365-371

Jones R, Haselkorn R (1989) The DNA sequence of the *Rhodobacter capsulatus ntrA*, *ntrB* and *ntrC* gene analogues required for nitrogen fixation. Mol Gen Genet 215: 507-516

Klipp W, Masepohl B, Pühler A (1988) Identification and mapping of nitrogen fixation genes of *Rhodobacter capsulatus*: Duplication of a *nifA-nifB* region. J Bacteriol 170: 693-699

Kranz RG, Haselkorn R (1985) Characterization of *nif* regulatory genes in *Rhodopseudomonas capsulata* using *lac* gene fusions. Gene 40: 205-215

Kranz RG, Haselkorn R (1986) Anaerobic regulation of nitrogen-fixation genes in *Rhodopseudomonas capsulata*. Proc Natl Acad Sci USA 83: 6805-6809

Kranz RG, Haselkorn R (1988) Ammonia-constitutive nitrogen fixation mutants of *Rhodobacter capsulatus*. Gene 71: 65-74

Magasanik B (1982) Genetic control of nitrogen assimilation in bacteria. Annu Rev Genet 16: 135-168

Masepohl B, Klipp W, Pühler A (1988) Genetic characterization and sequence analysis of the duplicated *nifA/nifB* gene region of *Rhodobacter capsulatus*. Mol Gen Genet 212: 27-37

Moreno-Vivián C, Schmehl M, Masepohl B, Arnold W, Klipp W (1989) DNA sequence and genetic analysis of the *Rhodobacter capsulatus nifENX* gene region: Homology between *NifX* and *NifB* suggests involvement of *NifX* in processing of the iron-molybdenum cofactor. Mol Gen Genet 216: 353-363

Novak PD, Maier RJ (1987) Inhibition of hydrogenase synthesis by DNA gyrase inhibitors in *Bradyrhizobium japonicum*. J Bacteriol 169: 2708-2712

Pollock D, Bauer CE, Scolnik PA (1988) Transcription of the *Rhodobacter capsulatus nifKDK* operon is modulated by the nitrogen source. Construction of plasmid expression vectors based on the *nifHDK* promoter. Gene 65: 269-275

Römermann D, Warrelmann J, Bender RA, Friedrich B (1989) An *rpoN*-like gene of *Alcaligenes eutrophus* and *Pseudomonas facilis* controls expression of diverse metabolic pathways, including hydrogen oxidation. J Bacteriol 171: 1093-1099

Ronson CW, Nixon BT, Albright LM, Ausubel FM (1987) *Rhizobium meliloti ntrA* (*rpo N*) gene is required for diverse metabolic functions. J Bacteriol 169: 2424-2431

Vignais PM, Colbeau A, Willison JC, Jouanneau Y (1985) Hydrogenase, nitrogenase, and hydrogen metabolism in the photosynthetic bacteria. Adv Microbial Physiol 26: 155-234

Willison JC, Ahombo G, Chabert J, Magnin JP, Vignais PM (1985) Genetic mapping of the *Rhodopseudomonas capsulata* chromosome shows non-clustering of genes involved in nitrogen fixation. J Gen Microbiol 131: 3001-3015

Willison JC, Magnin JP, Vignais PM (1987) Isolation and characterization of *Rhodobacter capsulatus* strains lacking endogenous plasmids. Arch Microbiol 147: 134-142

Willison JC, Vignais PM (1982) The use of metronidazole to isolate Nif^- mutants of *Rhodopseudomonas capsulata*, and the identification of a mutant with altered regulatory properties of nitrogenase. J Gen Microbiol 128: 3001-3010

Zhu YS, Hearst JE (1988) Transcription of oxygen-regulated photosynthetic genes requires DNA gyrase in *Rhodobacter capsulatus*. Proc Natl Acad Sci USA 85: 4209-4213

MOLECULAR BIOLOGY OF RESPIRATORY NITRATE REDUCTION (DENITRIFICATION) OF *PSEUDOMONAS STUTZERI*

W.G. Zumft [1], A. Viebrock, S. Blümle, C. Braun, and A. Jüngst

Introduction

Respiratory nitrate reduction by bacteria (denitrification) consists in the sequential transformation of an ionic N-oxide to dinitrogen. The process is catalyzed by four different metalloenzymes which, for the first time, were all purified to homogeneity from the same denitrifying bacterium, *Pseudomonas stutzeri* ZoBell (ATCC 14405). Full understanding of a biological system such as denitrification requires knowledge of the structure and function of the genes and their arrangement within the genome. Denitrification is plasmid-coded in the lithoautotrophic bacterium *Alcaligenes eutrophus* H16 (Schneider et al. 1988). The genus *Pseudomonas* harbors many species that were shown to carry plasmids with important metabolic functions, but it is not known whether denitrification would also be plasmid-coded. To detect plasmids in *P. stutzeri* standard procedures and those specifically designed to preserve megaplasmids were applied (Eckhardt 1978; Simon 1984; Nies et al. 1987). The plasmid preparations were subjected to pulsed field gel electrophoresis (CHEF variant), a technique which would facilitate the identification of megaplasmids because of an increased band-sharpening of DNA-molecules with several megabase pairs in size. No indications for plasmids were found, so we consider denitrification of *P. stutzeri* to be chromosomally encoded.

Nitrate Reduction

Respiratory nitrate reductase from *P. stutzeri* was solubilized by heat treatment from the membrane fraction of anaerobically nitrate-grown cells. The purification procedure involved chromatography on DEAE-Sepharose, gel filtration, and preparative polyacrylamide gel electrophoresis. The enzyme shares properties with the respiratory nitrate reductase from *Escherichia coli* and other denitrifying pseudomonads (Carlson et al. 1982; Ishizuka et al. 1984; Stewart 1988). It is an iron-sulfur protein and contains molybdenum in form of a bactopterin molecule of about M_r 710 (Krüger et al. 1987). Basic properties of this enzyme are summarized in Table 1.

Two enzyme forms were observed in nondenaturing electrophoresis with apparent M_r 140,000 (form I) and M_r 132,000 (form II). Both protein bands stained positively

[1] Department of Microbiology, University, Kaiserstr. 12, D-7500 Karlsruhe, FRG

for nitrate reductase activity. Trypsin treatment converted the two forms into a single species of M_r ca. 100,000, indicating that the faster migrating species was a proteolytic cleavage product of the form with higher molecular mass. The subunit composition of the two forms of nitrate reductase was heterogeneous. Form I had components of 112, 60, and 46 kDa, form II had only two components of 112 and 46 kDa. The cause for different subunit sizes was previously attributed to proteolytic modification of the small subunit during heat treatment (DeMoss 1977; DeMoss et al. 1981; Ishizuka et al. 1984).

By employing methods described below for identifying and cloning *nos* specific functions, we sought access to *nar* related functions and a strategy for identifying the structural genes for nitrate reductase. Mutants with defects in nitrate utilization were generated by random transposon Tn*5* insertion into genomic DNA of *P. stutzeri* (Zumft et al. 1985). Tn*5* was transferred by conjugation with *E. coli* SM10-20 via the mobilizable plasmid pSUP2021 (Simon et al. 1983) to a spontaneous streptomycin-resistant mutant of *P. stutzeri*. Kanamycin-resistant exconjugants were screened for chlorate resistance. By immunological and biochemical analysis three classes of mutants were identified, so far. One class was defective in nitrate respiration (Nar^- phenotype), but still able to assimilate nitrate (Nas phenotype); another class was lacking both activities (Nar^- and Nas^-); one pleiotropic strain was unable to respire or assimilate nitrate, but overproduced the respiratory enzyme about 5-fold.

Nitrite Reduction

The dissimilatory nitrite reductase of *P. stutzeri* is the tetraheme protein cytochrome cd_1. The enzyme is a homodimer of M_r 120,000, with one heme c and one heme d_1 per each subunit of 60 kDa (Table 1). The enzyme reduces *in vitro* nitrite to nitric oxide (NO) and catalyzes the oxidation of cytochrome c_{551} (Zumft et al. 1988 b). Heme d_1 is a dioxo-isobacteriochlorin with four methyl, two acetyl, two oxo groups, a propionate and an acrylate group as side chains (Chang et al. 1986). Heme c was identified as the site of entry of electrons (Wharton et al. 1973), the site of nitrite-binding probably being at heme d_1 (Pettigrew and Moore 1987). Cytochrome cd_1 of *P. stutzeri* shows the characteristic absorbance maxima of heme d_1 around 465 and 625 nm.

Mutants with defective respiratory nitrite utilization (Nir^- phenotype) were obtained by random transposon Tn*5* mutagenesis. Seven mutants (MK201-207) were selected that did not form gas from nitrate and were unable to grow anaerobically with nitrite (Zumft et al. 1988 a). Mutant MK202 showed no immunochemical cross reaction with anti-cytochrome cd_1 antiserum. Mutants MK203 and MK205 did not have the absorbance maxima of heme d_1 and the purified enzyme of strain MK203 lacked the heme d_1 moiety (Zumft et al. 1988 a). Mutant MK201 overproduced cytochrome c_{552} (400%),

Table 1. Properties of the enzymes of anaerobic respiration (denitrification) from *Pseudomonas stutzeri* ZoBell ATCC 14405

Property	Nitrate reductase[a]	Nitrite reductase[a]	Nitric oxide reductase[b]	Nitrous oxide reductase[c]
EC nr.	1.7.99.4	1.9.3.2	1.7.99.[d]	1.7.99.6
Reaction	$NO_3^- \rightarrow NO_2^-$	$NO_2^- \rightarrow NO$	$NO \rightarrow N_2O$	$N_2O \rightarrow N_2$
Location[f]	CM	PPL	CM	PPL
Mol. mass (kDa)				
Gelfiltration	–	120	180[e]	120
PAGE[g]	140, 132	–	–	–
Subunits (kDa)				
SDS PAGE	112, 60	61	38, 17	62-74
Quart. structure	αβ	α_2	αβ	α_2
Metal atoms/M_r	~ 0.6 Mo/140 8-10 Fe/140	2 Fe/120	~ 6 Fe/55	~ 8 Cu/140
S^{2-}	6-8/140	none	none	–
Prosthetic groups	Fe_nS_n Mo-pterin	heme *c* heme d_1	heme *b* heme *c*	Cu complex
pI	5.6	5.4	–	5.04
Absorbance maxima				
Oxidized form (nm)	280, 290(s) 410, 420(s)	278, 415 465, 535, 625	280, 411.5 537	280, 350,480 540, 780
Reduced form (nm)	420	417, 460 522, 554, 625	420.5, 522.5 552.5, 560(s)	640
Spec. activity[h]	30	4.2	40[i]	60[j]
pH Optimum	7.2	–	4.8	9.8[j]
Electron donors				
Artificial[k]	MV, BV	PMS	PMS	BV, MV, MB
Physiologic	cyt *b*?	cyt c_{551}	?	?

[a] Unpublished data; [b] Heiss et al. 1989; [c] Coyle et al. 1985; Riester et al. 1989; [d] no valid entry at present; [e] M_r of Triton X-100 micelle; [f] CM, cytoplasmic membrane; PPL, periplasm; [g] polyacrylamide gel electrophoresis; [h] units in µmol substrate reduced per min per mg protein with artificial electron donors; [i] in the presence of phospholipids; [j] base-activated form; [k] MV, methyl viologen; BV, benzyl viologen; PMS, phenazine methosulfate; MB, methylene blue

but functional cytochrome cd_1 was present. Mutants MK206 and MK207 had low cytochrome cd_1 and cytochrome c_{552} contents and a low nitrite reductase activity (10%). Together these mutants establish the essentiality of cytochrome cd_1 for nitrite respiration in *P. stutzeri*. A defect in heme d_1 precludes anaerobic growth on nitrite; hence no alternative respiratory nitrite reductase is operative in this organism. All mutants continued to reduce NO *in vivo* and *in vitro* and, thus, provide proof that cytochrome cd_1 and the NO-reducing system of this bacterium are separate entities.

Genomic DNA of the Nir⁻-mutants was partially digested with *Sau*3A into fragments of 30 - 50 kbp which were cloned into the cosmid vector pJA1. Cosmids, coding for kanamycin resistance, were digested with *Eco*RI and subcloned into the vector pBR325. The *Eco*RI-fragments, carrying Tn*5*-insertions, were then physically mapped. Three different *Eco*RI-fragments were isolated. Strains MK201 and MK202 were mutagenized in the same fragment (8.6 kbp), mutants MK203 and MK205 in a 10.5 kbp fragment, and mutants MK206 and MK207 in a fragment of 11 kbp. To isolate the structural gene of cytochrome cd_1, a phage λgt11 expression library (Huynh et al. 1985) was constructed with genomic DNA of *P. stutzeri* (see below) and screened with antiserum against cytochrome cd_1. The protease-deficient *E. coli* strain Y1090 was used for infection to prevent proteolytic cleavage of expressed proteins. Eleven clones were isolated that had inserts with sizes from 1750 (clone A2) to 3400 base pairs. The insert of clone A2 hybridized with the inserts of seven other clones. The identity of the structural gene of nitrite reductase was confirmed by hybridizing the biotinylated insert of clone A2 with the cloned 8.6 kbp *Eco*RI-fragment of mutant MK202. As final proof the restriction map of the *Eco*RI-fragment of λ clone A2 was constructed in pBR325. This map was identical to that of the flanking regions of the Tn*5*-insertion site of mutant MK202.

Nitric Oxide Reduction

Nitric oxide reductase catalyzes the transformation of NO to N_2O. This reduction is mechanistically important because of the formation of the N-N bond. The enzyme of *P. stutzeri* represents the first NO reductase that was purified to apparent electrophoretic homogeneity (Heiss et al. 1989). The enzyme has two subunits of molecular masses 38 and 17 kDa, which contain heme *b* and heme *c*, respectively (Table 1). The electron paramagnetic resonance spectrum shows high and low-spin ferric heme signals. The enzyme also possesses nonheme iron, though apparently no iron-sulfur groups. The reduction of NO to N_2O in the presence of phenazine methosulfate-ascorbate requires phospholipids to proceed with the maximum rate of 40 μmol per min per mg protein.

For identification of the NO reductase structural genes (*norBC*), a phage λgt11 expression library of partially *Sau*3A and *Alu*I digested genomic DNA of *P. stutzeri* was constructed. Antisera against native NO reductase and the separated subunits were raised in rabbits and purified with cell-free extract of *E. coli* Y1090 by affinity chromatography. About 1 x 10^5 p.f.u. of the gene library were screened. Immunologically positive λgt11 clones were detected by a non-radioactive labelling system in 15 plaques. The DNA from these clones was isolated and electrophoretically analyzed after digestion with *Eco*RI. The clones showed inserts ranging in size from 1200 to 1500 base pairs. The relationship of the cloned DNA-fragments was shown by restriction analysis and Southern hybridization. The fragments were subcloned in pBR325 for this purpose and clone pNO303 was used as the hybridization probe. Work is in progress to introduce a kanamycin resistance gene cartridge into the cloned gene regions, transfer the cartridge by marker exchange into the wild type, and thus generate mutants with Nor^- phenotype.

Nitrous Oxide Reduction

Nitrous oxide reductase of *P. stutzeri* is a Cu-containing enzyme with approximately 8 Cu atoms per 140 kDa molecular mass (Table 1) (Coyle et al. 1985). The Cu chromophore is unusual. It consists of a mixed valence species and a Cu_A center similar in its spectroscopic properties to cytochrome *c* oxidase (Kroneck et al. 1988; Riester et al. 1989; Jin et al. 1989; Scott et al. 1989). The enzyme is located in the periplasm (Minagawa and Zumft 1988). There is circumstantial evidence that the apoenzyme is first exported and Cu is subsequently inserted to yield the catalytically active enzyme. Work with Cu-deficient cells (Matsubara et al. 1982) and Tn*5*-induced mutants (Zumft et al. 1985) indicated that specific gene products are required for the incorporation of Cu into the apoenzyme. The inactive form of N_2O reductase, which lacks Cu, is activated by added Cu(II) in the absence of *de novo* protein synthesis. The protein from mutants synthesizing only a chromophoreless apoenzyme, cannot be activated (Minagawa and Zumft 1988). The outer membrane of *P. stutzeri* JM300 carries a Cu-containing, channel-forming protein that is also required for the biosynthesis of catalytically active N_2O reductase (Lee et al. 1989).

Analysis of the genes for N_2O respiration (designated *nos*) has been advanced farthest of all other denitrification genes. The entire *nos* region identified by random Tn*5* mutagenesis of *P. stutzeri* was cloned by plasmid integration and excision (Viebrock and Zumft 1988). The almost complete sequence of the 8 kbp *nos* gene cluster is now available. Translation of the sequence in the six possible phases revealed several open reading frames (ORF) (Viebrock and Zumft 1988; and unpublished data). One ORF encodes a 638-residue polypeptide and represents the structural gene (*nosZ*) for N_2O reductase. This was confirmed by internal amino acid sequences of purified N_2O reductase. Heterologous expression of *nosZ* in *E. coli* resulted in the synthesis of apo-N_2O reductase.

Three contiguous open reading frames (*nosDFY*) were identified downstream from *nosZ*, which code for proteins of 436, 308, and 276 amino acids, respectively. These proteins are involved in Cu-processing for N_2O reductase, since insertion of Tn*5* into any one of these ORFs leads to the synthesis of a copper-free enzyme The *nosDFY* genes show an organizational relationship to bacterial transport systems for small molecules. The deduced product of the *nosF* gene belongs to the superfamily of prokaryotic and eukaryotic nucleotide-binding proteins (Higgins et al. 1988). The translation start of *nosF* overlaps with the end of *nosD* and the start of *nosY* overlaps with the end of *nosF*. The lack of intergenic sequences suggests that these genes form part of the same transcriptional unit. The presumptive initiation codons of the three genes are preceded by potential ribosome-binding sites. No transcriptional terminator signal seems to be present at the end of these genes. A further ORF starts 28 bp downstream from *nosY* and proceeds beyond the region characterized by insertional mutagenesis. Work is in progress to determine whether this gene is also required for N_2O respiration. The sixth ORF with protein-coding potential within the *nos* regions ends 77 bp upstream of *nosZ*. Mutants with a Tn*5* insertion in this ORF show nil or low levels of expression of N_2O reductase. The function of this gene is not known, although the phenotype is suggestive of a regulatory role.

Acknowledgments. Supported by the Deutsche Forschungsgemeinschaft.

References

Carlson CA, Ferguson LP, Ingraham JL (1982) Properties of dissimilatory nitrate reductase purified from the denitrifier *Pseudomonas aeruginosa*. J Bacteriol 151: 162-171

Chang CK, Timkovich R, Wu W (1986) Evidence that heme d_1 is a 1,3-porphyrin-dione. Biochemistry 25: 8447-8453

Coyle CL, Zumft WG, Kroneck PMH, Körner H, Jakob W (1985) Nitrous oxide reductase from denitrifying *Pseudomonas perfectomarina*. Purification and properties of a novel multicopper enzyme. Eur J Biochem 153: 459-467

DeMoss JA (1977) Limited proteolysis of nitrate reductase purified from membranes of *Escherichia coli*. J Biol Chem 252: 1696-1701

DeMoss JA, Fan TY, Scott RH (1981) Characterization of subunit structural alterations which occur during purification of nitrate reductase from *Escherichia coli*. Arch Biochem Biophys 206: 54-64

Eckhardt T (1978) A rapid method for the identification of plasmid desoxyribonucleic acid in bacteria. Plasmid 1: 584-588

Heiss B, Frunzke K, Zumft WG (1989) Formation of the N-N bond from nitric oxide by a membrane-bound cytochrome *bc* complex of nitrate-respiring (denitrifying) *Pseudomonas stutzeri*. J Bacteriol 171: 3288-3297

Higgins CF, Gallagher MP, Mimmack ML, Pearce SR (1988) A family of closely related ATP-binding subunits from prokaryotic and eukaryotic cells. BioEssays 8: 111-116

Huynh TV, Young RA, Davis RW (1985) Constructing and screening cDNA libraries in λgt10 and λgt11. In: Glover DM (ed) DNA Cloning, vol I. IRL Press, Oxford, pp 49-78

Ishizuka M, Toraya T, Fukui S (1984) Purification, properties and limited proteolysis of nitrate reductase from *Pseudomonas denitrificans*. Biochim Biophys Acta 786: 133-143

Jin H, Thomann H, Coyle CL, Zumft WG (1989) Copper coordination in nitrous oxide reductase from *Pseudomonas stutzeri*. J Am Chem Soc 111: 4262-4269

Kroneck PMH, Antholine WA, Riester J, Zumft WG (1988) The cupric site in nitrous oxide reductase contains a mixed-valence [Cu(II),Cu(I)] binuclear center: A multifrequency electron paramagnetic resonance investigation. FEBS Lett 242: 70-74

Krüger B, Meyer O, Nagel M, Andreesen JR, Meincke M, Bock E, Blümle S, Zumft WG (1987) Evidence for the presence of bactopterin in the eubacterial molybdo-enzymes nicotinic acid dehydrogenase, nitrite oxidoreductase, and respiratory nitrate reductase. FEMS Microbiol Lett 48: 225-227

Lee HS, Hancock RE, Ingraham JL (1989) Properties of a *Pseudomonas stutzeri* outer membrane channel-forming protein (NosA) required for production of copper-containing N_2O reductase. J Bacteriol 171: 2096-2100

Matsubara T, Frunzke K, Zumft WG (1982) Modulation by copper of the products of nitrite respiration in *Pseudomonas perfectomarinus*. J Bacteriol 149: 816-823

Minagawa N, Zumft WG (1988) Cadmium-copper antagonism in the activation of periplasmic nitrous oxide reductase of copper-deficient cells from *Pseudomonas stutzeri*. Biol Metals 1: 117-122

Nies D, Mergeay M, Friedrich B, Schlegel HG (1987) Cloning of plasmid genes encoding resistance to cadmium, zinc, and cobalt in *Alcaligenes eutrophus* CH34. J Bacteriol 169: 4865-4868

Pettigrew GW, Moore GR (1987) Cytochromes *c*. Biological Aspects. Springer, Berlin Heidelberg New York

Riester J, Zumft WG, Kroneck PMH (1989) Nitrous oxide reductase from *Pseudomonas stutzeri*. Redox properties and spectroscopic characterization of different forms of the multicopper enzyme. Eur J Biochem 178: 751-762

Schneider B, Nies A, Friedrich B (1988) Transfer and expression of lithoautotrophy and denitrification in a host lacking these metabolic activities. Appl Environ Microbiol 54: 3173-3176

Scott RA, Zumft WG, Coyle CL, Dooley DM (1989) *Pseudomonas stutzeri* N_2O reductase contains Cu_A-type sites. Proc Natl Acad Sci USA 86: 4082-4086

Simon R (1984) High frequency mobilization of gram-negative bacterial replicons by the *in vitro* constructed Tn*5*-Mob transposon. Mol Gen Genet 196: 413-420

Simon R, Priefer U, Pühler A (1983) A broad host range mobilization system for *in vivo* genetic engineering: Transposon mutagenesis in gram-negative bacteria. Bio/Technology 1: 784-791

Stewart V (1988) Nitrate respiration in relation to facultative metabolism in enterobacteria. Microbiol Rev 52: 190-232

Viebrock A, Zumft WG (1988) Molecular cloning, heterologous expression, and primary structure of the structural gene for the copper enzyme nitrous oxide reductase from denitrifying *Pseudomonas stutzeri*. J Bacteriol 170: 4658-4668

Wharton DC, Gudat JC, Gibson QH (1973) Cytochrome oxidase from *Pseudomonas aeruginosa* - reaction with copper protein. Biochim Biophys Acta 292: 611-622

Zumft WG, Döhler K, Körner H (1985) Isolation and characterization of transposon Tn*5*-induced mutants of *Pseudomonas perfectomarina* defective in nitrous oxide respiration. J Bacteriol 163: 918-924

Zumft WG, Döhler K, Körner H, Löchelt S, Viebrock A, Frunzke K (1988 a) Defects in cytochrome cd_1-dependent nitrite respiration of transposon Tn*5*-induced mutants from *Pseudomonas stutzeri*. Arch Microbiol 149: 492-498

Zumft WG, Viebrock A, Körner H (1988 b) Biochemical and physiological aspects of denitrification. In: Cole JA, Ferguson SJ (eds) The Nitrogen and Sulphur Cycles. Cambridge University Press, Cambridge, pp 245-279

Part 6 Distribution of Storage and Metabolism, Environmental Stress

INFLUENCE OF NITROGEN SOURCE ON CARBON DISTRIBUTION IN PLANTS EXHIBITING THE C_3 AND C_4 PHOTOSYNTHETIC PATHWAYS

O.A.M. Lewis, M. Cramer, and T. Van Der Leij [1]

Introduction

It is well known that in plants such as wheat and barley which possess the C_3 photosynthetic pathway, ammonium-grown plants are reduced in size compared to nitrate-fed plants, the roots being particularly responsive (Haynes and Goh 1978). In maize plants, which exhibit the C_4 photosynthetic pathway, this effect is less apparent and the roots of ammonium plants are often larger than those of nitrate plants (Lewis et al. 1989).

A possible explanation for these different growth responses is the difference in the proportion of carbon allocated to nitrogen compounds and to carbohydrates participating in root energy storage and structural activities which results from the feeding of the different nitrogen forms. These differences arise from the fact that in both wheat and maize, nitrate assimilation occurs mainly in the shoot, whereas *nutrient* ammonium assimilation is restricted to the root (Lewis and Chadwick 1983; Murphy and Lewis 1987). Furthermore C_4 plants photosynthesize more rapidly than C_3 plants (Fig. 1), consequently a larger carbon supply is available for allocation to the different metabolic pools in maize plants than in wheat plants. The experimentation reported here has been designed to answer the following four key questions:

i. Does root-based ammonium assimilation divert too high a proportion of root carbon away from structural material to nitrogen compounds, thus retarding root growth?

ii. Do the energy requirements of root-based ammonium assimilation elevate root respiratory C loss above the rate occurring in nitrate-assimilating plants?

iii. In ammonium-fed plants, does the supply of nitrogen to the shoot exclusively in carbon-rich compounds, critically deplete the root of carbon required for root extension?

iv. Does the C_3 response to nitrate and ammonium feeding differ in any of the above respects from the C_4 response?

Materials and Methods

Zea mays L. var. Hazera and *Triticum aestivum* L. cv. Gamtoos were grown to 21

[1] Department of Botany, University of Cape Town, Rondebosch, Cape Town, South Africa

days in sand culture (for ^{14}C labelling) and hydroponically (for root respiration rate determinations) in controlled environment chambers set to the following conditions: for maize a 14 h day with irradiance of 1400 μmol m^{-2} s^{-1}, day/night temperature of 35°C/20°C and relative humidity of 50%/60%; for wheat an irradiance of 400 μmol m^{-2} s^{-1}, day/night temperature of 25°C/18°C and relative humidity of 50%/50%. Long Ashton growth media modified to contain nitrate-only or ammonium-only as the 4 mM nitrogen source (Hewitt 1966) were used in all experiments, the pH of the hydroponic solutions being maintained between 5.5 and 6. The sand culture plants were fed every day.

^{14}C Labelling. The plants to be labelled were enclosed in sealed 0.7 x 0.2 m perspex canopies through which air containing 50 μC $^{14}CO_2$ was circulated for 15 min. Thereafter the canopies were removed and the plants allowed to continue growing until harvested 9 h or 24 h after the commencement of isotope application. All treatments were triplicated, each individual unit consisting of 2 plants in the case of maize, and 15 plants in the case of wheat. On harvesting, the plants were divided into root and shoot, homogenized and extracted for 24 h at 0°C with 80% (v/v) ethanol-water. Thereafter the extract and residue were separated into the following fractions as described in Lewis et al. (1986) and Lewis et al. (1987); a) free (ethanol-water soluble) nitrogen compounds (mainly amino acids), b) bound (HCl-hydrolyzed) nitrogen compounds (protein), c) free (ethanol-water soluble) non-nitrogen compounds (sugars), d) HCl-hydrolyzed non-nitrogen compounds (starch etc.), e) HCl-insoluble non-nitrogen compounds (cellulose of cell walls).

The procedure for the estimation of the radioactivity of the cellulose fraction was modified so that the cell wall material was oxidized and prepared for counting in a Packard Tricarb Sample Oxidizer Model 306, yielding more reproducible results than the previously employed method. The radioactivity of each fraction was determined as described in Lewis et al. (1986).

Measurement of Photosynthetic Rate. Photosynthetic rates of the leaves of plants fed the two nitrogen sources were measured using a portable ADC infrared gas analyzer connected to a Parkinson's leaf chamber. Five determinations on separate plants of each treatment were made and averaged.

Collection and Analysis of Xylem Sap. In wheat plants the shoots were severed and sap expelled from the root and basal region of the plant by the use of a Scholander pressure bomb. In maize plants xylem sap was collected by severing the shoots of the plants and the xylem sap expelled from the stump was collected by Pasteur pipette during the next hour.

Transpiration Rate Determination. The rate of transpiration, an indirect measurement of the flow rate of xylem sap from root to shoot, was gravimetrically determined for each nutrient regime over an 8 h period.

Root Respiration Rate Determination. Root respiration rates were determined in individual plants growing in nutrient solutions enclosed in sealed 1 litre containers through which an airstream was constantly drawn at a rate of 400 ml min^{-1}. Respiration rates were calculated from the CO_2 content of the emerging airstream monitored using an ADC Model 220 Bench infrared gas analyzer. Six replicated determinations were averaged.

Results and Discussion

Photosynthetic Rates. Maize plants gained carbon at a rate approximately 3 times greater than that of wheat plants under their respective growth conditions (Fig. 1). This indicates that maize plants (C_4) have far more newly fixed carbon at their disposal for distribution to various metabolic pools than do wheat (C_3) plants.

Carbon Allocation Patterns. The distribution of newly assimilated carbon from shoot to root is more rapid in maize plants than in wheat plants (Fig. 2); in wheat plants the roots of ammonium-grown plants appear to sequester carbon at a faster rate than nitrate-grown plants, while the reverse is true in maize (Fig. 2).

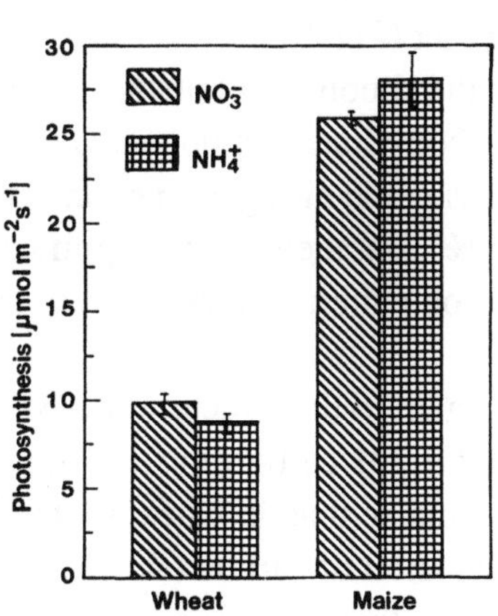

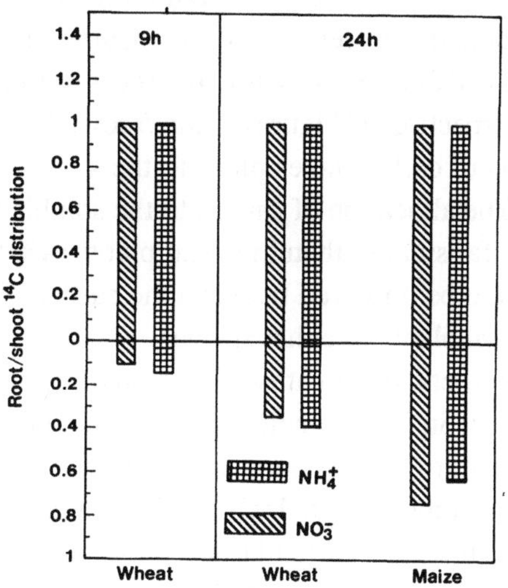

Fig. 1 *(left).* Photosynthetic rates of wheat and maize plants fed 4 mM nitrate and 4 mM ammonium

Fig. 2 *(right).* ^{14}C distribution between shoot and root in wheat (9 h and 24 h harvest and maize (24 h harvest) plants fed nitrate or ammonium

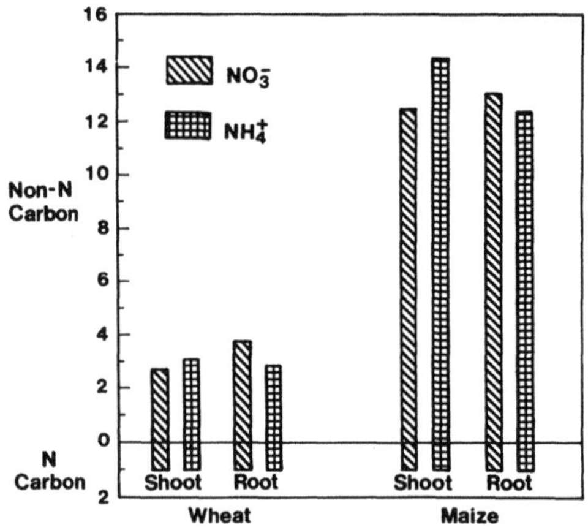

Fig. 3. ^{14}C distribution ratios between nitrogen and non-nitrogen components in 21-day-old wheat and maize plants fed nitrate or ammonium

In maize, the diversion of newly assimilated carbon is far greater to the carbohydrate component of the whole plant compared with the nitrogen component than is the case in wheat (carbohydrate carbon: nitrogen carbon 13:1 for maize, 2.5:1 for wheat). In the root of ammonium-grown wheat the nitrogen component sequesters a significantly greater fraction of the newly assimilated carbon than does the root of nitrate-grown wheat, presumably because of the need to supply carbon skeletons for ammonium assimilation, thus reducing the amount of carbon available for growth. In maize roots this difference is not apparent, presumably because of the large excess of carbon available for growth in the roots of this C_4 plant (Fig. 3).

An inspection of Figures 4 and 5 reveals that in the shoots of all plants, the greatest diversion of carbon takes place to the structural carbohydrate component (i.e. the cell walls). The allocation of carbon to the soluble and insoluble nitrogen fractions in maize plants is far smaller than in wheat plants (4% vs 26%), but there are no significant differences to be observed between the carbon allocation patterns of ammonium and nitrate-grown plants of either species.

In the roots, the greatest diversion of carbon is again to the structural carbohydrate component in both species (ca. 40%). The allocation of carbon to the soluble and insoluble nitrogen fractions is also much smaller in maize than in wheat roots (3% vs. 23%). In ammonium-fed wheat plants a greater diversion of carbon takes place to the nitrogen fractions, particularly the free-amino fraction, than in nitrate-fed plants at the expense of the structured carbohydrate fraction. This reflects the greater demand for carbon skeletons by nitrogen assimilation in the roots of ammonium-fed plants.

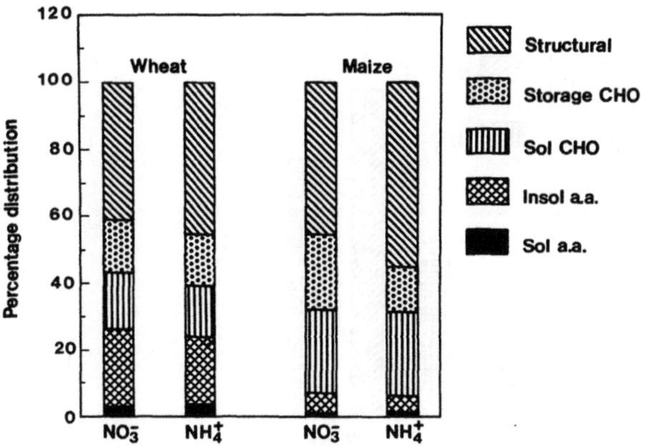

Fig. 4. Percentage distribution of ^{14}C between various nitrogen and non-nitrogen fractions of wheat and maize shoots fed nitrate or ammonium

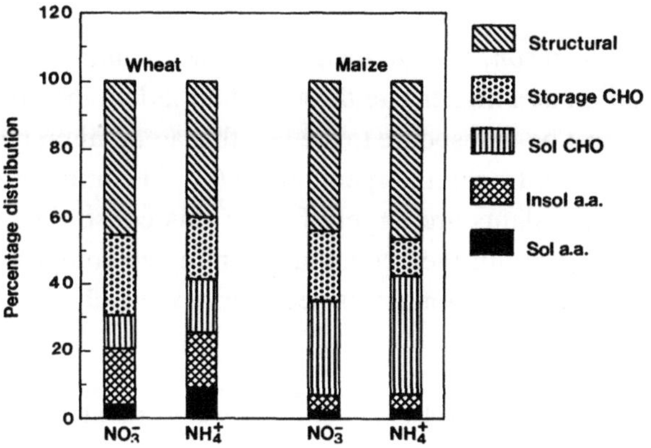

Fig. 5. Percentage distribution of ^{14}C between various nitrogen and non-nitrogen fractions of wheat and maize roots fed nitrate or ammonium (24 h harvest)

No such difference is discernible in maize plants, but this may be masked by the very low allocation of carbon to nitrogenous compounds compared to carbohydrate.

Root Respiration Rates. The respiration rate of ammonium-grown wheat plants is 34% higher than that of nitrate-grown plants (Fig. 6). This must reflect the difference between the energy expenditure of ammonium uptake + assimilation in the root compared with that of nitrate uptake. In maize the overall respiratory rate is some 100% higher than in wheat but no significant differences could be detected between the respiratory rates of ammonium and nitrate plants (Fig. 6). This could be due to the relatively low proportion of energy expended on nitrogen uptake and assimilation in maize compared with wheat.

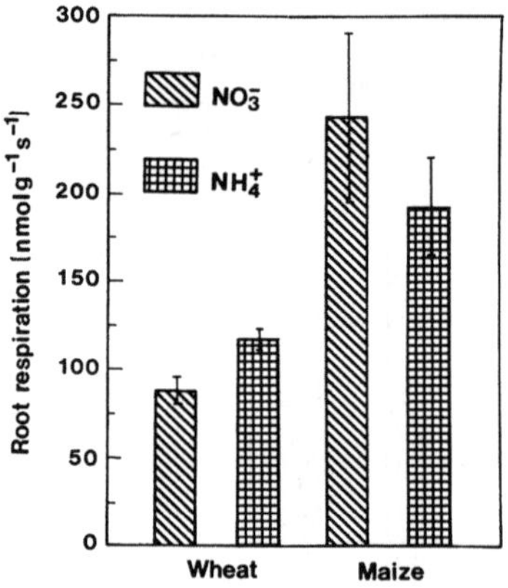

Fig. 6. Respiratory rates of maize and wheat plants fed nitrate or ammonium

Xylem Translocation of Carbon. In both maize and wheat, ammonium-grown plants appear to transport carbon at a higher rate through their xylem than do nitrate-grown plants (Table 1). This must be in response to the fact that ammonium plants are obliged to supply their shoots with nitrogen in organic form (i.e. in combination with carbon skeletons) whereas nitrate plants are largely free of this constraint. Because of the lower allocation of carbon to nitrogen compounds in maize compared to wheat, the loading of ^{14}C onto the xylem sap of maize is necessarily lower than that for wheat.

Table 1. Rates of ^{14}C translocation in xylem sap of maize and wheat plants fed 4 mM nitrate or ammonium (Bq g^{-1} dry matter h^{-1} ±SE).

	Maize	Wheat
NO_3^- fed	242 ±40	1303 ±305
NH_4^+ fed	342 ±93	1640 ±180

Conclusions

The restricted root growth in ammonium-fed wheat plants can be ascribed to the cumulative effect of increased respiration, greater allocation of carbon to nitrogenous compounds and increased export of carbon from root to shoot than that occurring in nitrate-fed plants. That these effects are not found in maize roots is probably due to the far greater availability of newly assimilated carbon for both ammonium assimilation and cell wall production due to the possession by this plant of the C_4 photosynthetic pathway (i.e. the additional demands of root ammonium assimilation are insignificant in the total carbon budget of the C_4 root).

Acknowledgments. The authors are grateful to the Foundation for Research Development for financial support, and to Mrs. A. Jenssen and Mr. I. Newton for expert technical assistance.

References

Haynes RJ, Goh KM (1978) Ammonium and nitrate nutrition of plants. Biological Reviews 53: 465-510

Hewitt EJ (1966) Sand and Water Culture Methods Used in the Study of Plant Nutrition. Techn. Commun 22 (revised), Commonwealth Bureau of Horticultural and Plantation Crops, East Malling, Commonwealth Agricultural Bureaux, Farnham Royal, England

Lewis OAM, Chadwick S. (1987) An ^{15}N investigation into nitrogen assimilation in hydroponically-grown barley (*Hordeum vulgare* L. cv. Clipper) in response to nitrate, ammonium and mixed nitrate and ammonium nutrition. New Phytologist 95: 635-646

Lewis OAM, Fulton B, Von Zelewski AAA (1987) Differential distribution of carbon in response to nitrate, ammonium and nitrate+ammonium nutrition in wheat. In: Ullrich WR, Aparicio PJ,Syrett PJ, Castillo F(eds), Inorganic Nitrogen Metabolism, Springer, Berlin Heidelberg New York, pp 240-246

Lewis OAM, Soares MIM, Lips SH (1986) A photosynthetic and ^{15}N investigation of the differential growth response of barley to nitrate, ammonium and nitrate+ammonium nutrition. In: Lambers H, Neetson JJ, Stulen I (eds), Fundamental Ecological and Agricultural Aspects of Nitrogen Metabolism in Higher Plants. Martinus Nijhoff, Dordrecht

Lewis OAM, Leidi EO, Lips SH (1989) Effect of nitrogen source on growth response to salinity stress in maize and wheat. New Phytologist 11: 155-160

Murphy AT, Lewis OAM (1987). Effect of nitrogen feeding source on the supply of nitrogen from root to shoot and the site of nitrogen assimilation in maize. New Phytologist 107: 327-333

ROLE OF NITRATE IN GROWTH OF HIGHER PLANTS WITH EMPHASIS ON REGULATION OF NITRATE ACCUMULATION

I. Stulen, M.W. ter Steege and P.J.C. Kuiper [1]

Introduction

Growth of higher plants is strongly dependent on mineral nutrition and changes therein. Experiments carried out in our laboratory in which the level of macronutrients in the solution was changed, both from high to low and from low to high concentrations, showed that nitrate probably is the most important component of the nutrient solution as to causing the observed positive and negative effects on growth (Lambers et al. 1981; Stulen et al. 1981). In *Plantago* the negative effect of a switch from high to low nutrient conditions on shoot growth was immediate; shoot growth did not decrease as a result of shortage of mineral nutrients. Apparently shoot growth was regulated in a manner which did not involve actual concentrations of minerals in the plant tissue itself, for growth stopped before any decrease in nitrate concentration in the tissue was observed (Lambers et al. 1981; Stulen et al. 1981).

Strong evidence for the involvement of cytokinins in growth regulation was obtained in further experiments with *Plantago* by D. Kuiper. Growth reduction, caused by lowering the mineral nutrition, could be retarded by addition of benzyladenine (Kuiper and Staal 1987). Further experiments in which plants of an inbred line of *Plantago* were subjected to an alteration in mineral supply strongly suggested that nitrate was the effector. The observed responses of growth rate to increased nutrient availability were caused by the production of cytokinins in the root tips, and not by the nitrate concentration in the plant tissue (Kuiper 1988, Kuiper et al. 1988).

In other experiments, performed in our laboratory with spinach, also no direct relation between nitrate concentration in the plant and growth rate of either shoot or root could be obtained. In this case the cause for the lack of correlation between nitrate concentration in the plant and growth could be found in differences in demand for nitrate as an osmotic agent for growth. When spinach plants were grown at rather low light intensity, there was a shortage of organic intermediates. In this case, nitrate apparently served as an osmotic agent for growth and was accumulated in the vacuoles of the leaves. The concentration in the vacuoles increased during the night and decreased during the day. A fluctuation in nitrate concentration was not found in either petioles or roots. It was concluded that the demand for nitrate as an osmotic in the shoot regulated nitrate uptake by the root (Steingröver 1986).

[1] Department of Plant Biology, State University of Groningen, P.O. Box 14, 9750 AA, Haren (Gn), The Netherlands

To investigate whether this conclusion holds for other than the studied cultivar (Vroeg Reuzenblad), diurnal nitrate accumulation and fluctuation over the day were determined in three more spinach varieties (Glares, Subito and Spinoza), known to differ in the extent to which they accumulate nitrate.

Material and Methods

Seeds from three spinach cultivars (*Spinacia oleracea*, cv.Glares, Subito and Spinoza) were obtained from the Institute of Horticultural Plant Breeding (Wageningen, the Netherlands). The seeds were germinated in a greenhouse in sterilized soil at a temperature of 22°C during the day (14 hours) and 16°C during the night (10 hours). After 10 days the seedlings were transferred into a nutrient solution and grown in a growth chamber under low light conditions (170 $\mu E\ m^{-2}\ s^{-1}$) for 10 hours a day, low temperatures (19°C during the day, 15°C during the night) and a relative humidity of 60 - 70 %, at a nitrate concentration of 4 mM (for composition of the nutrient solution see Steingröver et al. 1986), for up to 30 days. At this time the plants had reached a growth stage as for commercial use.

The plants were harvested 21 and 30 days after sowing at the beginning and at the end of the light period. At each harvest 5 sets of three plants each were taken. The plants were divided into leaf blades, petioles and roots and stored at -80°C before nitrate concentration was determined. Nitrate was determined in a watery extract by means of HPLC (Maas et al. 1986). The results were analyzed with the Student T-test.

Results

When the plants were 30 days old, all three cultivars had reached the same total fresh weight (Table 1). Although total fresh weight of the three cultivars did not differ significantly, the partitioning of fresh weight over the different plant parts was not the same: cultivar Spinoza showed a significantly lower shoot to root ratio (Table 1).

Table 1. Fresh weight of spinach cultivars after a growth period of 30 days under the conditions described in "Material and Methods". The plants were harvested at the beginning of the light period

Cultivar	Fresh weight in g				Shoot/root ratio
	leaf blade	petiole	root	total	
Glares	7.86	2.02	2.79	12.67	3.9
Subito	7.87	2.41	2.91	12.69	4.2
Spinoza	6.91	1.76	3.04	11.71	3.1

Shoot morphology of Spinoza was also different; the leaves were more curled, and not expanded as much as those of the other cultivars.

In contrast to fresh weight in nitrate concentration marked differences were found between the cultivars (Table 2). For all three cultivars, nitrate concentrations in the plant (as a whole) were significantly higher (at the $p = 0.01$ level) in the morning than in the afternoon, at both harvests. Marked differences in nitrate concentration of the various plant parts were found between the cultivars (Table 2). On days 21 and 30 nitrate concentration was significantly higher ($p = 0.01$) in the leaves of Glares than in

Table 2. Nitrate concentration in three spinach cultivars, grown for 21 and 30 days. Plants were harvested at the beginning and at the end of the light period. Data are the average of three sets of three plants each (±SE)

Cultivar		Glares	Subito	Spinoza
		nitrate concentration in µmol g Fwt^{-1}		
morning	21 days			
leaf blade		29.9 ±5.0	20.1 ±0.6	19.5 ±3.1
petiole		74.3 ±2.7	72.3 ±1.3	73.4 ±4.8
root		65.4 ±2.4	62.8 ±1.4	69.5 ±1.1
whole plant		41.6 ±3.2	34.5 ±1.0	34.9 ±1.2
afternoon	21 days			
leaf blade		17.2 ±3.4	16.2 ±0.9	12.8 ±3.4
petiole		69.9 ±3.8	64.0 ±3.8	56.7 ±3.5
root		24.2 ±3.8	39.2 ±2.5	45.4 ±5.5
whole plant		27.8 ±3.7	27.1 ±1.2	24.8 ±1.6
morning	30 days			
leaf blade		31.9 ±2.9	22.8 ±1.6	11.5 ±1.6
petiole		98.1 ±2.1	80.2 ±7.5	70.5 ±1.9
root		63.5 ±1.8	57.0 ±3.3	55.8 ±5.2
whole plant		49.0 ±1.0	39.9 ±2.0	30.4 ±3.8
afternoon	30 days			
leaf blade		16.4 ±4.0	12.6 ±1.4	17.8 ±0.6
petiole		82.3 ±6.6	65.6 ±9.1	80.7 ±2.2
root		60.5 ±4.1	45.3 ±3.3	59.9 ±1.9
whole plant		34.6 ±5.1	27.7 ±2.1	37.1 ±0.2

Table 3. Relative changes in nitrate concentration over the day in three spinach cultivars. The afternoon harvest is given as % of the morning harvest

Cultivar	Glares		Subito		Spinoza	
harvest day	21	30	21	30	21	30
leaf blade	57.6	51.6	80.7	54.9	65.4	154.7
petiole	93.4	83.9	88.6	81.9	77.2	114.6
root	69.2	95.3	62.5	79.4	65.3	107.4

those of the other cultivars (Table 2). In the petioles nitrate concentration was comparatively highest for all three cultivars. On day 30 nitrate concentration in the petioles of Glares was by far the highest. Nitrate concentrations in the roots on day 21 did not differ significantly between the cultivars; on day 30 there was a significant difference only between Glares and Subito ($p = 0.05$).

Comparison of the diurnal changes in nitrate concentration in the plant parts is given in Table 3. For Glares diurnal changes were predominantly found in the leaf blades at both harvests, in a relative (Table 3) as well as absolute sense (Table 2). On day 21 for all three cultivars considerable changes in nitrate concentration were found in the root. On day 30 the change in root concentration over the day was still marked in Subito. On day 21 a considerable change in nitrate concentration in the petiole was found in Spinoza, but for this cultivar an increase in nitrate concentration instead of a decrease was found on day 30. This coincided with a difference in shoot morphology: the leaf blades were less expanded.

Discussion

The present experiments with three cultivars of spinach showed that plants that had reached the same fresh weight after a growth period of 30 days at similar conditions, showed marked differences in internal nitrate concentrations. Hence in this case, concentration in the plant and growth were clearly not correlated. A number of authors has drawn the conclusion that nitrate can act unspecifically as a general osmotic in cell expansion, especially under low-light conditions, in lettuce (Blom-Zandstra and Lampe 1985), spinach (Steingröver et al. 1986), tomato (Van de Dijk 1981) and ryegrass (Veen and Kleinendorst 1985).

Previous experiments with spinach were performed with the cultivar Vroeg Reuzenblad (Steingröver 1986). Nitrate concentration in this cultivar decreased during the day, which could be solely ascribed to a change in leaf blade nitrate (Steingröver 1986). In contrast to these findings with Vroeg Reuzenblad, the cultivars used in the present experiments showed changes in petioles and roots as well (Table 3). For Glares the diurnal changes were highest in the leaf blade. But for Subito and Spinoza, changes in nitrate concentration were less in the leaf blade than in the root (Subito) or in the petiole (Spinoza).

Steingröver (1986) showed by a compartmental analysis that changes in nitrate concentration in the leaf blades were caused by changes in nitrate accumulation in the vacuoles. The accumulation capacity of the vacuoles depends upon both the volume of the vacuole and the nitrate flux across the tonoplast (Glass et al. 1985, Clarkson 1986). The lower nitrate accumulation in the leaf blades of Subito and Spinoza, as compared to Glares, might be explained by smaller vacuoles and/or differences in transport over the tonoplast. Cv. Spinoza showed considerable changes in nitrate concentration in the petioles as well. This might be explained by assuming that regulation of nitrate transport from the xylem into the cytoplasm of the leaf blade cells is important in this cultivar. Steingröver (1986) concluded that xylem loading and unloading is important in nitrate accumulation in the plant as a whole. Cv. Subito showed striking diurnal changes in root nitrate concentration, in contrast to cv. Vroeg Reuzenblad, which showed no diurnal changes in nitrate concentration in the root at all (Steingröver 1986). We conclude that in cv. Subito accumulation in the vacuoles of the root may be decisive in nitrate accumulation in the plant as a whole.

Summarizing we conclude that the three spinach cultivars used in the present experiments differ in regulation of nitrate accumulation and distribution of nitrate over roots, petioles and leaf blades. We state that regulation of nitrate distribution can take place at several points in the chain of transport steps leading to nitrate accumulation, and that the point of major regulation differed for the cultivars investigated, viz:

For cv. Glares: accumulation in the leaf vacuoles;

For cv. Subito: uptake and transport processes in the root;

For cv. Spinoza: transport between petiole and leaf blade.

References

Blom-Zandstra G, Lampe JEM (1983) The effect of chloride and sulfate salts on the nitrate content in lettuce plants (*Lactuca sativa* L.). J Plant Nutr 6: 611-628

Clarkson DT (1986) Regulation of the absorption and release of nitrate by plant cells: A review of current ideas and methodology. In: Lambers H, Neeteson JJ, Stulen I (eds) Fundamental, Ecological and Agricultural Aspects of Nitrogen Metabolism in Higher Plants. Martinus Nijhoff, Dordrecht Boston Lancaster, pp 3-27

Kuiper D (1988) Growth responses of *Plantago major* L. ssp. *pleiosperma* (Pilger) to changes in mineral supply. Evidence for regulation by cytokinins. Plant Physiol 87: 555-557

Glass ADM, Thompson RG, Bordeleau L (1985) Regulation of nitrate influx in barley. Plant Physiol 77: 379-381

Kuiper D, Schuit J, Kuiper PJC (1988) Effects of internal and external cytokinin concentrations on root growth and shoot to root ratio of *Plantago major* ssp. *pleiosperma* at different nutrient conditions. Plant and Soil 111: 231-236

Kuiper D, Staal M (1987) The effects of exogenously applied plant growth substances on the physiological plasticity in *Plantago major* ssp. *pleiosperma*: Responses of growth, shoot to root ratio and respiration. Physiol Plant 69: 651-658

Lambers H, Posthumus F, Stulen I, Lanting L, Van de Dijk SJ, Hofstra R (1981) Energy metabolism of *Plantago lanceolata* as dependent on the supply of mineral nutrients. Physiol Plant 51: 85-92

Maas FM, Hoffman I, Van Harmelen MJ, De Kok LJ (1986) Refractometric determination of sulphate and other anions in plants separated by High-Performance Liquid Chromatography. Plant and Soil 91: 129-131

Steingröver E (1986) Nitrate accumulation in spinach. Thesis, Groningen

Steingröver E, Ratering P, Siesling J (1986) Daily changes in nitrate uptake, reduction and storage of nitrate in spinach, grown at low light intensity. Physiol Plant 66: 550-556

Van de Dijk SJ (1981) Nitrogen metabolism in higher plants in different environments. Thesis, Groningen.

Veen BW, Kleinendorst A (1985) Nitrate accumulation and osmotic regulation in Italian ryegrass (*Lolium multiflorum* Lam.). J Exp Bot 36: 211-218

Stulen I, Lanting L, Lambers H, Posthumus F, Van de Dijk SJ, Hofstra R (1981) Nitrogen metabolism of *Plantago lanceolata* as dependent on the supply of mineral nutrients. Physiol Plant 51: 93-98

NITROGEN ASSIMILATION OF PLANTS UNDER STRESS AND HIGH CO_2 CONCENTRATIONS

S.H. Lips, E.O. Leidi and M. Silberbush [1]

Introduction

Most crop plants reduce their biomass production when growing in relatively mild salinity (electric conductivity, EC, 5 - 13 dSi m^{-1}) This was assumed to be due either to the limited ability of these plants to adjust osmotically to their saline environment or to the energetic cost of the osmotic adaptation process (Yeo 1983). During the last years an increasingly large number of publications point out that this may not be the case (e.g. Termaat et al.1985) since plants readily adjust osmotically to these salinity levels. The inhibition of biomass production by salinity is probably due to a number of failing metabolic processes each of which develops at different levels of salinity according to their particular sensitivity to salt. Once we define the physiological characteristics of a given crop, salinity may be applied to enhance quality and quantity of fruit and seed production in the range of 0 - 10 dSi m^{-1}. This idea will be presented in relation to two central primary assimilation processes: uptake of nitrogen and photosynthesis.

Nitrogen Uptake and Assimilation under Saline Conditions

Plants keep a fairly constant concentration of total organic nitrogen under most environmental conditions (Silberbush et al. 1988; Leidi et al. 1990 a,b). The amount of nitrogen taken up by the plant determines the amount of biomass they will produce at standards predetermined genetically. Since Cl^- reduces nitrogen uptake, one may assume that salt-induced inhibition of nitrogen uptake limits growth. Nitrogen supplementation to some crops (e.g.: peanut, sunflower, cotton, maize, wheat) prevents or limits salt inhibition of growth. The sensitivity of different species to salt is reflected in the Cl^- sensitivity of their nitrate uptake mechanisms (Leidi et al. 1990 a,b). We compared (Fig. 1) the ratio of total N/Cl^- in the plants as affected by the ratio of NO_3^-/Cl^- in the nutrient medium with salt-sensitive peanut and with salt-tolerant cotton plants. Increasing medium salinity affects more rapidly the uptake of nitrogen by the sensitive than by the tolerant species. It should be possible, consequently, to improve nitrogen uptake by slightly increasing the NO_3^-/Cl^- ratio around the plant roots. When

[1] Center of Desert Agrobiology, J. Blaustein Institute for Desert Research and Dept. of Biology, Ben Gurion University of the Negev, Sede Boqer, 84993 Israel

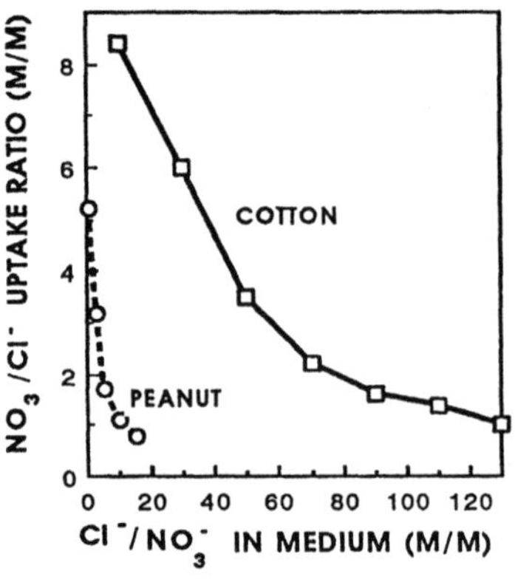

Fig. 1. Nitrate uptake as affected by Cl^-

peanut plants are grown in cultures with 0 - 100 mM NaCl, plant growth is rapidly inhibited (Fig. 2).

However, if plants growing in 60 mM NaCl, whose growth is inhibited by 40 - 60 %, receive in their nutrient medium (sand) increasing concentrations of KNO_3 (Fig. 3), the growth of these plants is renewed and reaches rates higher than that of the controls. It was observed that the overall content of chloride in the plants remained fairly constant while the amount of nitrate increased proportionally to growth (Fig. 4). We concluded that the salt sensitivity of the *Rhizobium* strain in peanut root nodules is very

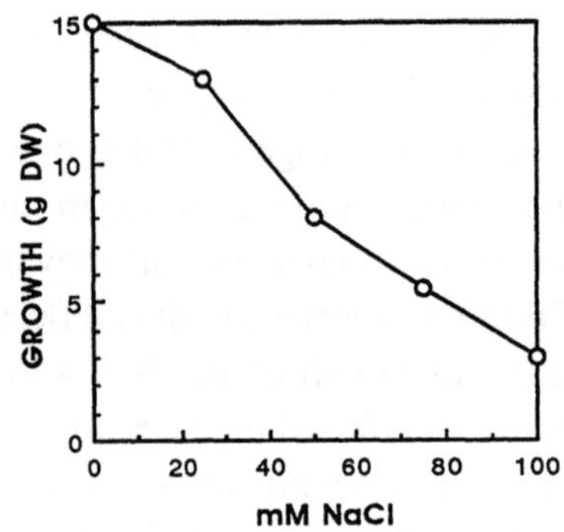

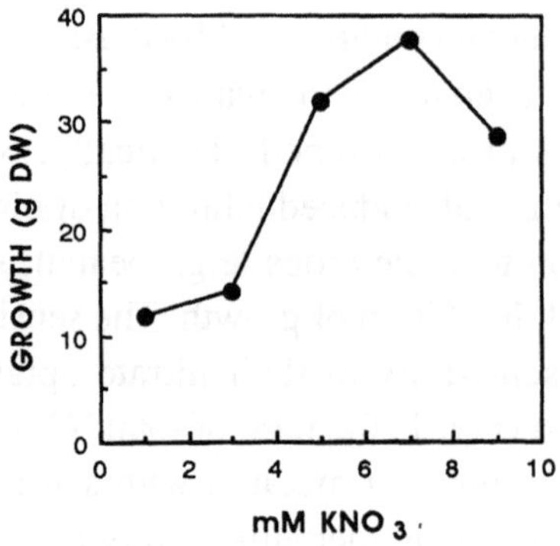

Fig. 2 *(left)*. Growth of peanut plants as affected by NaCl. No fixed nitrogen added, good nodulation

Fig. 3 *(right)*. Growth of peanut plants grown in NaCl as affected by KNO_3 concentration

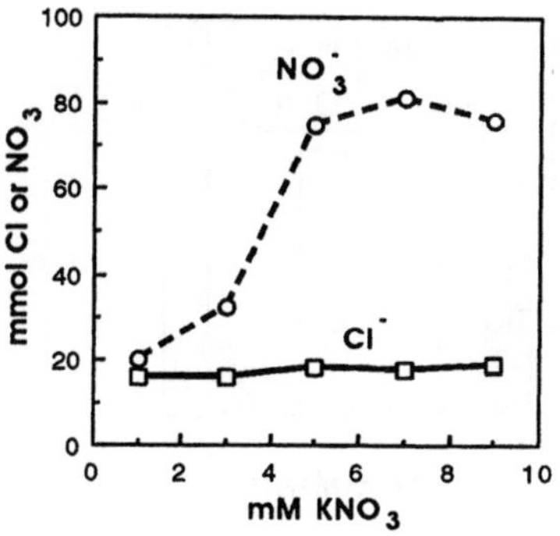

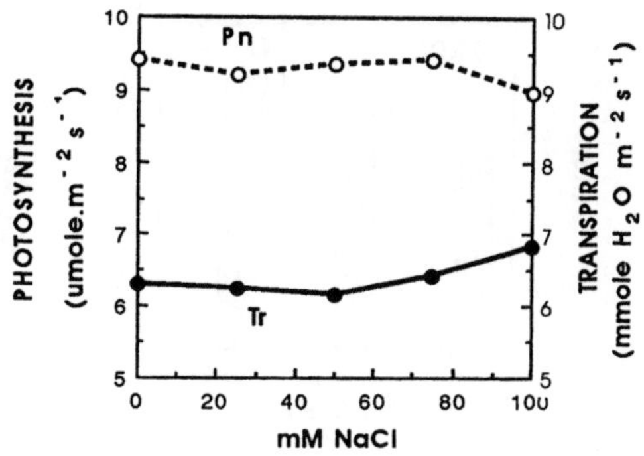

Fig. 4 *(left)*. Chloride and nitrogen content of peanut plants grown in 60 mM NaCl as affected by KNO_3 concentration

Fig. 5 *(right)*. Photosynthesis and transpiration rates of wheat as affected by NaCl concentrations

pronounced at even relatively low concentrations of NaCl. The addition of nitrate or ammonium bypasses the salt sensitivity of the root nodules, re-establishes adequate nitrogen supply to plants and the result is uninhibited growth.

Photosynthesis and Transpiration

Contrary to nitrate uptake mechanisms, photosynthesis and transpiration are stable when plants are irrigated with up to 80 mM NaCl (Fig. 5). This is the case for both salt-sensitive and tolerant species. The decreased growth of salt-sensitive plants is not due, at the salinity range indicated, to salt-induced damage to the photosynthetic apparatus. This is the case even when a considerable decrease in extracted chlorophyll is observed.

Study Case 1: Peanut Production

Since the growth capacity of peanut plants was sensitive to salinity but could be restored by increasing the supply of nitrogen (Fig. 2), we assumed that peanut production could also take place under field conditions when irrigated with saline water supplemented with nitrogen fertilizer (Silberbush et al. 1988 a,b). Flowering was greatly stimulated by salinity (Silberbush et al. 1988 a,b), even in plants with good

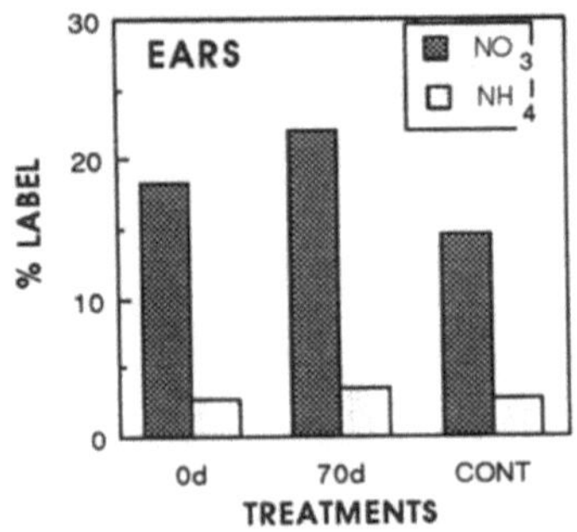

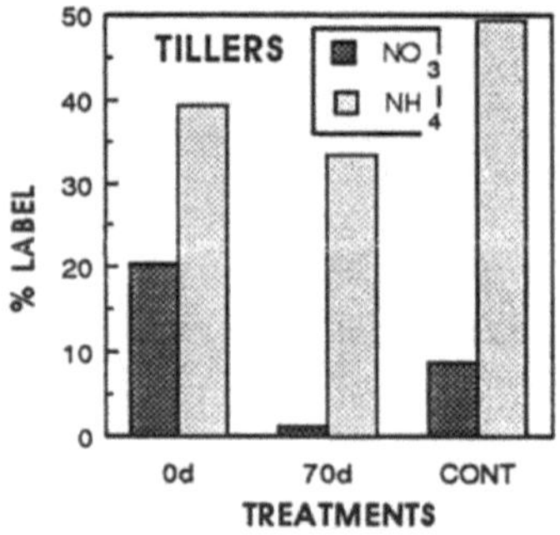

Fig. 6 *(left)*. ^{14}C-allocation to ears as affected by the type of nitrogen fed to wheat plants

Fig. 7 *(right)*. ^{14}C-allocation to young tillers as affected by the type of nitrogen fed to wheat plants

vegetative growth which did not show any stress symptoms. Salinity damage signs (chlorotic and necrotic leaf areas) appeared in the crop soon after flowering at the time of gynophore development. The end result of our efforts in the field at this stage was that we were obtaining large plants with enhanced flowering but lower yields of seeds. Gynophore development was very sensitive to soil salinity and most of them did not develop pods and seeds when in contact with more than 100 mM NaCl. What was the reason for the salt-sensitivity of peanut during the reproductive stage? We are presently studying two alternatives: (a) the capacity of the root to discriminate between Cl^- and NO_3^- decreases during gynophore development and (b) gynophores may take up ions from the soil in a way which introduces large amounts of Cl^- into the plant.

As a result of these observations, best results can be achieved today by using saline water (6-10 dSi m^{-1}) during vegetative growth until flowering and continuing irrigation with good quality water (up to 1 dSi m^{-1}) during gynophore development. The use of saline water irrigation during the vegetative stage enhances overall production of peanut seeds 70 - 90 % above yields obtained with peanut plants grown with good quality water during their entire life cycle.

Study Case 2: Wheat Production

The use of saline water irrigation during wheat production differs from that used with peanut. Nitrogen (especially ammonium nitrogen) stimulates tiller development while chloride inhibits it (Leidi et al. 1990 a,b). Growth inhibition of wheat by salinity is due to the decrease in the number of tiller produced and in the dry mass of single tillers. As a result, less and smaller grains are produced (Silberbush and Lips 1990 b). Adding nitrogen to the irrigation saline water releases some of the salt-induced inhibition of shoot growth but the effect is less dramatic than in peanuts during the vegetative stage.

With wheat we learned that the type of nitrogen supplied, however, seems to affect the extent of mobilization of assimilates to filling grains (Botella et al. 1989). Nitrate, but not ammonium nitrogen, stimulates assimilate translocation from leaves to ears in plants irrigated with saline water (Fig. 6). Supply of nitrate, under saline conditions, causes production of larger grains (Silberbush and Lips 1990 b). Translocation of recently photosynthesized ^{14}C-assimilates to the grains is stimulated by salt (Leidi et al. 1990 a). While nitrate-fed plants translocate assimilates to the grains, ammonium-fed plants allocated most of their carbon assimilates to developing young tillers (Fig. 7) (Botella et al. 1989).

From here one may conclude that wheat irrigated with good quality water and ammonium nitrogen during vegetative growth, followed by saline water (50-60 mM NaCl) and nitrate, may produce a better yield of grain than wheat grown on good quality water throughout its entire growth cycle. In this case the combined effects of nitrate nitrogen and salinity during the grain filling stage may result in an increased production of grain biomass.

Recent work by Berliner et al. (1986) indicates that placement of nitrate in deep layers of soil (about 0.80 - 1 m deep) promotes wheat root proliferation in these layers. Deep roots allow the plant to use larger amounts of soil moisture in areas where its supply is limited to seasonal rains.

Atmospheric CO_2 Concentration and the Effects of NaCl on Biomass Production

Salt-tolerant and salt-sensitive sunflower plants growing in atmospheric CO_2 concentrations (340 ppm) respond to relatively low concentrations of NaCl (10 - 25 mM) by reducing their growth to some extent. However, when these plants are placed in 1200 ppm CO_2, the same low concentrations of NaCl will stimulate growth to a large extent (Figs. 8 and 9). Since total nitrogen concentration in the leaves remains constant under all treatments, the salt sensitivity of the nitrate uptake mechanism could be also a function of carbohydrate supply to the roots. The concept of stress is, therefore, very much a relative function of environmental conditions.

We are presently studying various aspects of the possible mechanisms involved in this NaCl stimulation of growth, such as Na^+ and Cl^- distribution and compartmentation, K^+ requirements of the plants under high CO_2, etc.

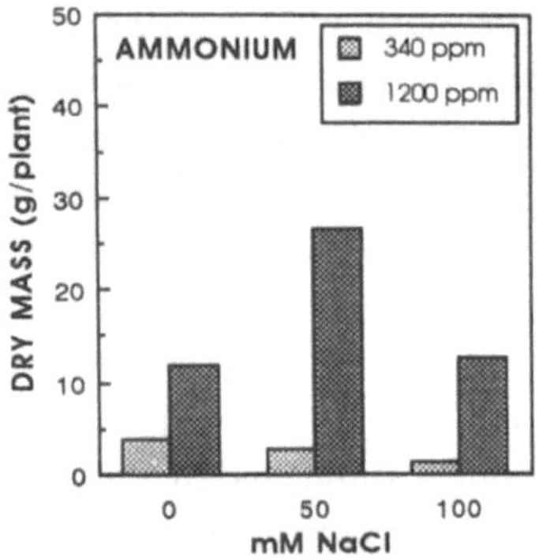

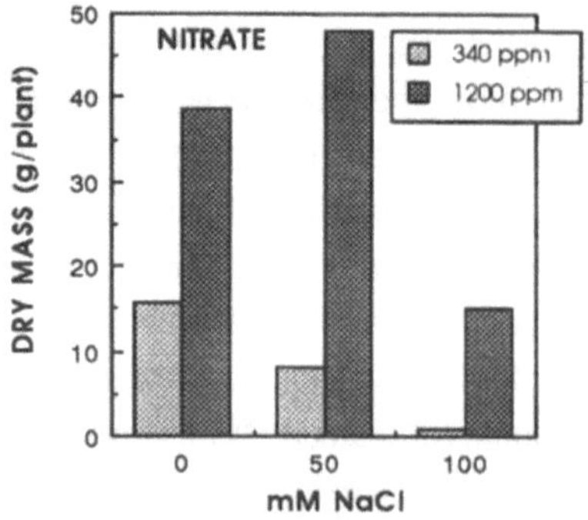

Fig. 8 *(left)*. Dry mass production of sunflower plants grown in ammonium nitrogen as affected by CO_2 and NaCl concentrations

Fig. 9 *(right)*. Dry mass production of sunflower plants grown in nitrate nitrogen as affected by CO_2 and NaCl

Determination by Salinity of Sink Priorities in Nitrate- and Ammonium-Fed Plants

Wheat plants well supplied with nitrogen develop new tillers even when grain of older tillers is at the filling stage. Nitrate-fed plants produce most or all of their tillers during a limited period of time, mostly during the vegetative stage. During the second period of growth, carbohydrate and nitrogenous reserves of the leaves are transported exclusively to the filling grains. In this way, a large percentage of tillers produces mature ears. Ammonium-fed wheat produces tillers during the entire life cycle of the plant. A smaller percentage of the total number of tillers produced will bear ears and not all of the latter will mature before the onset of plant senescence (Botella et al. 1989). Salinity stops development of young tillers and enhances translocation of assimilates to the seeds; both processes are also typical of nitrate-fed wheat plants. The combination of salinity and nitrate produces more efficient plants in the sense that assimilates will be invested totally in grain production and none in the generation of late tillers whose ears will not mature.

These observations confirm our work (Silberbush and Lips 1990 b) in which we reported the production of larger seeds by wheat plants irrigated with saline water and nitrate as compared with plants receiving saline water and ammonium.

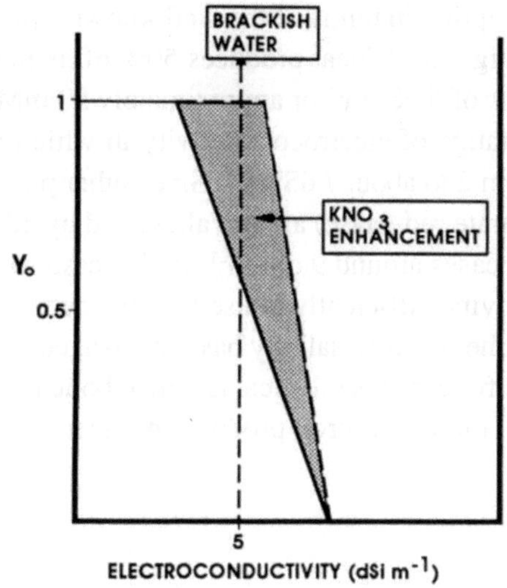

Fig. 10. Response of a salt-sensitive plant to increasing salinity and additional KNO_3

Crop Production and Salinity

Brackish or saline water with electroconductivities up to 13 dSi m^{-1} constitute frequently the main or only water resource for agriculture in extensive arid and desert areas of the world. Even in more humid regions, high water tables, inadequate use of irrigation or excessive pumping from wells in costal areas, may cause salinization of the medium in which crops are produced. At this lower range of salinity, nitrogen and potassium fertilization are especially important. Nitrogen can counteract, at this range of electroconductivity, much of the growth inhibition induced by salt when it is provided in adequate amounts, forms (ammonium or nitrate) and timed to specific development stages of the plants.

At higher salt concentrations, the increasing number of metabolic processes affected by salinity cannot be corrected or alleviated by enhanced nitrogen fertilization (Fig. 10).

Plants respond to salinity in different ways since different developmental processes are affected. At this stage of our awareness of the interaction between NaCl and different metabolic processes of the plants we may have to devise different strategies for each crop. Salinity not only inhibits physiological plant processes (nitrogen uptake, chlorophyll synthesis, tiller induction, N_2-fixation) but it stimulates others (flowering, nitrogen uptake under high CO_2, assimilate translocation to fruits and seeds). By alleviating some of the more salt-sensitive processes while stimulating others, we should be able to use saline water (4 - 10 dSi m^{-1}) to produce better yields.

One can express this assumption in terms of the well-known salt-sensitivity plot by Maas and Hoffman (1977) (Fig. 10). Wheat produces 50% of its potential grain yield (Y_0) when irrigated with water of 5 dSi m^{-1} or approximately 50 mM NaCl. When we add to this water KNO_3 the range of electroconductivity in which normal yields are obtained may be extended from 2 to about 7 dSi m^{-1}. Since other physiological systems (such as photosynthesis or nitrate reductase) are not alleviated by adding KNO_3, yield production will cease in both cases around 9 dSi m^{-1}. In this case we extended the full yield range of electroconductivity sufficiently to use locally abundant 5 dSi m^{-1} water, although the survival limit of the plants to salinity has not changed.

As our knowledge of the different physiological and metabolic processes affected by salinity increases, a universal model for crop production under saline conditions will presumably be developed.

References

Berliner P, Oosterhuis DM , Green GC (1986) The effect of depth placement of nitrogen fertilizer on the response of dry land wheat. Report to the Department of Agriculture of South Africa
Botella A, Cerda A, Lips SH (1989) Studies on the effect of salinity at different growth stages of wheat growing on ammonium or nitrate. Abstract, 2nd Advanced Course on Inorganic Nitrogen Metabolism, Seiano, Italy, p 132
Leidi EO, Silberbush M, Lips SH (1990 a) Effects of nitrogen source, pH and salinity on wheat growth. (I) Mineral balances in the plant. Physiol Plant (in press)
Leidi EO, Soares MIM, Silberbush M, Lips SH (1990 b) Salinity and nitrogen nutrition studies on peanut and cotton plants. Physiol Plant (in press)
Maas EV, Hoffman GJ (1977) Crop salt-tolerance - Current assessment. ASCE J Irrig Drain Div 103: 115-134
Silberbush M, Ben-Asher J, Lips SH (1988 a) Nutrient-salt interactions and plant growth. In: Advances in Desert and Arid Land Technology and Development, Vol 5, Harwood Academic Publishers, New York
Silberbush M, Golan-Goldhirsh A, Heimer Y, Lips SH (1988 b) Nitrogen concentration, ammonium/ nitrate ratio and NaCl interaction in the vegetative and reproductive growth of peanut plants. J Plant Physiol 132: 229-233
Silberbush M, Lips SH (1990 a) Potassium, nitrogen, ammonium/nitrate ratio, and sodium chloride effects on wheat growth. (1) Shoot and root growth and mineral composition. Physiol Plant (in press)
Silberbush M, Lips SH (1990 b) Potassium, nitrogen, ammonium/nitrate ratio, and sodium chloride effects on wheat growth. (2) Tiller and grain production. Physiol Plant (in press)
Termaat A, Passioura J, Munns R (1985) Shoot turgor does not limit shoot growth of NaCl-affected wheat and barley. Plant Physiol 77: 869-872
Yeo AR (1983) Salinity resistance: Physiologies and prices. Physiol Plant 58: 214-222

EFFECT OF UV-B (290-320 nm) RADIATION ON UPTAKE OF ^{15}N-NITRATE BY MARINE DIATOMS

G. Döhler [1]

Introduction

Since the first studies on nitrate uptake (Warburg and Negelein 1920) nitrogen metabolism of microalgae was investigated by several groups in more detail (see e.g. Syrett 1981; Ullrich 1983; Wheeler 1983). The use of the ^{15}N-tracer technique gave insights into cellular nitrogen metabolism (Collos and Slawyk 1980; Döhler and Roßlenbroich 1981; Dugdale and Goering 1967). Monochromatic light exhibits a regulatory effect on nitrate uptake in algae (Aparicio and Maldonado 1979; Calero et al. 1980; Döhler and Alt 1989). More recently, also the impact of ultraviolet-B radiation on metabolic processes, especially on the nitrogen metabolism of microalgae has been studied (Döhler 1985, 1988). It could be shown in field (Smith et al.1980; Worrest 1982; G. Döhler E. Hagmeier, E. Grigoleit, unpublished) and laboratory experiments (Döhler 1989, and references therein) that UV radiation can damage the pigmentation and the cellular metabolism of the organisms as well as the community composition of aquatic ecosystems.

Material and Methods

The marine diatoms *Ditylum brightwellii* (West) Grunow, *Lithodesmium variabile* Takano, *Odontella sinensis* Greville, *Synedra planctonica* and *Thalassiosira rotula* Meunier were grown at +18°C under normal air conditions (0.035 vol.% CO_2) and a light/dark cycle of 14:10 h (white light intensity: 8 W m^{-2}) as pure cultures. The artificial nutrient solution (35‰ salinity) developed by v. Stosch and Drebes (1964) was used. Algae harvested during exponential growth were exposed to UV-B radiation (400, 439, 464, 717 and 1184 J m^{-2} d^{-1}) in special quartz vessels or in an assimilation chamber, of which one side consisted of quartz. The different doses of UV-B (Philips lamps TL 40/12, cut-off filters WG 305, 3 mm thick) were obtained by changing the distances from the UV lamps to the vessels or by variation of the exposure time. Filters for visible monochromatic light (intensity 436 mW m^{-2}) were narrow-band interference filters from Balzers (Liechtenstein): 454 ± 8 nm, 544 ± 7 nm, 604 ± 10 nm, 624 ± 9 nm and 670 ± 9 nm. The diatoms were concentrated to a density of 3 or 6 x 10^4 or 1.4 x 10^6 cells ml^{-1} in a nitrogen-free medium. $K^{15}NO_3$ (96.8 atom %, final concentration

[1] Botanisches Institut der Universität, Siesmayerstr. 70, D-6000 Frankfurt a.M., FRG

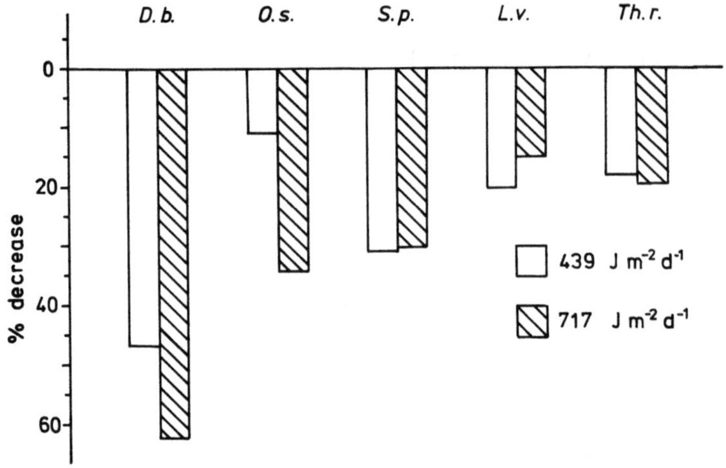

Fig. 1. Effect of UV-B radiation (439 and 717 $Jm^{-2}d^{-1}$) on ^{15}N-nitrate uptake after 10 min photosynthesis by several marine diatoms. Values are presented as % decrease related to non-UV-B irradiated cells (= 0). D.b. *Ditylum brightwellii*; O.s. *Odontella sinensis*; S.p. *Synedra planctonica*; L.v. *Lithodesmium variabile*; Th.r. *Thalassiosira rotula*. Further details in Material and Methods

0.5 or 1.0 mM) was applied after 15 min preillumination with white or monochromatic light. For synchronization of the marine diatom *Synedra planctonica* a light/dark regime of 6.5/17.5 h (5,000 lux) was used. The Dumas method served for sample preparation, and ^{15}N-analysis was performed with an atomic emission spectro-meter (Statron NOI-5; Zeiss, Jena). For further details see Döhler and Alt (1989), Döhler and Biermann (1985).

Results

Several species of marine diatoms were exposed to UV-B (5 h/d) during growth for 2 days. The effect of this treatment on the uptake of ^{15}N-nitrate after 10 min photosynthesis is presented in Figure 1. A depression of the uptake rates could be found in all marine diatoms tested. *Ditylum brightwellii* was a very sensitive species, whereas *Lithodesmium variabile* and *Thalassiosira rotula* were less affected by UV-B irradiance. These results clearly show species-dependence of the response to UV-B stress.

The impact of UV-B on uptake of ^{15}N-nitrate during irradiance with monochromatic light has been investigated with *Thalassiosira rotula* (Fig. 2). A reduced ^{15}N-accumulation in the diatom cells was found under blue (454 nm), orange (604 nm) and red (670 nm) light, whereas an increase could be measured under green (544 nm) and orange-

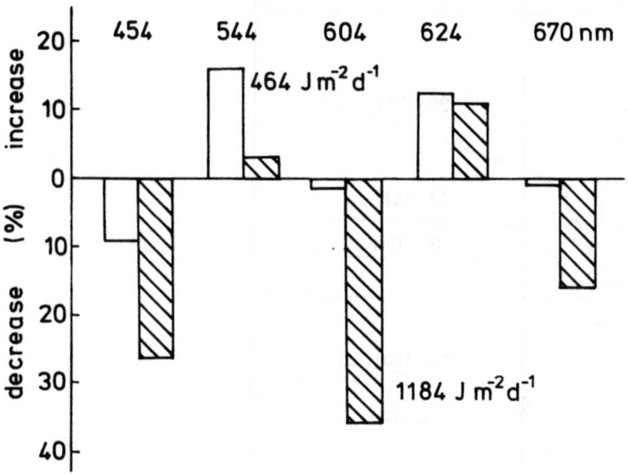

Fig. 2. Effect of UV-B radiation (464 and 1184 J $m^{-2}d^{-1}$) on ^{15}N-nitrate uptake of *Thalassiosira rotula* after 10 min exposure to UV-B and irradiance with monochromatic light. Values are % increase or decrease calculated to not UV-B treated cells (= 0). For further details see Material and Methods

red (624 nm) illumination of the cells after UV-B exposure for 10 min. A more pronounced depression of nitrate utilization was observed in the blue than in the red region of the spectrum. Rates of ^{15}N-nitrate uptake and ^{15}N-incorporation into the amino acids of non-UV-B treated cells were higher under blue than red light irradiation. A reduction of the ^{15}N-label in glutamic acid and an enhancement in glutamine were measured after additional UV-B irradiation of the diatoms independent of the used light quality.

In another series of experiments the effect of UV-B radiation on ^{15}N-nitrate uptake was studied with synchronized cells of *Synedra planctonica.* Algae were exposed to UV-B (2h; 400 J $m^{-2}d^{-1}$) during the 3rd division cycle at different times (Fig. 3). In the light period uptake of ^{15}N-nitrate increased whereas no or only a low uptake rate could be observed in the dark. Figure 3a presents data of ^{15}N-nitrate uptake by algae in the 3rd cycle but irradiated with UV-B during the 2nd cycle of cell division. The inhibitory effect was relatively low after UV-B treatment during the 20th - 22nd hour in the dark and 4th - 6th hour in the light period (Fig. 3 b). During DNA synthesis (8th - 10th hour) a marked depression of ^{15}N-nitrate utilization was found. The UV-B damage was low at low rates of biosynthetic processes (e.g. protein synthesis; 0 - 4 hours); this could be shown by ^{15}N-labelling experiments in the protein fraction.

The effect of UV-B radiation on the activity of nitrate reductase and a possible reactivation were investigated with algae harvested during the light period (Fig. 4). A total inhibition of nitrate reductase could be detected after UV-B irradiation during the hours 0 - 2 and 4 - 6 in the division cycle of *Synedra* and a slight reactivation was observed 30.5 hours after UV-B exposure. On the other hand, a reduction in nitrate reductase activity by 34% was determined after UV-B radiation during the 2nd - 4th hour of the

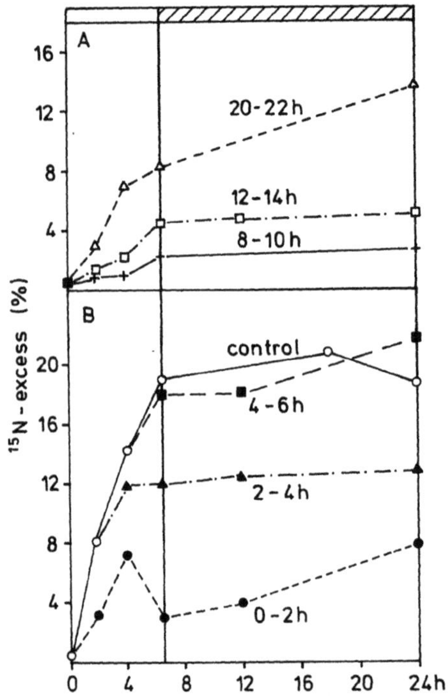

Fig. 3. Effect of UV-B radiation (400 J $m^{-2}d^{-1}$) on ^{15}N-nitrate uptake during the 3rd division cycle of *Synedra planctonica*. Algae irradiated with UV-B during the dark period in the 2nd cycle (**a**) or in the light period of the 3rd cycle (**b**). The exposure time to UV-B was varied over different periods of the cycle (0 - 2; 2 - 4; 4 - 6; 8 - 10; 12 - 14 and 20 - 22 hours after onset of cycle). Values as % ^{15}N in excess. Further details in Material and Methods

cycle. Summarizing, our results show that, also in marine diatoms, the response to UV-B radiation is dependent on the growth stage of the cells. A detailed description of the studies with synchronized *Synedra planctonica* will be published elsewhere (Döhler and I. Biermann, unpublished).

Discussion

Photosynthetic uptake of ^{15}N-nitrate was depressed after UV-B radiation in all marine diatom species tested under white light during the experiments (Fig. 1). The species-dependent response to UV-B irradiance can be explained by the adaptation to the different environmental conditions of the habitats. A different response of several phy-

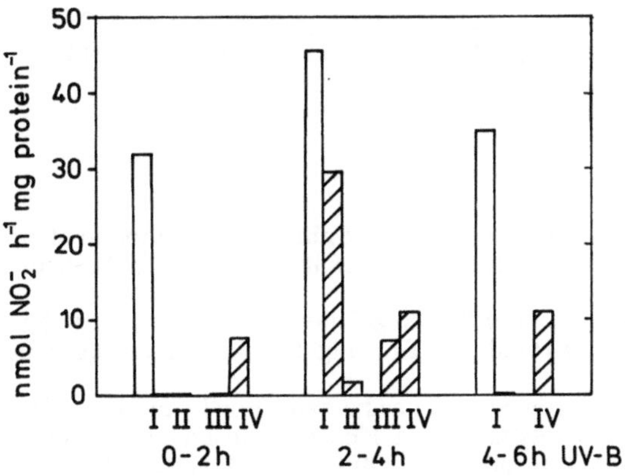

Fig. 4. Effect of UV-B radiation (400 J $m^{-2}d^{-1}$) on the activity of nitrate reductase of synchronized *Synedra planctonica* exposed to UV-B at different times during the light period. White columns, control (not UV-B treated); striped columns, after 2h UV-B radiation; measurements carried out with algae immediately (I), 6.5 (II), 24 (III) and 30.5 h (IV) after UV-B exposure

toplankton species has been also observed by other researchers (Worrest 1982, and references therein). The reduction of ^{15}N-nitrate uptake by UV-B can be attributed to several sorts of damage: (1) UV-B might directly affect the membranes via destruction of protein and lipid arrangements and by that the transport system of nitrate (e.g. carrier protein, ATPase). A significant reduction of digalactosyl diglycerol synthesis was found in synchronized *Ditylum brightwellii* (Döhler and Biermann 1988). (2) A reduction in ATP supply via inhibition of photosysnthetic reaction centre II (Kulandaivelu and Noorudeen 1983) and/or carbon skeletons available for amino acid biosynthesis due to reduced photosynthetic CO_2 fixation or respiration; Döhler (1988) found an inhibition of total $^{14}CO_2$ fixation by marine diatoms. (3) Alterations in the regulation mechanism of nitrate uptake, e.g. via variations of amino acid and/or ammonia pools; an increase of the pool size and biosynthesis of glutamine was observed (Döhler 1988). (4) A direct impact of UV-B radiation on the key enzymes of carbon and nitrogen metabolism; a total inhibition of nitrate reductase could be detected (see Fig. 4). On the other hand, a reduction in protein synthesis was found by using ^{15}N-leucine and ^{3}H-leucine as well as by investigating the *in vitro* translation products with m-RNA of UV-B irradiated *Odontella sinensis*. The different influence of UV-B radiation on ^{15}N-nitrate uptake of *Thalassiosira rotula* cells in dependence of the wavelength (Fig. 2) can be attributed to damages to chlorophyll a and/or flavins. On the other hand, a different regeneration from the UV-B damages dependent on the wavelenght might be a possible explanation. Further studies will have to elucidate this problem.

Acknowledgments. This work was supported by a grant (0704556 B/O) from the Bundesministerium für Forschung und Technologie, Bonn. Thanks are expressed to Dipl. Biol. Irene Biermann for her contribution.

References

Aparicio PJ, Maldonado JM (1979) Regulation of nitrate assimilation in photosynthetic organisms. In: Hewitt EJ, Cutting CV (eds) Nitrogen Assimilation of Plants. Academic Press, London New York, pp 207-215

Calero F, Ullrich WR, Aparicio PJ (1980) Regulation by monochromatic light of nitrate uptake in *Chlorella fusca*. In: Senger H (ed) The Blue Light Syndrome. Springer, Berlin Heidelberg New York, pp 411-421

Collos Y, Slawyk G (1980) Nitrogen uptake and assimilation by marine phytoplankton. In: Falkowski G (ed) Primary Productivity in the Sea. Plenum Publishing Corp, New York, pp 195-211

Döhler G (1985) Effect of UV-B radiation (290 - 320 nm) on the nitrogen metabolism of several marine diatoms. J Plant Physiol 118: 391-400

Döhler G (1988) Impact of UV-B radiation on nitrogen metabolism of marine phytoplankton. In: Öztürk M (ed) Proceedings, Plants and Pollutants Izmir, pp 57-87

Döhler G, Alt MR (1989) Assimilation of ^{15}N-ammonia during irradiance with ultraviolet-B and monochromatic light by *Thalassiosira rotula*. CR Acad Sci Paris 308 (III) 513-518

Döhler G, Biermann I (1985) Effect of salinity on ^{15}N-ammonia and ^{15}N- nitrate assimilation of *Bellerochea yucatanensis* and *Thalassiosira rotula*. Biochem Physiol Pflanzen 180: 589-598

Döhler G, Biermann Th (1988) Variations in the pattern of lipids during cell cycle of the marine diatom *Ditylum brightwellii* (West) Grunow and after exposure to UV-B radiation (290-320 nm). Hoppe-Seyler's Z Biol Chem 369: 19-20

Döhler G, Roßlenbroich HJ (1981) Photosynthetic assimilation of ^{15}N-ammonia and ^{15}N-nitrate in the marine diatoms *Bellerochea yucatanensis* (von Stosch) and *Skeletonema costatum*. Z Naturforsch 36 c: 834-839

Dugdale RC, Goering JJ (1967) Uptake of new and regenerated forms of nitrogen in primary productivity. Limnol Oceanogr 12: 196-206

Kulandaivelu G, Noorudeen AM (1983) Comparative study of the reaction of ultraviolet-C and ultraviolet-B radiation on photosynthetic electron transport. Physiol Plant 58: 389-394

Smith RC, Baker KS, Holm-Hansen O, Olson R (1980) Photoinhibition of photosynthesis in natural waters. Photochem Photobiol 31: 585-592

Stosch HA von, Drebes G (1964) Entwicklungsgeschichtliche Untersuchungen an zentrischen Diatomeen, IV. Die Planktondiatomee *Stephanopyxis turris*, ihre Behandlung und Entwicklungsgeschichte. Helgoländer wiss Meeresunters 11: 209-257

Syrett PJ (1981) Nitrogen metabolism of microalgae. Can Bull Fish Aquat Sci 210: 182-210

Ullrich WR (1983) Uptake and reduction of nitrate: Algae and fungi. In: Läuchli A, Bieleski RL (eds) Inorganic Plant Nutrition. Encyclopedia of Plant Physiology, NS vol 15A. Springer, Berlin Heidelberg New York, pp 376-397

Warburg O, Negelein E (1920) Über die Reduktion der Salpetersäure in grünen Zellen. Biochem Z 110: 66-116

Wheeler PA (1983) Phytoplankton nitrogen metabolism. In: Carpenter EJ, Capone DG (eds) Nitrogen in the Marine Environment. Academic Press, New York, pp 309-346

Worrest RC (1982) Review of literature concerning the impact of UV-B radiation upon marine organisms. In: Calkins J (ed) The Role of Solar UV Radiation in Marine Ecosystems. Plenum Press, New York, pp 429-458

Subject Index

MIX
Papier aus verantwortungsvollen Quellen
Paper from responsible sources
FSC® C105338

If you have any concerns about our products,
you can contact us on
ProductSafety@springernature.com

In case Publisher is established outside the EU,
the EU authorized representative is:
Springer Nature Customer Service Center GmbH
Europaplatz 3, 69115 Heidelberg, Germany

Printed by Libri Plureos GmbH
in Hamburg, Germany

FIRST THOUGHTS